Space Rescue

Ensuring the Safety of Manned Spaceflight

David J. Shayler

Space Rescue

Ensuring the Safety of Manned Spaceflight

David J. Shayler
Astronautical Historian
Astro Info Service
Halesowen
West Midlands
UK

Front cover illustrations: (Main image) Early artist's impression of the land recovery of the Crew Exploration Vehicle. (Inset) Artist's impression of a launch abort test for the CEV under the Constellation Program.

Back cover illustrations: (Left) Airborne drop test of a Crew Rescue Vehicle proposed for ISS. (Center) Water egress training for Shuttle astronauts. (Right) Beach abort test of a Launch Escape System.

SPRINGER–PRAXIS BOOKS IN SPACE EXPLORATION
SUBJECT *ADVISORY EDITOR*: John Mason, B.Sc., M.Sc., Ph.D.

ISBN 978-0-387-69905-9 Springer Berlin Heidelberg New York

Springer is part of Springer-Science + Business Media (springer.com)

Library of Congress Control Number: 2008934752

Printed in Germany

Cover design: Jim Wilkie
Project management: Originator Publishing Services, Gt Yarmouth, Norfolk, UK

Printed on acid-free paper

Contents

For those who develop the methods of ensuring a crew is safe during their mission. For the designers of systems and procedures, the training teams, safety teams and controllers and to the flight crews for their dedication and belief that if something does go wrong there is hopefully time and facilities to do something about it, and live to tell the tale.

Earth and Moon system.

Author's preface

In the late 1960s as America and the Soviet Union pushed to become the first nation to place its citizens on the Moon I became fascinated with what the media was calling 'the moon race'. When learning more about the missions of Apollo and Soyuz in 1969 I also read that along with the triumphs there were also tragedies on both sides. Apollo 1 had claimed three American astronaut lives in a pad fire in January 1967 and three months late the Soviet's Soyuz 1 had killed its single cosmonaut occupant during landing. I still have clear memories of a school memorial service for the Apollo 1 astronauts. I also recall following daily the near fatal flight of Apollo 13 in 1970 and the next year the loss of three cosmonauts returning in Soyuz 11 after the first successful space station mission. At the time of these setbacks, along with the details of future developments in manned spaceflight, media reports also featured articles on how in forthcoming programs it may be possible to rescue stranded crewmembers from a stricken spacecraft.

At my local cinema I had recently enjoyed the 1969 feature film *Marooned* based on a 1964 book by Martin Cadan. Here American astronauts were stranded in Earth orbit and a rescue mission had to be mounted to save them with help coming from the Soviets as well. For Christmas 1970 I received the *Daily Mirror Book of Space* in which a chapter entitled 'What would happen if . . .?' was devoted to the prospect of space rescue in light of the Apollo 13 incident. With information on the escape systems of past manned spacecraft, details of training for emergency landings and the prospect of a rescue craft for the Skylab space station astronauts my interest in providing spacecraft crew with adequate methods of escape from a hostile environment increased.

Over the years I gathered material on the various escape systems and real space accidents where use of these escape systems was not always possible. From these studies evolved the book *Disasters and Accidents in Manned Spaceflight* published in this series in 2000. What was not included in detail due to book extent limitations were the various methods of escape from spacecraft available for the crewmember, or the myriad of escape and rescue systems and procedures developed for use by a stricken crew. Many of these plans never left the drawing board, others have been incorporated on spacecraft for decades, while others could simply not be employed due to the rapid sequence of events of the accident concerned.

Any spaceflight crewmember would tell you that they would not fly on a mission if they did not expect to return home safe and sound. Despite that, most people who strap themselves to a rocket for a quick and rapid ride into space are usually fully aware of the risk and danger ahead. True, this might not be so apparent to those whose spaceflight "career" may not be long-term or their training in-depth, but most believe that they will have a good mission and come home to talk about it . They know of the safety and rescue systems available, though all of them hope not to have to put their training to the test and rely on emergency procedures to complete their mission.

This book then is the other side of the space accidents story: the rescue and safety of manned spaceflights both on the pad, during launch, in flight and getting home. With dramatic and vivid images of Challenger and Columbia, painful memories of Apollo 1, Soyuz 1 and Soyuz 11, the near misses of Gemini 8 in 1966, Apollo 13, the 1975 Soyuz launch abort and 1983 Soyuz pad abort and countless incidents recorded on a number of manned spaceflights. Contingency and emergency training and provisions for the use of rescue systems on rare occasions will continue to be a major part of human exploration of space. As we venture deeper into space the option of a quick return to Earth may not be always the best or safest route to follow. For future space explorers, however, the option of a rescue system is comforting, but using it is another matter.

David J. Shayler
Director
Astro Info Service Ltd
www.astroinfoservice.co.uk
Halesowen, West Midlands, U.K.
Summer 2008

Acknowledgments

This book has been compiled from resources and references gathered over a period of nearly 40 years and is a compression of that research, supplementing the earlier work on space accidents and disasters and other titles by the author and the expanded Praxis Space Science Series listed in the Bibliography at the end of this book.

This research was only possible by the continued support and cooperation of a number of individuals and agencies. Much appreciated thanks go to fellow researchers Colin Burgess, Michael Cassutt, John Charles, Phil Clark, Rex Hall, David Harland, Brian Harvey, Bart Hendrickx, Gordon Hooper, James Oberg, Andy Salmon, Asif Siddiqi, Bert Vis and Andy Wilson. Early research direction was provided by Robert Edgar in the 1970s and the assistance of Tim Furniss is also recognised

Special thanks are also due to Mark Wade and his *Encyclopaedia Astronautix* online reference site and to the staff and resources of the British Interplanetary Society, London.

The assistance of the Public Affairs departments at NASA, ESA and the Russian Space Agency is also acknowledged, as are the periodicals *Aviation Week and Space Technology*, *Flight International* and NASA's JSC Roundup publications.

I must also express my sincere thanks to former US astronauts and Russian cosmonauts who provided additional insights into their personal careers including aspects of survival training and related their own experiences during preparations for and flying on various spaceflights over the years. Special thanks are also extended to former Skylab and Shuttle astronaut Paul J. Weitz for his generous foreword.

I must thank members of my immediate family for their help, support and encouragement in the compilation of this volume, which as with all previous titles has never been an easy process. Thanks once again go my brother Mike for his early editorial input to this title.

I express thanks to Clive Horwood of Praxis Books and to the team at Springer-Verlag for their continued support of my work, and to Jim Wilkie for his continued skills in developing covers for my titles. Thanks also to Neil Shuttlewood and staff at Originator Books for their editorial and typesetting skills

Writing a book, just like any spaceflight, is a team effort though only one name may appear on the cover. The survival of the author to rescue early scribbles and turn them into the finished published book is dependent on his rescue team behind the scenes just as an space explorer requires trustworthy equipment and procedures to rescue them from a tricky situation and a support team on the ground to guide them though the difficulties.

To my support team many thanks once again at the end of a long and at times difficult and frustrating journey.

Special appreciation and love to my mother Mrs. Jean Shayler, who in approaching her 80th solar orbit has continued to be a solid foundation of support and encouragement in everything I have done. Thanks, Mom!

Foreword by Paul J. Weitz

Humans have an innate urge to explore—to see what is just over the ridge, what is on the other side of the river, or what is beyond that next planet. As Captain Picard said "... to boldly go where no one has gone before." (Artificial gravity and warp drive—life is good on the *Starship Enterprise*.) I am convinced that we will continue to do just that; therefore concerns for crew safety in spacecraft design must continue to be addressed.

The principal driver in the design of any spacecraft is mission success. This results in some level of redundancy and fail-safe implementation. When the spacecraft is also tasked to accommodate humans, the resultant life support requirements add significant complications both in added complexity and systems reliability. Some features, such as a fully capable crew escape module, in essence a spacecraft within a spacecraft, while very desirable, adds significant weight, and makes for a more demanding design, test and checkout process.

Paul J. Weitz.

During Apollo and Skylab, we underwent desert survival training in southeastern Washington state and jungle survival training in Panama. We also practiced emergency egress procedures from a command module in the water. When I flew STS-6 in

Weitz monitors Skylab's Apollo Telescope Mount console.

1983 we wore flight coveralls rather than pressure suits, so the extent of our emergency procedures was to practice getting out of the Orbiter after a survivable landing through the side hatch or an overhead hatch and lowering ourselves to the ground. Of course, after the Challenger accident, pressure suits were again worn and the crews had to train on bailout procedures using the extendable pole.

As both an astronaut and a manager, I have found that the appropriate approach is to design into the spacecraft and ground support systems the highest possible reliability so that the occurrence of a requirement for escape and rescue is minimized. However, no endeavor involving travel, whether on the surface of the Earth or far above it, can ever be designed and implemented to provide 100% safety and assurance of survival. The challenge then is to provide a vehicle that approaches as nearly 100% as possible.

Paul J. Weitz,[1] *Captain USN, Retired*
NASA Astronaut 1966–1994
Pilot Skylab 2 (1973)
Commander STS-6 (1983)

[1] Paul Weitz was selected for the Apollo Program in 1966 and was a leading candidate for assignment as CMP to Apollo 20 before its cancellation in 1970. Reassigned to Skylab he flew a 28-day mission to the Space Station in 1973 and commanded the first flight of Challenger (OV-099) in 1983. He served in various management roles until his retirement from NASA in 1994.

Other works by David J. Shayler in this series

Disasters and Accidents in Manned Spaceflight (2000), ISBN 1-85233-225-5
Skylab: America's Space Station (2001), ISBN 1-85233-407-X
Gemini: Steps to the Moon (2001), ISBN 1-85233-405-3
Apollo: The Lost and Forgotten Missions (2002), ISBN 1-85233-575-0
Walking in Space (2004), ISBN 1-85233-710-9

With Rex Hall
The Rocket Men (2001), ISBN 1-85233-391-X
Soyuz: A Universal Spacecraft (2003), ISBN 1-85233-657-9

With Rex Hall and Bert Vis
Russia's Cosmonauts (2005), ISBN 0-38721-894-7

With Ian Moule
Women in Space: Following Valentina (2005), ISBN 1-85233-744-3

With Andy Salmon and Mike Shayler
Marswalk: First Steps on a New Planet (2005), ISBN 1-85233-792-3

With Colin Burgess
NASA's Scientist Astronauts (2006), ISBN 0-387-21897-1

With Tim Furness and Mike Shayler
Praxis Manned Spaceflight Log 1961–2006 (2007), ISBN 0-387-34175-7

Figures

(all illustrations are via the AIS Photo Library and are courtesy NASA, unless otherwise stated)

Tables

Abbreviations and acronyms

AA	Ascent Abort
AA-OSF	Associate Administrator Office of Space Flight
AAP	Apollo Applications Program (Skylab)
ACM	Attitude Control Motor
ACRV	Assured Crew Return Vehicle
AFB	Air Force Base
AFRSI	Advanced Flexible Reusable Surface Insulation
AGS	Abort Guidance System
AIAA	American Institute of Aeronautics and Astronautics
AK	Apogee Kick
ALT	Approach and Landing Test (Shuttle)
AOA	Abort Once Around
APU	Auxiliary Power Unit
ARD	Abort Region Determinator
ARG	Anthropomorphic Rescue Garment
ascan	astronaut candidate
ASIS	Abort Sensing and Implementation System
ASTP	Apollo-Soyuz Test Project
ATB	Abort Test Booster
ATO	Abort To Orbit
BP	BoilerPlate
BPC	Boost Protective Cover
BSLSS	Buddy Secondary Life Support System
CAIB	Columbia Accident Investigation Board
Capcom	Capsule communicator
CB	NASA Astronaut Office, JSC, Houston, Texas
CEM	Crew Escape Module
CEV	Crew Exploration Vehicle

cg	center of gravity
CLV	Crew Launch Vehicle
CM RCS	Command Module Reaction Control System
CM	Command Module; CrewMember
CNES	French National Space Center
CSCS	Contingency Shuttle Crew Support
CSM	Command and Service Module
CWC	Contingency Water Container
DESFLOT	DEStroyer FLOTilla (USN)
DOD	Department Of Defense
DOI	Descent Orbit Insertion
DOM	Orbital maneuvering engines (Russian)
EDS	Emergency Detection System
EGRESS	Emergency Global Rescue Escape and Survival System
ELS	Earth Landing System
EMS	Entry Monitor System
ENCAP	ENcapsulated CAPsule
EO	Emergency separation (Russian)
EOM	End Of Mission
EST	Eastern Summer Time
ET	External Tank
ETR	Eastern Test Range, Florida
EVA	ExtraVehicular Activity
FCO	Flight Control Officer
FCOD	Flight Crew Operations Directorate (JSC)
FD	Flight Day; Flight Director
FDO	Flight Dynamics Officer
FE	Flight Engineer
FIRST	Fabrication of Inflatable Re-entry Structures for Test
FSS	Fixed Service Structure
FTA	Flight Test Article
FTO	Flight Test Office
GE	General Electric
GET	Ground Elapsed Time
GNCS	Guidance Navigation and Control System
IAF	International Astronautical Federation
ICTV	Instrumented Cylindrical Test Vehicle
IVA	IntraVehicular Activity
JSC	Johnson Space Center, Houston, Texas
KBOM	Design Bureau of General Machine Building (Russian)
KSC	Kennedy Space Center, Florida
LC	Launch Complex
LES	Launch Entry Suit; Launch Escape System
LM	Lockheed Martin; Lunar Module
LOK	*Luniy Orbitalny Korabl* (Lunar Orbital Spaceship)

LREE	Lifting RE-Entry
MCC	Mission Control Center
MDF	Mild detonating fuse
MDS	Malfunction Detection System
MECO	Main Engine Cut-Off
MET	Modularized Equipment Transporter
MIT	Mishap Investigation Team
MMT	Mission Management Team
MOL	Manned Orbiting Laboratory
MOOSE	Man Out Of Space Easiest
MOSES	Manned Orbital Shuttle Escape System
MPS	Main Propulsion System
MS	Mission Specialist
MSC	Manned Spacecraft Center, Houston (since 1973 JSC)
MV	Return maneuver (Russian)
MWS	Mini-WorkStation
NAA	North American Aviation (forerunner of NA Rockwell/ Rockwell International)
NACA	National Advisory Committee for Aeronautics
NASA	National Aeronautics and Space Administration
NIIKhSM	Scientific Research Institute of Chemical and Building Machines (Russian)
OBPR	Office of Biological and Physical Research (NASA)
OFT	Orbital Flight Test
OKB	Design bureau (Russian)
OKP	Cosmonaut training program (Russian)
OMS	Orbital Maneuvering System
OPF	Orbiter Processing Facility (KSC)
OPS	Oxygen Purser System
OT	Orbit Trajectory
OVEWG	Orbiter Vehicle Engineering Working Group
PA	Pad Abort
PARD	Pilotless Aircraft Research Division (NASA)
PCM	Pitch Control Motor
PDI	Powered Descent Initiation
PEAP	Personal Egress Air Pack
PGNS	Primary Guidance and Navigation System
PLSS	Portable Life Support System
PM	Propulsion Module
POS	Portable Oxygen System
PPA	Pilot Protective Assembly
PRE	Personal Rescue Enclosure
Prox ops	Proximity operations
PTV	Parachute Test Vehicle
RCC	Reinforced Carbon–Carbon

RCS	Reaction Control System
RMS	Remote Manipulator System
ROCAT	ROcket CATapult
RSLS	Redundant Set Launch Sequencer
RTLS	Return To Launch Site (abort)
SCRAM	Station Crew Return Alternative Module
Scuba	Self-contained underwater breathing apparatus
SLF	Shuttle Landing Facility (KSC)
SOPE	Simulated Off-the-Pad Ejection
SPS	Service Propulsion System
SRB	Solid Rocket Booster
SSME	Space Shuttle Main Engine
STG	Space Task Group
STS	Space Transportation System
TAL	Transoceanic Abort Landing
TEI	Trans-Earth injection
TKS	Transport Supply Ship (Russian)
TMA	Transport, modification, anthropometric
TsPK	Cosmonaut Training Center, Moscow (Soyuz)
TVC	Thrust Vector Control
USAF	United States Air Force
VA	Return Apparatus (Russian)
VAB	Vehicle Assembly Building
VHF	Very High Frequency
VTOL	Vertical Take-Off and Landing
WETF	Weightless Environment Training Facility
WSMR	White Sands Missile Range

Prologue

The idea of a crew stranded in space without hope of rescue is a nightmare not only of space explorers but also of those on the ground: controllers, management, administrations and of course the families. Though this is a tragedy no one would wish to see, it has been the subject of science fiction for decades. Indeed, as feature films expanded their genre into science fiction so did the prospect of a stricken crew with little or no hope of rescue. The gripping storylines held audiences captive as the plot unfolded to a not always happy conclusion. From landmark books and films, such as the 1968 *2001: A Space Odyssey*, to recent films depicting the human exploration of Mars a rescue scene is usually an integral part of the storyline.

In novels, too, space rescue is an ongoing theme. The landmark title was of course *Marooned* by Martin Caidin (1964) which featured a stranded crew in Earth orbit and the efforts to recover them. The original text was written in 1963 at the time of Mercury and Vostok, well before Gemini or Soyuz flew, and technical accuracy was important to the author. The film rights were sold in 1964 and after a five-year development resulted in the 1969 film *Marooned*. The screenplay and story was updated to include the Apollo spacecraft and a crew (call sign Iron Man One) visiting the Orbital Workshop (later Skylab). After some weeks in space the crew were scheduled to come home, but their main engine did not fire and with precious little oxygen aboard the race was on to try to launch a USAF lifting body–type vehicle to assist in their rescue. The story also includes the intervention of a specially launched lone cosmonaut on a Soyuz to try to help the stricken crew. The story of Iron Man One and the rescue of the astronauts is the main theme of the book, which also recorded the stress and pressure of extended spaceflight, international cooperation and the unforgiving environment of space. The year following the release of the feature film, Apollo 13 hit the headlines and space rescue became a reality. Ironically, when Skylab eventually flew in 1973 a problem with the second crew's CSM thrusters almost saw the launch of a rescue crew to bring the astronauts home, but this in the event was not required and they landed safely after 59 days in space.

In 1981 on the eve of the launch of the first Space Shuttle, Columbia, on STS-1 David C. Onley penned *Shuttle: A Shattering Novel of Disaster in Space*. Here again the technology of the day was the format of the story. In Onley's book the Space Shuttle Columbia was to be air-launched by a manned hypersonic jet transport vehicle called *Yorktown*. This idea was originally proposed for the real Shuttle where a large fly-back booster would take the Orbiter to the fringe of space before returning to the launch site, and the Orbiter would continue into space. Changes in the design due to technical difficulties and cost meant that the fly-back booster was replaced by twin solid rocket boosters and an External Tank, but in Onley's book the original idea for the Shuttle became the platform to create a novel of space accident, potential disaster and hopeful rescue. The *Yorktown* suffers a major failure resulting in the only option available where Columbia takes the stricken booster into orbit to save the crew. The story then unfolds on how another Shuttle is launched to rescue the two stranded crews.

Two years later a TV movie starring Lee Majors (of *Six Million Dollar Man* fame) was aired which also looked at a hypersonic aircraft stranded in space and the use of a Shuttle to rescue the crew. Called *Starflight: The Plane that Couldn't Land* (*Starflight One* on video) the "starring" shuttle was once again called Columbia.

In this scenario the *Starflight* hypersonic inaugural flight has a two-hour duration from Los Angeles to Sydney (Australia), and after launch it encountered debris from a communication satellite rocket that developed a fault and exploded and lay in the flight path of the hypersonic aircraft. When the scramjets of the hypersonic aircraft do not clear the debris and damage is suffered resulting in an inability to shut off the engines, *Starflight* heads for orbit, a place it was not designed to operate in. With fuel depleted the aircraft is stuck in orbit. Columbia is launched to rescue some of the passengers, and though the turnaround between flights stretches the imagination by relaunching the Orbiter twice more in a matter of hours, it cannot be launched again before atmospheric drag pulls the hypersonic aircraft to what seems inevitable destruction. So, using atmospheric drag and skip re-entry and protected by a second Orbiter riding a plow wave, the stricken aircraft follows behind and lands safety.

Three years later the real Challenger Shuttle was lost, and, ironically, 20 years after Columbia was the cameo in *Starflight* it was lost in the tragedy of STS-107. Although the idea of a rescue mission was investigated *after* the mission using Atlantis as a model, this hypothetical rescue mission recalled some suggestions made during the 1981 novel and 1983 TV film.

The idea of a (US) National Orbital Rescue Service (NORS) was proposed amongst others by the Martin Company in the mid-1960s. In 1965 Martin proposed a NORS to commence in a prompt and multi-phased manner spearheaded by consultant Earl Cocke, former National Commander of the American Legion. The Martin proposal would be developed over a ten-year period for estimated costs of $ (1965) 50 million per year which included a provisional system which would operate whilst a more permanent system was developed. This was passed to the National Aeronautics and Space Council which in turn forwarded it to Cyrus Vance, the then Secretary of Defence in Washington for Presidential Review. NASA Administrator James Webb felt that it was too early to attempt to develop a practical space rescue infrastructure, though it should be studied for future operations (Welsh, 1965; Vance, 1965).

In briefing Bill Moyers, Special Assistant to the President, Cyrus Vance noted that in light of the Martin proposal , if the USAF Manned Orbiting Laboratory proceeded as then planned then crew safety would far exceed that of earlier programs. The Gemini would be attached and available as a "space lifeboat" with a six-hour period orbital capability before selecting preferred de-orbit and landing sequences. In addition, MOL would have the same design redundancies as Gemini or Apollo, extensive quality testing and the availability of an astronaut abort mode for each phase of the flight.

As for a separate space rescue service (i.e., separate from basic flight hardware), this would only be used if a quick launch capability was possible. In addition rendezvous and docking could be assured under uncertain conditions, as the rescue vehicle was of a higher reliability than the vehicle to be rescued. These techniques were being developed for Gemini, Apollo and MOL; however, until these could be demonstrated, a separate program for space rescue could not proceed with reasonable and genuine objectives.

Cyrus Vance also noted: "It is possible we may strand an astronaut in orbit some day. It is very likely that astronauts will be killed, though stranding them is one of the less likely ways. The nation must expect such a loss of life in the space program. There have been several deaths already in our rocket development. We would be untruthful of we were to present any different image to our citizens."

It was clear that as long as the manned space program evolved at a sufficiently high pace of operation and capability that a space rescue program may be warranted and that the Department of Defense would probably play a larger role in regular operations including rescue operations. Comparisons were made with commercial airlines rescue and commercial ocean traffic rescue operations. In these cases every possible realistic precaution was taken to reduce the probability of catastrophic failures, and to ensure an effective and viable rescue infrastructure is in place to retrieve passengers from any major accident wherever possible. The extensive deployment of aircraft and ships by the DoD deployed across the world for each US manned spaceflight is typical of this practice. Vance also suggested that rescue from submersibles was only possible to (at that time) a depth of 400 feet, which reflected a very small percentage of the ocean and any accident at that depth could result in a similar stranding of a crew to that proposed in space.

Cyrus Vance stated he saw no advantage for a specific study into the idea of a separate space rescue service at that time but acknowledged that the safety of the crew is paramount at all times and that on the military program more investment in crew safety was likely along the lines of other dangerous aircraft testing programs.

Over 40 years later the idea of a Space Rescue Service remains unrealized. The growth and scope of the human spaceflight program has yet to reflect the long-range plans of the 1960s but when, as it surely will, the manned space program blossoms into extensive orbital and lunar operations supporting Martian exploration then perhaps a

Space Rescue Service will be created to support constant space operations on several fronts.

Ever since men and women first ventured out of the atmosphere, space hardware reliability and redundancies, as well as provision for contingency operations, alternative mission plans and survival training have featured in every program and continue to do so today. With the high-profile push for the Moon in Apollo and the openness of the program to the public, NASA conducted several studies into the potential and effects of a Saturn V exploding during its ascent, in full view of the world's media and the expected thousands witnessing the launch. A review of these studies was presented by Dwayne Day (2006) in *Space Review*. As Day highlighted, the Saturn 5 was the largest launch vehicle, had the heaviest load of fuel and had the fewest options for abort, all of which created a possible danger to the public.

The emplacement of LC39 was the result of some of these studies into allaying the effects of an explosion. However, after it was located it was determined that the effect of damage to surrounding areas was less from a Saturn explosion than to the crew using the escape tower to clear the fireball, which was expected to last for 33 s to 39 s and reach 1,370°C with heat radiating some 600 meters. It was from these studies that the design profile of the Apollo launch escape system was determined. Also studied were the Saturn colliding with the launch tower, the operation of the escape system in such a situation and what might affect the normal performance of the launch vehicle. This was determined as a $\pm 15\%$ variation in the hold-down force, 4% deviation in thrust, misalignments of the engines, a change in the vehicle's center of gravity, excessive wind shear, the failure of an engine and an engine "hardover". As the first Saturn V lifted off the pad unmanned in 1967 the onlookers in the launch control center could close protective covers should the vehicle suddenly explode; however, it did not and neither did the other 12 Saturn V vehicles launched over the next six years.

Despite this record of success, studies into what might happen and the resulting effects had to be evaluated and planned for. This is reflected in other aspects of manned spaceflight by provision of escape and rescue on the pad, during ascent in space and in getting back to Earth. It is launch and landing that are the most visible and dramatic times of a flight and options for rescue and recovery depend on circumstances and time. In space the categories of in-flight emergency increase and multiply as the distance from Earth increases. These include the loss of a ferry vehicle, a collision penetration of a pressurized module, extreme solar activity, failure of the attitude control system, separations of vehicle stages or modules, failure of controls and displays, loss of onboard propulsion, the failure of primary power supply, a major structural failure, a fire in the spacecraft, physiological and psychological issues with crewmembers, leaks, loss of communications, electrical failures, the failure of the Environmental Control System and the creation of toxic gases.

To date, providing support against these incidents has been the challenge of every spacecraft design team. In addition, providing an effective method of launch, space and landing escape or rescue is an integral part of ensuring mission success with every launch. The complexity and depth of this support and provision of hardware and procedures is one of the unsung aspects of human spaceflight.

REFERENCES

Memo to the President, May 21, 1965, from E.C. Welsh, Executive Secretary National Aeronautics and Space Council.

Memo for Mr. Bill Moyers, May 29, 1965, Special Assistant to the President, from Cyrus Vance, Secretary of Defence, Washington (declassified September 12, 1979).

Saturn's fury: Effects of a Saturn 5 launch pad explosion, Dwayne Day, *Space Review*, April 3, 2006. Available at *www.thespacereview.com/article/591/1*, last accessed January 14, 2008.

STS-107: Rescue or repair?

1

STS-107: Rescue or repair?

In early 2003 the Space Shuttle Columbia, pride of the fleet and the first to fly in orbit, was lost with all seven crewmembers in a tragic re-entry accident at the end of a highly successful two-week research mission.

During the course of the investigation into the accident, the evidence pointed to a sequence of events that originated during the ascent to orbit and the eventual re-entry into the atmosphere, but which were not evident either to the crew nor the controllers in time to prevent the tragic end to the mission. As the investigation continued, questions were raised about the possibility of a rescue, either by the astronauts themselves or by a second vehicle, had the full extent of the damage been known at the time. Both independent NASA and Accident Board–related studies were conducted, looking at hypothetical rescue scenarios based on the facts gleaned from real flight data. These studies were somewhat redundant, as they occurred after the fact, relied on a number of assumptions and were, as the Report put it, "without regard to political or managerial considerations." This, then, was a paper study about what might have been possible if the full implications of the damage to Columbia had become apparent to the crew and ground teams *early enough in the mission to allow something to have been done about it.*

COLUMBIA'S LAST FLIGHT

The mission of STS-107 was the latest in a series of scientific research missions, using the facilities of the Orbiter to support research and experiments planned for the International Space Station. When the mission was first proposed, it was part of a plan for several Shuttle-based research missions, using the Spacelab or Spacehab modules not only to perform useful scientific research in orbit prior to full scientific facilities becoming available on the Space Station, but also to train scientists and controllers how to prepare experiments and hardware for the Space Station.

Of course, the plans did not anticipate the delays caused by the International Space Station program itself. In fact, STS-107 was delayed 13 times over three years, from its original launch date in the third-quarter of 2000 to its final lift-off in January 2003. Funding restrictions caused the cancellation of a follow-up mission, with its funding instead allocated to STS-107. As the mission approached, it became clear that the flight would be one of the last scientifically based non-ISS Shuttle missions, although its on-orbit success generated renewed calls for a second mission as quickly as possible after STS-107 came home—which of course it never did. For a while after the accident it looked as though STS-107 would be the final non-ISS Shuttle mission, until a combination of NASA, public, political and scientific support salvaged a place on the flight manifest for the fourth Hubble Servicing mission, which is currently planned for the end of 2008 or early 2009.

Crew of STS-107. *Clockwise from bottom center:* Husband, Chawla, Brown, McCool, Anderson, Ramon, Clark.

Shuttle Columbia was chosen for STS-107 as it was capable of flying extended-duration missions, and was structurally strong enough to support flights with the Spacehab or Spacelab modules. STS-107 was manifested with a wide-ranging program of research in life sciences, the physical sciences, space and Earth sciences, and education. There were 70 scientists involved in the research program from across the US, Europe, Canada, Australia, China, Japan and Israel.

The crew for the flight were announced in July 2000. Crew selection followed the now standard procedure of assigning an 'Orbiter crew' (Commander Rick Husband, Pilot William McCool and the ascent and entry Flight Engineer, designated MS2, Kalpana Chawla, who would assist the Commander and Pilot during ascent and entry and monitor the Orbiter's systems during the mission). The "science crew" consisted of Payload Commander Michael Anderson (Mission Specialist 3), along with David Brown (MS1), Laurel Clark (MS4) and Israeli Payload Specialist Ilan Ramon. All seven astronauts would operate the range of scientific experiments and research hardware, working in two shifts. Red Shift comprised Husband, Chawla, Clark and Ramon, while the Blue Shift featured McCool, Brown and Anderson.

As this was purely a research mission, with no special requirements for visiting ISS, rendezvous or docking, payload deployment or retrieval, nor planned Extra Vehicular Activity (EVA), none of these activities featured in the crew's training program. However, as with all crews, a contingency EVA team was trained to cope with unplanned excursions outside, such as manually closing the payload bay doors. For STS-107, Anderson (EV1) and Brown (EV2) trained for these contingencies. McCool trained as Intravehicular Activity (IVA) crewmember, to assist in the preparation, chorography, and post-EVA activities from within the crew module (*Columbia Accident Investigation Report*, I).

Columbia's 28th mission

On 16 January 2003, Columbia was launched on its 28th mission into space, the 113th of the program. Two minutes and seven seconds later, the twin Solid Rocket Boosters separated from the External Tank. Six and one-half minutes after that, the Main Engines shut down normally, followed by the normal separation of the External Tank. A two-minute burn of the Orbital Maneuvering Engines placed Columbia in a 175-mile orbit inclined at 39° to the equator. During the first few hours in orbit, the crew configured the spacecraft for orbital flight and began activating both the Spacehab module and the suite of experiments and research hardware. Their 16-day mission around the Earth had begun.

During the mission, the crew received updates on the status of their vehicle. There were no reports from the crew of anything untoward affecting the performance of the spacecraft or the science program. Analysis of post-launch imagery was being conducted as the flight progressed and it became clear that something (it later turned out to be foam) had fallen from the ET and struck the left wing of Columbia during the early stages of the ascent. On January 17, during Flight Day (FD) 2, and unbeknown to the crew, the flight controllers or anyone else at the time, a small object floated away from the Orbiter. Subsequent analysis after the accident suggested that this object was related to the impact suffered during ascent. The object's detachment was discovered following a review of Air Force space command radar tracking data after the accident.

During FD 8 (January 24), Mission Control in Houston emailed Husband and McCool with the news that foam had struck the wing during the ascent. Flight Control indicated that this was of no concern regarding the insulation tiles or the reinforced carbon–carbon (RCC) located on the leading wing edge. The message also indicated that, as similar events had been recorded on other launches, there was no cause for concern during re-entry. Video of the debris strike was also emailed to the astronauts, who then shared this information with the rest of the crew. The mission progressed as per the flight plan. On FD 13 (January 28), the astronauts and ground controllers observed a moment of silence to honor the memory of the lost crews of Apollo 1 (January 27, 1967) and Challenger (January 28, 1986). No-one could have foreseen that STS-107 would soon be tragically added to this painful roll call.

The early science results coming from the mission looked very promising, and talk of a follow-up mission began to gather momentum. Around the clock each day, the crew performed a variety of chores associated with their research program. Mostly located in the Spacehab module the research included nine commercial payloads encompassing 21 separate investigations. There were also four payloads from the European Space Agency with 14 investigations, 18 payloads from NASA's Office of Biological and Physical Research (OBPR) supporting 23 investigations and another investigation connected to ISS Risk Mitigation. Three additional experiments were located in Columbia's payload bay and another six on a Hitchhiker Pallet. Much of the scientific data from these experiments were downlinked to the ground during the flight. In the recovery of debris from the accident, some of the experiments were found, although samples and data collected for return at the end of the mission were mostly lost.

On February 1, Columbia began her flight home. Over East Texas at about 9 AM EST, some 16 minutes prior to the planned landing at the Kennedy Space Center in Florida, all data and signals from the flight were lost. The mission had lasted 15 days 22 hours and 20 minutes and America and the world would soon mourn the loss of another seven brave pioneers of space exploration.

Columbia is lost

It soon became evident to flight controllers that the vehicle and her crew had perished. Following standard procedures, the control room was locked and data were archived. The effort to preserve data and materials was part of a Contingency Action Plan that had been established following the loss of Challenger in 1986. NASA declared a 'Shuttle Contingency' and senior members of the agency began contacting the astronauts' families, the President and members of Congress. President Bush in turn contacted the President of Israel to inform him of the loss of Illan Ramon. President Bush later addressed the nation with the chilling words: "The Columbia is lost. There are no survivors." By then, NASA Administrator Sean O'Keefe had instigated the International Space Station and Space Shuttle Mishap

Interagency Investigation Board. Within minutes of confirmation of the accident, the Board began coordinating the search for and recovery of any debris that had survived re-entry and the break-up. Reports of debris impacting the ground came flooding in to the 911 Emergency Call service as pieces of Columbia fell out of the sky. A huge search and recovery team was organised, and over the next four months a total of 85,000 pieces of Orbiter debris (representing 38% of the dry weight of the vehicle) were recovered over an area of approximately 2.3 million acres, spread over 2,000 square miles. Shipped to KSC, the debris was examined to identify its location on the Columbia and to determine any damaged areas. The information gathered from the recovered debris, retrieved flight and ground support data, testimonies, and reviews of mission preparations and operations was reviewed by the Investigation Board that worked for seven months following the accident.

The physical cause

According to the findings of the Columbia Accident Board based on their examination of available evidence:

> "... the physical cause of the loss of Columbia and its crew was a breach in the Thermal Protection System on the leading edge of the left wing. The breach was initiated by a piece of insulating foam that separated from the left bipod ramp of the External Tank and struck the wing in the vicinity of the lower half of Reinforced Carbon–Carbon panel 8 at 81.9 seconds after launch. During re-entry, this breach in the Thermal Protection System allowed superheated air to penetrate the leading edge insulation and progressively melt the aluminum structure of the left wing, resulting in a weakening of the structure until increasing aerodynamic forces caused loss of control, failure of the wing and break-up of the Orbiter."

After reviewing the launch video, it was determined that the falling foam was the most likely cause of the initial damage. Radar images of Columbia in orbit and evidence from eye witnesses (including video of debris shedding during the re-entry and break-up) were combined with archived flight data and evidence from recovered hardware. Tests that simulated the impact of foam on the RCC demonstrated that such damage could breach the area sufficiently to duplicate the sequence of events that ended Columbia's last mission so tragically.

A more in-depth account of the loss of Columbia, the events leading up to the accident, its investigation, and the effects on the program, are described in detail in several other titles recorded in the bibliography and is not the aim of this current volume. However, it is planned for inclusion in the revised edition of *Disasters and Accidents in Space*, the companion work to this volume. The focus of this section is to review the options available for rescuing the crew had it become clear how severely damaged Columbia was early enough to mount a rescue mission.

RESCUE OR REPAIR?

During the investigation into the accident, several NASA officials indicated that there was nothing that could have been done to save the crew, even if they had known about the damage. It was decided to investigate that opinion in more depth, so the Investigation Board asked NASA to evaluate hypothetical scenarios in which the damage had been identified early enough in the mission either for the crew to repair the damage or a rescue mission to be mounted. In these scenarios, NASA would be working with the knowledge that there had been damage to the left wing and could focus their studies accordingly. In a real-time situation on STS-107, even if it was known the vehicle had been damaged, the extent and location of such damage may not have been apparent enough to identify a possible course of action in time. Therefore, a significant amount of assumption had to be included in the assessments (*Columbia Accident Investigation Report*, II).

Extending the mission

In order to allow time to conduct a repair or a rescue, it would be essential to determine how long the use of onboard consumables could be extended to make either crew repair or a rescue launch viable. Evaluating the data from the vehicle and the crews' metabolic rate during their mission, it was determined that the maximum on-orbit lifetime (while maintaining acceptable levels of CO_2 concentration) would be 30 days. The limiting factor in this estimate was the availability of lithium hydroxide (LiOH), which is used to remove carbon dioxide from the atmosphere within the crew module. This meant that, in theory, Columbia could have remained safely in orbit up to February 15, two weeks longer than its intended mission. All other consumables, such as oxygen, hydrogen, nitrogen, food and water, and propellant levels, were deemed sufficient to support a mission in excess of 30 days (i.e., beyond February 15). This, of course, assumed that the damage was discovered very early in the mission so that the use of consumables could be stretched to the maximum. It was also determined that an inspection EVA by the crew would have helped to identify the scope of the problem and to decide whether repair or rescue was the better option.

Inspection EVA

In evaluating the two options, NASA had to estimate an alternative timeline for the early days of the mission; one in which the damage was identified early and the crew were able to inspect and report the extent of the damage in sufficient time to consider and implement the chosen option. This "alternative timeline" assumed that on FD 1, images of the ascent would be recorded and sent for analysis. During FD 2, photographic evidence from the launch camera would have indicated that the Orbiter had possibly been damaged during ascent. On FD 3, NASA would have requested "national assets" to examine the Orbiter (DoD radar/satellite imagery). In order to simulate both options to the full, the team decided to conclude that the imagery was too inconclusive to reveal the extent of the damage and they would therefore have initiated a crew contingency EVA to inspect the suspected area.

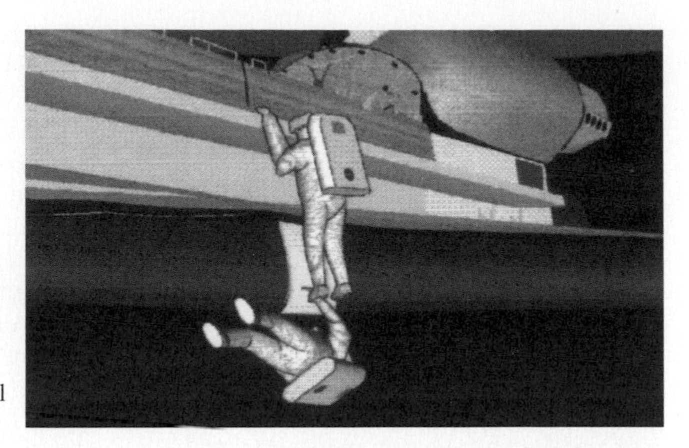

Impression of a hypothetical inspection EVA.

The decision for Anderson (EV1) and Brown (EV2) to perform a contingency EVA would have been made on FD 4, and the crew informed. The astronauts would then have unpacked and prepared their EMUs for the excursion the following day, assisted by McCool (IV). NASA determined that such a contingency EVA on FD 5 was feasible and well within the capability of the crew and equipment. The lack of an RMS system with its integrated cameras aboard STS-107 meant that surveys of the vehicle outside the payload bay would have been more difficult, but not impossible. The team then decided that this inspection would have revealed potentially catastrophic damage to the wing's leading edge (which was thought to be the cause of the eventual loss of the vehicle). This would have led to the decision to instruct the crew to conserve consumables, while round-the-clock processing of a rescue Shuttle mission would have been initiated. At the same time, the possibility of repair would come under evaluation. From FD 6 until FD 26, evaluations and preparations for both the rescue and repair options would have been followed. By FD 26 (February 10), a decision on which option to follow would have to be made to allow time to implement the decision. The other determining factor beyond FD 30 would be the level of oxygen aboard the vehicle. This, it was determined, would be the absolute limit for the crew to either leave Columbia for a rescue vehicle, or to attempt a return to Earth on FD 31 (February 16).

For the contingency inspection EVA, EV1 would have translated along the outer edge of the port payload bay door until reaching the vicinity of RCC Panel 8. From here, the upper surfaces of the wing's leading edge could be visually inspected. If no obvious damage was spotted, it may still have been possible for EV2 to inspect the lower surfaces of the wing. In this situation, the study suggested that EV1 (wearing towels taped to his boots in order to protect the wing's surfaces) could hang on to the door edge and lower his feet on to the wing. His left foot would be placed in front of the wing's leading edge and his right foot on the upper wing surface. There is plenty of room to do this as there is approximately 1.2 m from the edge of the open payload bay door to the upper surface of the wing. Using EV1 as a human bridge, EV2 could then follow his colleague out of the airlock along the wing edge, moving over him and

Representation of a rescue mission launch.

hanging on to visually inspect the lower surfaces of the RCC leading-edge structure and surface area. There would be no detailed digital or photographic recording, as the mission did not carry EVA-compatible equipment. All reports would have had to have been verbal.

There were several safety issues regarding this plan, including the crew having to ensure they had no direct contact with any damaged surfaces due to the risk of sharp edges piercing their pressure suits or of them inadvertently adding further damage to the area. Though both Anderson and Brown were trained in a variety of EVA contingency procedures, there was no specific training to prepare for this type of operation. In addition, as STS-107 had no scheduled EVAs, the amount of support material was extremely limited. With no SAFER devices carried, the possibility of the EV crewmembers becoming separated from the Orbiter was heightened.

As part of the review into this possible activity, a team of two specialist EVA flight controllers and two EVA-experienced astronauts were tasked to assess the idea at the virtual reality lab located at JSC. Their findings revealed that such an operation, though untried and untrained for, was possible. The level of difficulty was classed as moderate, with a low risk of crew injury and a low to moderate risk of further area damage. For the objective of providing conclusive information regarding the severity of the damage, the outcome was judged to be high.

Atlantis to the rescue

If the decision was made instead to rescue the crew of STS-107, the assessment reviewed the possibility of launching a second Shuttle to rendezvous with Columbia and return the crew. Timing was of the essence and the next vehicle in the flow was Atlantis, which was being prepared for STS-114. During FD 4 (January 19) of the STS-107 mission, Atlantis was in the Orbiter Processing Facility being prepared for a launch to ISS on March 1. The twin SRBs and ET had already been mated in the VAB on January 7 and there remained approximately ten days of routine Orbiter processing prior to moving Atlantis over to the VAB. The RMS was not installed and there were no payload elements in the payload bay. The planning for STS-114 processing allowed for a rollover on January 29, with any cargo elements being installed on the pad by February 17.

A review of options by senior government and contractor management at KSC yielded a hypothetical timeline in which to launch Atlantis to rescue the stranded crew. If notification to proceed with the rescue launch had come on FD 5 (January 20), then a round-the-clock processing cycle could have been implemented quickly, while still preserving safety in the processing and ensuring optimum time in space to effect the rescue. After all standard vehicle checks had been completed, a rollout to the VAB would have been possible on January 26 (Columbia FD 11), with a VAB flow of four, rather than the standard five, days. Atlantis would have been moved to the pad to support a launch on February 10 (Columbia FD 26). Several standard tests would not have been completed due to the restricted time span, but their omission was deemed to be, at worst, a low risk level. The weather would of course have been a critical launch factor for the rescue mission, and evaluation of the actual weather data for that period indicated that no launch criteria would have been violated. Other changes implemented would have included updates to flight and mission control software.

Atlantis would have carried the "core" mid-deck stowage for extra LiOH canisters for Columbia. Notebook computers would have been carried for rendezvous and proximity operations. In addition, four EMUs would have been required; two for use by the Atlantis EVA crew and two for the transfers. Two SAFERs and wireless video helmet units would also have been included for the Atlantis crew and two portable foot restraints could have been installed on each side of the rescue vehicle's payload bay. There would also have been a telescopic boom, stowed on the forward bulkhead.

Rescue mission crewing

In order to support the return of seven astronauts from STS-107, Atlantis would have had to carry a minimum crew complement. With rendezvous, proximity operations and EVA tasks involved in the rescue operations, however, that minimum would have been no fewer than four astronauts: a Commander, a Pilot and two Mission Specialists. The two MSs would serve as the rescue EVA crew. The four would fly on the flight deck, with the MS2 serving as Flight Engineer for ascent and entry. The Commander and Pilot would perform the rendezvous and proximity operations, which would have involved about eight or nine hours of manual flying.

For such an intensive and high-pressure mission, it would have been essential to choose a crew that was capable of quickly adapting to the micro-gravity environment (no Space Adaptation Syndrome history). The rescue crew would also have to be selected quickly but be chosen for their past experience in EVA and rendezvous and proximity operations. At the time of STS-107, there were seven Commanders, seven Pilots and nine EVA-experienced astronauts who had such capabilities and experience, so the rescue crew could have been formed quickly using any combination of these.

THE SPACE SHUTTLE RESCUE BALL

During the early 1970s, NASA studied the requirements of the Shuttle Extra Vehicular Activity suit for planned spacewalking operations, as well as an Intra Vehicular Activity garment for use inside the spacecraft. These studies included an Emergency IVA suit concept to protect the crew from rapid decompression, but a combination of budget restrictions, weight issues and available room inside the vehicle meant the concept never proceeded beyond a study. At the end of the decade, as the launch of the first Orbiter loomed closer, NASA began a study for a crew survival system based on IVA and EVA suit developments; one which could be used in the event of emergency evacuation and crew transfer from one Orbiter to another. This consisted of a Portable Oxygen System (POS) and a Personal Rescue Enclosure (PRE), commonly known as the "rescue ball" (Thomas and McMann, 2006).

The fabric enclosure measured 0.86 m in diameter and was fitted with a small, round observation window. The occupant, wearing normal flight overalls and a POS, would enter the sphere by opening the zipper and would be zipped inside by another

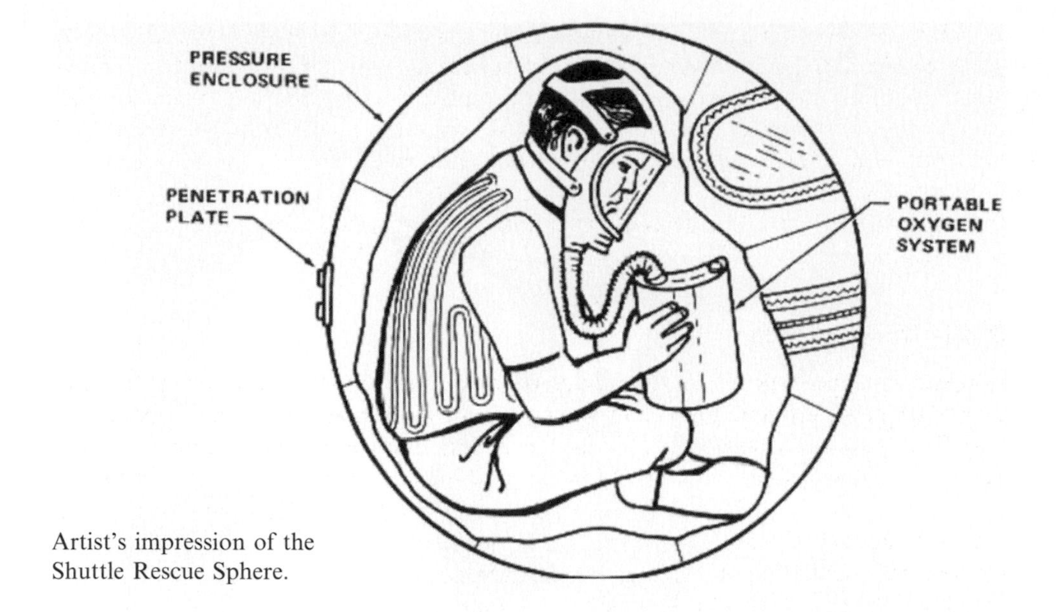

Artist's impression of the Shuttle Rescue Sphere.

crewmember. To ensure the POS could be used to its maximum during the EVA transfer if required, the sphere could be pressurized and supplied with oxygen from the Shuttle. During rescue, the PRE and its occupant would be transferred by an EVA crewmember to a second Shuttle or Space Station. The idea was that, in the event of a Shuttle becoming stranded in orbit, a second Shuttle could be launched and the stranded crew transferred using the EVA/PRE system. They could then return to Earth in the rescue craft or remain on a Space Station pending the arrival of a recovery vehicle.

Up to five of these PRE units could be carried on a Shuttle thanks to its compact design, along with three extra POSs and the two EVA suits that are carried on all missions. An "environmental safe haven" would thus be available for a seven-person crew in the event of an on-orbit emergency. The likely scenario following the successful transfer of the rescued crew would have seen the abandoned Orbiter returned on an unmanned destructive re-entry, if possible.

During 1976 and 1977, further studies were conducted on an alternative idea for rescuing a crewmember from a stranded Orbiter in space. The Anthropomorphic Rescue Garment (ARG) idea was studied by both the David Clark Company and ILC under a NASA contract. This featured a suit-like enclosure, again using a POS, but with a more limited degree of mobility and self-help facilities than the rescue sphere could offer.

Such systems were abandoned during the early years of the program, as the operational realities of the program revealed the true lead time it would take to prepare a second vehicle, launch it and rendezvous with a stricken vehicle. Artistic impressions of what such a rescue system may have looked like were released, but the actual enclosures were never flown on a real mission. Had these been available

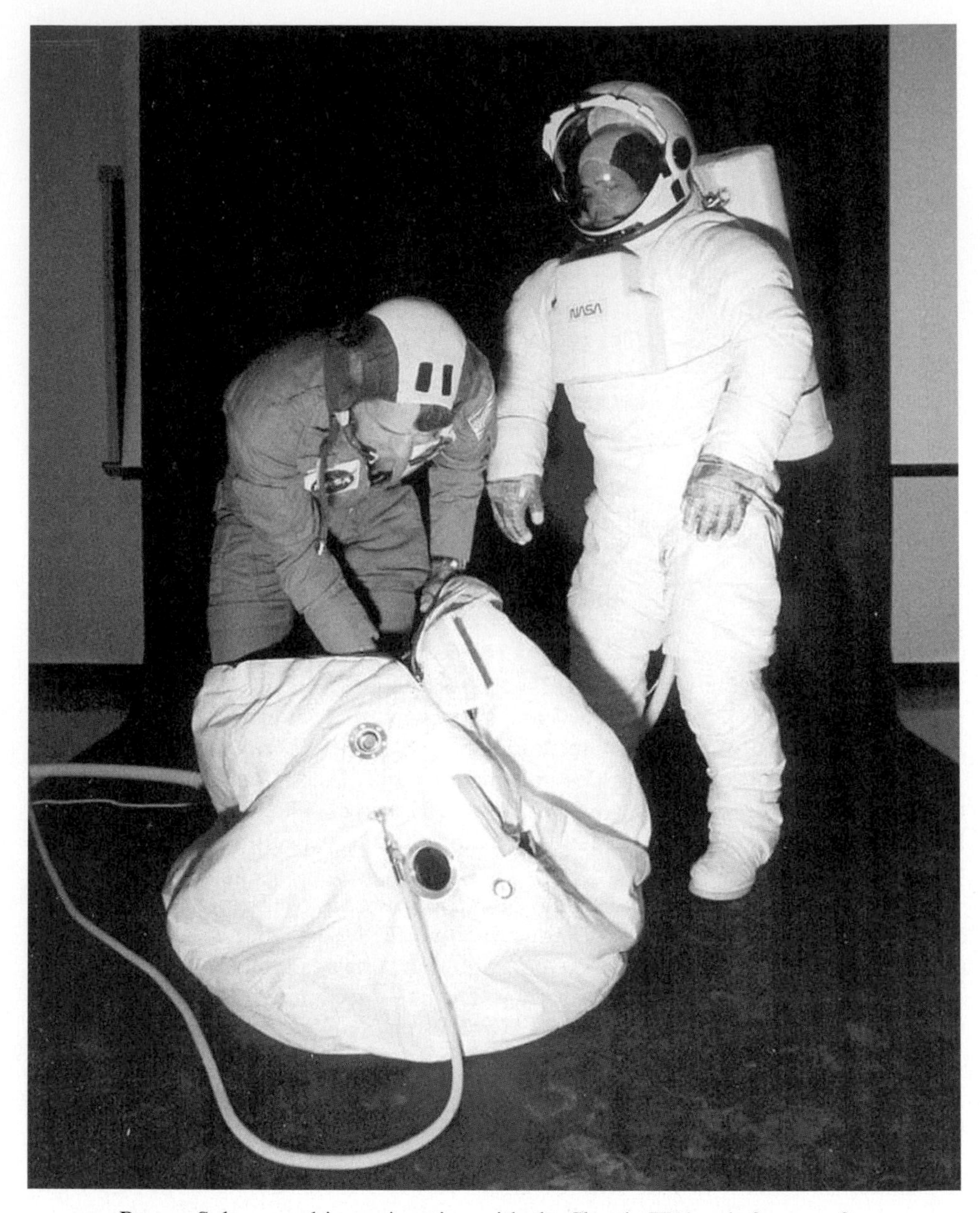

Rescue Sphere used in conjunction with the Shuttle EVA suit for transfer.

operationally, they could certainly have proven useful for the rescue scenarios envisaged during the Columbia investigations. By 2003, however, these methods were over 20 years out of date, and the idea of keeping a Shuttle in a constant state of launch readiness for a potential rescue mission was both expensive and complicated. In the near 30 years of the Shuttle program, there have been two tragic accidents, and

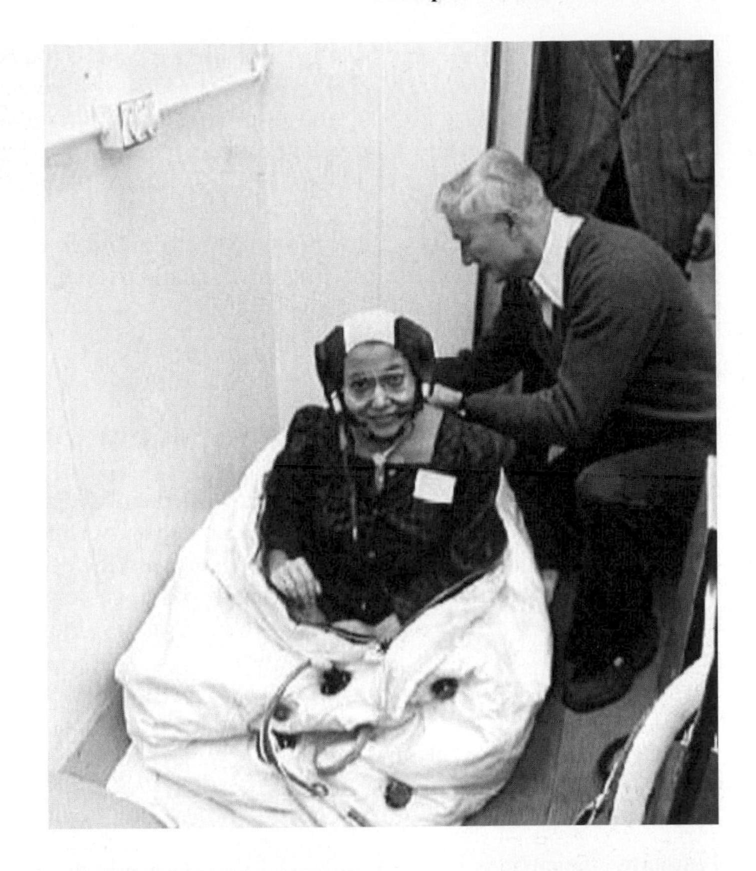

A tight squeeze (top). Confined inside the sphere (bottom).

the rescue vehicle would only have been of use for one of them—providing the need for it had been identified early on. Challenger was lost too early in the mission for any possibility of rescuing the crew, and events from Columbia suggest that although hypothetically a rescue may have been possible, the reality is that it would have taken a combination of mitigating circumstances, a sequence of specific events and a lot of luck to succeed.

Though PREs have never flown on a Shuttle mission, they continue to be used during the selection process for astronaut candidates, in an evaluation of their reaction to confined spaces and stress situations.

HYPOTHETICAL COLUMBIA RESCUE MISSION PROFILE

Any rescue mission for STS-107 would have depended upon reliability, experience and above all timing. The rescue flight would have required rendezvous with Columbia on FD 1. With no training beforehand and the eyes of the world watching, the chosen crew would have had to be able to pull together from the word go, adapt to micro-gravity with no difficulty and perform their duties under the most intense pressure.

Three days before the planned launch of Atlantis, the crew onboard Columbia would have initiated a 22.5 m/s translation maneuver, designed to raise the orbit of their Shuttle to 342 × 250 km. This would give Atlantis a wider range of launch windows in which to mount the rescue mission.

Template comparisons

In the review of a hypothetical rescue mission, it was determined that the ideal launch time for Atlantis would be on February 9, 2003 at 22:05 PM EST, allowing for a rendezvous with Columbia on February 10. Other launch windows were available on February 11 and 12, which would support a rendezvous on February 13 and still allow a margin of 36 hours before the LiOH onboard the stricken Orbiter was depleted.

The restricted time to prepare for and execute the rescue mission resulted in what was termed an "aggregate risk". If everything was processed as planned and equipment worked as designed, then in a best-case scenario a rescue attempt could have been attempted. There was of course risk inherent in shortening the processing time and abbreviating the training of the rescue crew. The review by the accident investigation team determined the levels of such risks as low to moderate.

On a "nominal mission" processing flow, Orbiter processing took about ten days in the OPF before moving to the VAB, not including the scientific or mission-specific payload requirements. In the rescue template, this was abbreviated to seven days. The Vehicle Assembly Building (VAB) flow, usually about 5 days, was reduced to 4 days in the rescue scenario. At the pad, the shortest turnaround had been 14 days, but for the rescue mission, 11 days was all that was allowed for. The risk assessment for the

OPF, VAB and pad flows were listed as moderate, but this depended on no failures hindering the flow cycle.

While the hardware is being processed, the software to support the mission also has to be developed, tested, trained on and implemented. Nominally, this takes up to six months prior to a mission. Flight software for the planned STS-114 mission that would have followed STS-107 was already completed, so the planned 7–8 days of verification for the rescue mission was determined to be low risk. For system integration, a nominal mission profile takes about six months to load, with four to five months for developing and loading thermal profiles, and up to ten months for drawings to be produced. As all work was completed on these for STS-114, the estimated eight days for deltas and verifications for the proposed rescue template was deemed a low risk. The software for Mission Control in Houston was also developed for STS-114 and was also a low-risk assessment for the proposed rescue mission. The Certificate of Flight Readiness nominally takes 12 to 15 weeks prior to a mission, but for the rescue proposal, no more than two weeks would have been available. This was listed as a moderate risk by the investigation board. Additional software "patches" would have to be implemented during ascent to ensure the ET did not impact on inhabited land masses. The ET impact area would have been north of New Zealand's North Island and south of the island of Tonga in the Pacific.

The third element was the composition and preparation of the rescue crew. Nominal mission training was suggested for a Commander/Pilot team. The MS2/FE role was not specified in this assessment, but was probably included. In previous crew assignments, when an Orbiter flight crew was assigned independent to a science crew, this comprised a Commander, a Pilot and an MS2, who were normally named together. Their training as a core unit continued in parallel. Training for an EVA crew (EV1, EV2 and probably IV1), when EVAs were actually planned for a given mission, varied between 40 to 50 weeks for a nominal mission cycle. In other words, a full year of training was required to prepare a Shuttle flight crew for their next mission. In the rescue profile, only two weeks would be available to train the crew. This was listed as moderate risk for the Commander/Pilot/MS2 cycle and moderate to high for the EVA team. Previous experience would therefore be essential, so the crew would probably have included either recently flown or forthcoming crew combinations to take full advantage of either peak training or in-flight experiences. The assignments were of course never suggested or proposed, as the whole idea of mounting an Atlantis rescue flight to Columbia occurred after the event, but either the recently returned STS-113 crewmembers or forthcoming STS-114 crewmembers would most likely have been the prime considerations for any STS-107 rescue mission.

Atlantis rescue mission

The template of the rescue mission was based on two significant assumptions. The first was that the damage to the Columbia had been identified, but in reality of course the damage to the leading edge of the wing was not known, and could not be positively identified with any accuracy. Post-incident investigations, modeling and tests have led to the most likely chain of events that resulted in the fatal damage to the leading edge of the left wing. This was the basis for evaluating whether a rescue or repair may have been mounted had sufficient information been available during Columbia's flight. The review also presumed that this damage could have been detected visually, either by the crew or by "national assets".

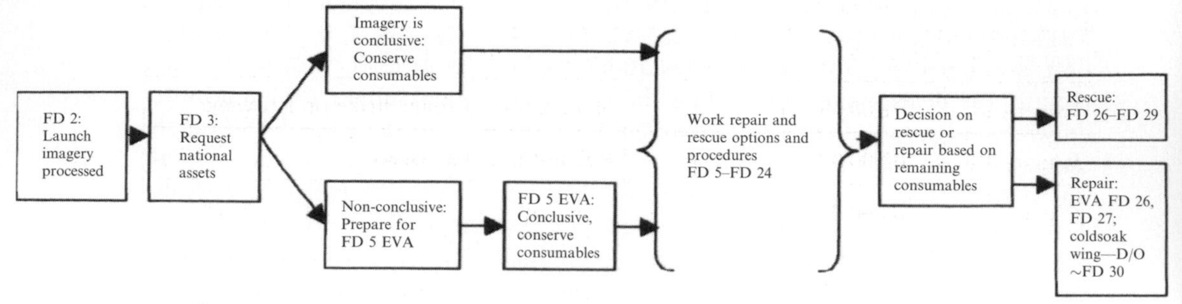

The second assumption was that NASA and the country would be willing to launch a second Shuttle to rescue the Columbia, knowing full well that the damage had been caused during the launch process. There would be no time to redesign the area to prevent a repeat of the STS-107 launch incident before the rescue mission itself had to lift off. Such work was completed in the months following the loss of Columbia and prior to the launch of STS-114 in 2005. Ground and onboard inspections of the leading edge of the Orbiter wing and underside thermal protection are now a feature of each mission and some facilities for added inspection and possible repair have been included in the profile of each mission flown since 2003. If Atlantis had been launched to rescue the Columbia crew, it would have done so without such safeguards and under perhaps the highest risk of the entire scenario. Launch safety would likely have been a telling factor in the discussion about whether a rescue mission was viable at all. The rescue mission template did include an EVA inspection of the leading edge of the Atlantis wing area to see if similar damage had been sustained by the second vehicle. But if it had, then that would probably have been a far more fatal blow to the American space program than the loss of the Columbia crew was. The possibility of both the Columbia crew and the rescue crew being stranded in orbit would have been too great for the program to bear.

The decision to launch the rescue mission would also have to factor in a night launch (thus restricting ascent imagery, a factor in limiting the first return to flight missions to daytime launches from 2005). The processing flow and training would have to be done quicker than ever before and far outside demonstrated templates, and many techniques and procedures would have seen their first use during the flight. The risk to a second vehicle and crew would have to be accepted, by the agency, the government and the public (as well as the rescue astronauts and their families), before the incident to the first vehicle had been fully investigated or understood. As the template indicated: "The timing of decisions and the information for their basis is critical and highly optimistic to allow a rescue mission to have been mounted, both in time to save the Columbia crew and safely ensure the return of the rescue crew, whether with the Columbia astronauts or not."

Table 1.1. Timeline of a hypothetical rescue mission.

Calendar date	*Columbia—mission elapsed time (MET) at 10:39* AM	*Columbia flight day*	*Events*
January 16	00/00:00	1	Columbia launch—10:39 AM EST
January 17	01/00:00	2	Notification of foam strike on left wing
January 18	02/00:00	3	Request national assets
January 19	03/00:00	4	Plan inspection EVA; notify KSC to begin processing Atlantis
January 20	04/00:00	5	Perform inspection EVA; major powerdown begins; LiOH conservation
January 21	05/00:00	6	
January 22	06/00:00	7	Last day to notify KSC for vehicle processing (to make 2/14 7:40 PM FD 1 rendezvous window)
January 23	07/00:00	8	
January 24	08/00:00	9	
January 25	09/00:00	10	
January 26	10/00:00	11	Atlantis rollover; OPF to VAB
January 27	11/00:00	12	
January 28	12/00:00	13	
January 29	13/00:00	14	
January 30	14/00:00	15	Atlantis rollout; VAB to pad
January 31	15/00:00	16	
February 1	16/00:00	17	
February 2	17/00:00	18	
February 3	18/00:00	19	
February 4	19/00:00	20	
February 5	20/00:00	21	

(*continued*)

Table 1.1 (*cont.*)

Calendar date	*Columbia—mission elapsed time (MET) at 10:39* AM	*Columbia flight day*	*Events*
February 6	21/00:00	22	
February 7	22/00:00	23	
February 8	23/00:00	24	
February 9	24/00:00	25	First launch window—11:09 PM
February 10	25/00:00	26	Second launch window—10:40 PM
February 11	26/00:00	27	Third launch window—10:05 PM EST; rendezvous on February 13
February 12	27/00:00	28	
February 13	28/00:00	29	
February 14	29/00:00	30	Last FD 1 rendezvous window 8:40 PM EST
February 15	30/00:00	31	LiOH depleted—morning

Courtesy CAIB.

When Atlantis reached orbit, the crew would have flown a standard rendezvous profile, with a +R bar approach profile in which Atlantis would approach from below the Columbia. Such an approach would give a much longer period for proximity operations to be flown between the two vehicles, and would be easier to maintain. Thus, for an experienced rendezvous commander, minimal training would be required. This type of approach had been used for the Shuttle–Mir docking missions (1995–1998), and the ISS missions from STS-88 in 1998 up to STS-102 in 2001. With the active Atlantis approaching the passive Columbia from below, the Earth would be behind the Atlantis crew and would therefore not affect the visibility of approach and proximity operations as the rescue vehicle closed in.

The activities at Columbia (called Proximity Operations) were deemed 'straight-forward' by the template documentation, though it was acknowledged that any activity would have been at an unprecedented duration. Onboard Columbia, Husband and McCool would have orientated their vehicle wing forward, with the payload bay facing the Earth and under active attitude control mode. Meanwhile, the Commander and Pilot of Atlantis would have orientated their Orbiter with its nose forward and payload bay facing that of Columbia. This would have resulted in a "90 degree clocking" of the two Orbiters, allowing for close approach without the added hazard of one vertical tail impacting the other as two vehicles closed in.

Experience from Mir and ISS missions had shown that two large vehicles could be flown very close to each other and held at a distance of about 3 m. In addition, previous flights had demonstrated that it was possible to capture payloads either by EVA crewmembers or by RMS operations. On ISS flights, the Commander normally holds the Orbiter at a distance of about 9 m, to allow for dampening any rotational and positional errors prior to closing in for docking. The Shuttle program therefore had several examples to draw upon of an Orbiter flying close to another large object or keeping station there for several minutes before grappling the target with the RMS. For the template rescue profile, however, there would have been a significant difference to the proximity operations. It was estimated that between 8–9 hours would be required to fly the vehicles in "Prox Ops", far longer than ever attempted before, and an additional strain on the concentration of the Atlantis "Orbiter crew". This was one of the reasons for choosing a four-person rescue crew. The MSs would be able to assist in at least the early rendezvous and proximity operations before committing themselves to the rescue EVA.

In the rescue scenario investigated by the Columbia Review Board, a retro reflector would have been installed on the top of the Spacehab module of Columbia during the first rescue EVA operation. The Trajectory Control System installed on the Atlantis would have been used with state-of-the-art rendezvous tools to help in the day/night/day cycles. It was also suggested that as soon as some of the Columbia crew had been transferred to Atlantis, they might have been able to help with the long station-keeping exercise until the transfer had been completed. This would depend on their physical (and physiological) condition.

Rescue EVA

It was determined that the rescue EVA crew would have followed a 10.2 psi cabin pressure pre-breathing protocol, possibly starting at their ingress into the Orbiter on the pad to maximize their pre-breathe time and to initiate EVA preparations as soon as Atlantis was safely in orbit. Onboard Columbia, the two EMUs would be prepared for the transfer and sized to fit the first two crewmembers (CM1 and CM2) to save time in the actual transfer operation. As Atlantis approached, two of the Columbia crew would already be inside the airlock in the EMUs, ready for depressurization. At 6 m separation of the payload bay sills of the two Orbiters, the rescue EVA could begin, with the Atlantis EVA crew preparing as much of the equipment and completing as many pre-exit procedures as possible prior to station keeping.

With both airlocks depressurized, the EVA would begin, with the initial priority given to transferring replacement LiOH to Columbia. EV2 would have attached a portable foot restraint on the payload bay sill of Atlantis. Then, using the EVA boom to increase their reach, EV2 would have assisted EV1 in transferring the additional LiOH canisters and two further EMUs to Columbia. EV1 would assist CM1 and CM2 to exit from the airlock of Columbia and would then have placed two spare EMUs and the LiOH canisters inside the now vacant airlock before closing the outer hatch. The re-pressurization of Columbia's airlock would begin as soon as possible and once the inner hatch was opened, the next two crewmembers (CM3 and CM4)

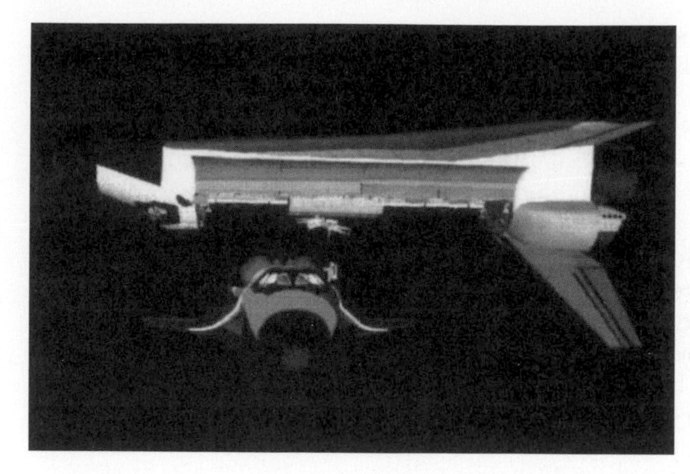

Proximity operations between two Orbiters.

would have donned the new EMUs and entered the airlock. The transfer sequence of Columbia's crew would probably have been determined beforehand, so that the sparc EMUs could be pre-sized to accommodate CM3 and CM4 as far as possible to save precious time.

Meanwhile, CM1 and CM2 would be transferred to the airlock of Atlantis using the extension boom, accompanied by EV1. After the first transfer, EV1 and EV2

Hypothetical rescue EVA operations.

Table 1.2. EVA transfer sequence.

Sequence	*Atlantis*	*Columbia*
A	EV1 and EV2 exit and transfer to OV-102 with two spare EMUs	CM1 and CM2 in STS-107 EMUs in OV-102 airlock
B	CM1 and CM2 EVA transfer to airlock of OV-104	CM3 and CM4 don spare EMUs from Atlantis
C	CM3 and CM4 EVA transfer to empty airlock of OV-104 the original EMUs from Columbia returned by Atlantis EV crew	CM5 dons one of the original OV-102 EMUs
D	CM5 transfers to OV-104; one spare EMU transferred to OV-102	CM6 and CM7 don one OV-102 EMU and a spare EMU
E	CM6 and CM7 transfer to OV-104	
F	EV1 and EV2 return to OV-104 airlock	

would conduct a SAFER inspection of the thermal protection system of Atlantis and install a portable TCS laser reflector on Columbia for any subsequent rendezvous operations. They may also have conducted additional inspections and photo-documented the damage to the Columbia if safe and prudent to do so. The first two members of the Columbia crew would be taking off their suits, now safely in Atlantis, and preparing for the next transfer. The next transfer would see CM3 and CM4 move across to Atlantis, assisted by one of the Atlantis EV crewmembers, while the other EV crewmember moved the now vacant Columbia EMU suits across to be used by the next transfer crewmembers. In any transfer of empty EMUs, they would have to be powered up and pressurized to prevent the water coolant from freezing up, adding to the difficulty in manipulating the units. With only some of the Columbia crew being EVA-trained, the Atlantis EV crew would have to act as primary EVA crewmembers for all activities.

During the third transfer, it was envisaged that only one crewmember should be rescued, leaving the last two onboard Columbia to assist each other in suit-up and departure for the final transfer. Though a nominal standard EVA pre-breathe protocol would be used by the astronauts onboard Columbia as far as possible, it would probably have to be modified due to the time constraints, providing this was approved by the flight surgeons.

The donning of suits by CM6 and CM7 would have been difficult, as no IV crewmember would be present. It was suggested that using Columbia suits for this operation would have been the best option, and the donning techniques would have been similar to those developed for the first Shuttle flights, which carried only two crewmembers.

The rescue mission profile anticipated that it would actually require two separate EVAs, unless everything went exceptionally well and pre-breathing times were reduced as far and as safely as possible. The complete duration for the EVA rescue operation was expected be 8.5 to 9 hours, pushing the safe limits of a single EVA operation by the Atlantis crew. This was not entirely new ground, however. During STS-49 in 1992, the efforts to capture the Intelsat VI satellite included an 8-hour 29-minute EVA, the first three-person EVA in history. The longest Shuttle-based EVA set a new world record and occurred during the STS-102 mission to ISS in 2001. It lasted 8 hours 56 minutes. Neither of these involved the difficulties inherent in the STS-107 rescue EVAs.

Atlantis return and Columbia disposal

One factor of the rescue that had to be taken into consideration was calculation of the return weight and centre of gravity of the rescue vehicle. Atlantis would come home with 11 crewmembers, six EMUs and the core mid-deck stowage. It is likely that little or no data and items from Columbia's science program would be returned with the Atlantis in order to save weight. The estimated mass of Atlantis would be 94,873.6 kg with the center of gravity 27,462.4 mm, well within certification requirements. There would be no need for OMS or RCS ballasting and enough propellant would be available for the Atlantis crew to employ nominal de-orbit targeting profiles.

The final crewmember of Columbia (presumably Commander Husband) would have to configure a small number of switches prior to departure, to allow controllers in Houston's Mission Control to command Columbia's de-orbit burn automatically. Though a number of operations can be performed remotely by MCC, they do not have the capability to land the vehicle. Processes such as starting the APUs, deploying air data probes or lowering the landing gear and "steering" the vehicle towards a prescribed landing target would be beyond their control, so Columbia would have followed a destructive entry path over the South Pacific.

Both the OMS and RCS systems would be pressurized. The OMS engines would have been armed and the onboard computer systems would have been configured by the crewmember to allow flight controllers to command the necessary actions to take Columbia out of orbit and burn up, along with all its scientific instruments, Spacehab, experiments and most of the results. A valuable and historic vehicle would have been lost, but a successful rescue mission would have saved the crew and NASA would once again recall the heady days of sending men to the Moon and returning the Apollo 13 crew to Earth. At least on paper and in theory.

If such a mission could actually have been staged and successfully completed, a number of notable firsts would have been written into the pages of space history:

1. An inspection EVA, if attempted, would have seen the first excursion in the wing area of the Orbiter. There would have been unknown communication issues, and the crew would have had to tether around Freon panels and translate along the outer edge of the open payload bay door without the benefit of EVA handrails or pre-inspection of any sharp edges.

2. A rescue EVA would have seen the two Atlantis EVA astronauts completely isolated outside the spacecraft during the period between the EVA transfers of Columbia's crewmembers, with both EVA airlocks sealed. Had a problem occurred with one of the Atlantis crew EMUs then a contingency operation would have been very difficult, if not impossible, to complete.
3. In the mission profile, a record 11-person return on one Orbiter (with not all crewmembers in seats) would have been completed. The ground command of a de-orbit burn and unmanned re-entry and destruction of an Orbiter would be necessary. In addition, extended Prox Ops of around 9–10 hours would be required, as well as the first time two manned Shuttles would have been in orbit at the same time.

Though each element was possible in a best-case scenario, all operations would have required 100% success rates, and would have to be carried out on time and without added incidents. This was a major assumption of the profile evaluation.

A REPAIR EVA OPTION?

This option was a far more hypothetical suggestion and depended on an inspection to reveal the damage to the wing leading edge. The actual repair would have had to withstand all the problems of re-entry and allow the crew to control the vehicle to an altitude where they could bail out using onboard equipment. In this scenario, a second Shuttle would not have been launched to Columbia.

One of the major considerations for the repair option was the assumption that adequate materials could be found onboard Columbia with which to repair the damage to the wing. In the evaluation of the investigation board, three categories of materials were considered.

The materials considered had to be capable of surviving re-entry heating and stress levels, and to seal the damaged area sufficiently for the vehicle to complete a safe re-entry. The first material identified for this was the Orbiter's thermal protection system, which would have been used by taking components from other, less critical areas of the vehicle. This would have included RCC to replace that lost or damaged on the wing, and this presented the first problem, as this material was beyond the areas accessible to the crew by EVA. Additionally, the TPS on the less critical areas of the vehicle simply could not survive the searing temperatures encountered by the leading wing edges of the Shuttle during atmospheric entry.

An alternative considered was to temporarily interrupt the flow of hot gases around the area of the wing spar. Several TPS materials were considered, but were soon rejected due to their low thermal mass. Aluminum could have been inserted and there were sufficient quantities of the metal available, but the problem here would have been how to effectively secure it so that it retained its position during entry. Consideration was also given to various non-essential materials, which may have been used to temporarily seal off the wing's damage enough to allow the crew to evacuate the vehicle in mid-air and descend by parachute. Using adhesive material to bond the repair materials to survive re-entry was also considered, though it was found that none of the adhesive material on Columbia could have survived the levels of re-entry heating.

Limited options

An effective seal of the damaged area might have provided the protection needed to get Columbia home safely. To achieve this would have necessitated finding material capable of surviving the re-entry heat and stress and some method of retaining it in place. Since no adhesives were available to stick the repair in place, only a tight friction fit would have been sufficient for the purpose, although nothing was identified as capable of restraining a loose tile or filling a hole in the RCC panel. A reshaped tile may have been friction-fitted into the gap, however.

Other options considered focused on interrupting the flow of the hot gases across the wing spar. The difficulties with this method again came down to choosing the correct type of material to withstand the heat and stress of re-entry, keeping it there for the duration of re-entry, and getting at the area to repair it in the first place. Stowage bags inserted into the hole, with the mouth facing outwards, would allow items to be stuffed inside to expand the bag and thus restrain it against the spar beams on the edge of each RCC panel. Titanium or other suitable metals salvaged from inside the Orbiter, it was suggested, could have kept the expanded bag in place inside Panel 8 at least until the bag burned through.

Another investigation focused on using water ice to disrupt the flow of hot gases. Onboard Columbia was enough hose to stretch from a test port in the airlock to RCC Panel 8. It could have been used to fill a Contingency Water Container (CWC) placed into the area in a similar fashion to the stowage bag proposal. There were up to four CWCs carried on Columbia and some, if not all, could have been inserted, filled with water and sealed. Over a period of 3–6 days, the water would have turned to ice, forming a solid blockage in the damaged area. If free water had been sprayed inside Panel 8, some method would have had to be found to keep it there until it froze, a much less controllable method than the bagged water proposal.

Of all these options, by far the most suitable would have been to harvest TPS tile fragments and bring them inside the Orbiter for reshaping. A second EVA would then have been conducted to push the material into the gaps. Demonstrations of this idea on the ground revealed that although the tiles could be reshaped and force-fitted into the gap, the number needed and the accuracy of shaping required would have made it very difficult for the crew to control or determine the success of their labors. The investigation team did not conduct any testing to determine how much friction was required to restrain a tile in place, nor how large a gap could be filled this way. They did determine that for a 15.2 cm hole, a combination of titanium and water ice could fill the cavity between the spar and panel and could then be sealed in with AFRSI.

In order to complete the repair, Columbia would have been placed in a cold soak attitude, allowing the structural temperature to drop by some 18°C over the next three to six days, prior to re-entry. On a nominal mission, this type of profile would not be considered, as it would seriously impact on other systems in the area, such as the main landing gear, the aerodynamic surfaces and the payload bay doors. But since the crew would have bailed out rather than landing on the ground, the condition and temperatures of the tires and landing gear would not have been critical. Studies indicated that cold soak alone would not have been successful in maintaining the structural integrity of the wing, but combined with the repair options, it may have held the vehicle together long enough to reach a point where the crew could bail out.

Significant work on entry options was conducted by Flight Director Leroy Cain, who presented his Entry Options Tiger Team report to the Orbiter Vehicle Engineering Working Group (OVEWG) of the Columbia Investigation Board on April 22, 2003. The Tiger Team evaluated the effectiveness of jettisoning a considerable amount of payload bay cargo and cold-soaking the wing as a possible method of decreasing re-entry heating and thus the overall heating load. The challenges in achieving this would have included difficult EVAs involving cutting power and fluid cables, cutting and separating the Spacelab transfer tunnel and handling large-mass objects. The difficulty of these tasks was not specifically determined, but it was concluded that jettisoning whatever could be jettisoned during the two repair EVAs could have helped to decrease the entry heat loads. The Tiger Team also looked at either changing the angle of attack to 45° instead of the nominal 40°, or following a lower drag profile. It was clear that while either method would reduce the heating load on the left wing, it would increase the load on other area of the TPS. Analysis of the temperature decrease suggested a reduction of over 148°C. This alone would not be sufficient, but combined with other options would have considerably reduced the stress and heat on the damaged area. An additional software patch would have to have been added to the entry guidance software.

The success of each of these proposals depended upon the accuracy of the repair and the duration that it would delay spar heating and burn through. It was impossible to determine this delay with sufficient accuracy to predict that any of the methods would have allowed Columbia to survive long enough for the crew to bail out. Many problems were identified, adding to the uncertainty of the repair's effectiveness. These included the size of gaps between inserted tiles, keeping the tiles in place, material redistribution within the RCC cavity, and the shifting of material once the hot gas entered the area and melted the ice.

It was not a prospect on which anyone would have liked to gamble with the lives of Columbia's crew. Perhaps a combination of repair, cold-soaking the wing, de-orbiting from the lowest perigee, jettisoning available cargo bay payloads and flying a different angle of attack would have sustained Columbia's integrity long enough to allow the crew to bail out. But there were simply too many variables and too few certainties.

The option of bailout in respect of Columbia is discussed from p. 329 where the options for crew escape from a returning Shuttle are detailed.

Repair EVA techniques

Any attempt to repair the left wing would have required two separate EVAs. The first would have focused on gathering the material required and preparing equipment for the second EVA. That second excursion would be the repair procedure itself.

The first EVA was assessed to have a moderate to high degree of difficulty, due to the type and location of materials to be collected and how hard it would be to retrieve them successfully. Between the EVAs, the crew would modify the mid-deck ladder as an aid for the second EVA. To protect the Orbiter's wing from direct contact and additional damage, towels or foam would be wrapped round the top of the ladder and a Mini Workstation (MWS) attached to the upper rung for use by EV1 during the repair.

For the second EVA, the crewmen would transport the ladder and any other tools and equipment required along the port payload bay door. At the worksite, they would have to secure the ladder by inverting it, with the protection against the leading edge of the wing. Using EVA retention devices, the ladder would then be secured against the payload bay door and then tensioned to pull it gently against the leading edge and ensure its rigidity. EV1 would then translate along the ladder, assisted by EV2, and then attach his EMU to the MWS, firmly securing himself to the ladder close to the worksite. With equipment and tools located near the worksite, EV1 would be firmly stabilized to begin the repairs.

The evaluation of the repair EVA determined it to be high risk. The risk to the crew themselves would probably have been only moderate, but the risk of further damaging the wing while repairing the original breach was high. The success of the operation was deemed moderate to low, and depended upon the extent of the damage and the effectiveness of the techniques used at the time.

ALTERNATIVE OPTIONS

In addition to the EVA and rescue mission options, the review considered alternatives. These included regeneration of LiOH, the use of other vehicles to re-supply Columbia, or using ISS as a safe haven until an Atlantis rescue mission could be launched at a later date.

LiOH regeneration

To test whether the LiOH supply onboard a Shuttle could be extended without re-supply, research was carried out at Ames Research Center. This research was conducted following the loss of Columbia and was intended to determine whether changes of temperature could supply additional LiOH and thus extend the orbital life of the vehicle. When LiOH is exposed to CO_2 it turns into lithium carbonate (Li_2CO_3). The research at Ames revealed that at high temperature (677°C) and using a vacuum, Li_2CO_3 could be converted into LiO. Further research continued using low temperatures. As the maximum temperature of any part of the payload bay of an

Orbiter is 121°C, they theorized that extended exposure to a vacuum could convert some of the Li_2CO_3 to LiO, which in turn could be hydrated to create LiOH. The tests, aimed at future contingency options, suggested that taking LiOH canisters by EVA into a warmer area of the payload bay might provide additional LiOH.

Alternative vehicles

There was also some discussion over using other launch systems to send supplies to Columbia to extend their orbital duration, thus extending the time available to mount a rescue mission. As Columbia was orbiting at a 39° orbital inclination, launch sites situated outside this region would not be able to reach the stricken Orbiter. This ruled out the use of Soyuz Progress vehicles from Baikonur.

On 15 February, an Ariane 4 successfully launched an Intelsat Comsat from French Guiana in South America. Could the vehicle have been used to send supplies to Columbia instead? The investigation board concluded that this would have been possible, but it would have been very difficult to coordinate and operate in the three-week time frame dictated by the constraints onboard Columbia. The challenge would have included developing a new supply kit and attaching it to a new housing (which would have had to be built), developing a separation system and reprogramming the Ariane vehicle. Exactly how the crew would have rendezvoused and transferred the supplies was not immediately clear. Though considered unlikely for Columbia, such an option was pursued as a possible alternative after the publication of the Columbia Accident Board findings.

As for reaching ISS, with Columbia in a 39° inclination orbit, the Orbiter could not be maneuvered sufficiently to reach the 51.6° inclination of ISS, as it lacked significant translational capability. Columbia had only 136.5 m/s of available propellant and the maneuver to ISS would have required about 3,840 m/s capability. Even if Columbia could have rendezvoused with ISS, it did not have the option of docking, so the crew would still have had to transfer by multiple EVA using EMUs. The Russian Orlan suits were not compatible with the Shuttle airlock. Supplies and equipment could have been sent to ISS by Progress or Soyuz and the crew could have been recovered by Soyuz. Rendezvous and proximity operations with ISS would have given the station crew the chance to evaluate the damage to Columbia and once that was known, it is unlikely that a second Shuttle would have been launched to return the STS-107 crew for fear of a recurrence of the problem. That being the case, it would have required four one-man rescue Soyuz launches to ISS in order to return the seven Columbia crew and leave the ISS resident crew with their own Soyuz rescue vehicle. The logistics and hardware availability for such an operation would have been questionable in light of the actual difficulties encountered in re-supplying ISS during the Shuttle's downtime between 2003 and 2005.

REALITY CHECK

The Columbia Accident Investigation Board (CAIB) had to make two large assumptions in generating the repair and rescue proposals. First, that undisputed proof was available to NASA's management that a catastrophic breach would be discovered by the third or fourth day of flight. There was no such evidence. Second, NASA had to fully commit a second Orbiter and crew to a rescue flight without clear proof that it would not experience the same fate, and that was extremely unlikely. The scenarios set out in the repair and rescue options were made with the benefit of hindsight after the accident. So although the investigation suggested that repairs and/or rescue was technically and theoretically feasible, all the options involved extraordinary difficulty and great risk and held no firm guarantee of total success (i.e., the safe return of the seven astronauts).

The evidence of the foam strike came on the day following the launch during reviews of long-range-tracking camera footage. An engineering analysis was ordered to help determine the severity of the event. This concluded that it had probably not caused catastrophic damage and therefore requests for more detailed imagery to examine the impact area on Columbia while in orbit were halted. Shuttle program manager Ronald Dittemore stated on the day of the accident that managers did not ask for additional imagery, as past experience had shown that the images probably would not have helped much. There was also little the crew could have done if a major problem had been discovered. When the door from the drag parachute compartment at the base of the vertical tail fell off during the launch of STS-95 in 1998, pictures were analyzed to see what remained to affect thermal heating loads. But the pictures were not that clear and did not provide much more information. From the data the managers received, it appeared that the technical analysis was sufficient and there was not much more that could have been done, so again further imagery was not requested or pursued.

The CAIB commissioned the repair/rescue study after several board members became concerned that repeated comments from NASA, as well as contractual engineers and program managers, had indicated that there was nothing that could have been done even if the extent of the breach was known. Despite the availability of photos and data from the foam strike, there was a debate about whether a rescue of the crew could have been mounted or not. This continued throughout the CAIB investigation. Studies indicated that even after off-loading over 15 tons of equipment, Columbia would not have survived. The CAIB investigations examined whether a repair or a rescue mission could have been mounted and what effect such activities may have had on the end result.

In today's space program, the cutting edge engineering and pioneering spirit that was a feature of the early years of the space age have been somewhat blunted by bureaucracy, health and safety, and budgets. The idea that Atlantis could have rescued the crew of Columbia depends upon several "what ifs" and a sequence of events based on the most suitable circumstances. If any one of these failed or simply did not exist, then the whole idea of rescue and repair would have been in jeopardy. Ruling out any form of rescue or repair sealed the fate of Columbia. This, however, was a different situation from Apollo 13, and a different NASA, operating in a different era.

It appeared, from the data supplied to the board during the investigation, that the option was a rescue mission rather than a repair. But if sufficient information was available about the extent of the damage, perhaps a more refined decision could have been made as to whether a repair or rescue was in fact possible. Whether either would have succeeded, we will never know. Both options were technically possible and both had their risks and unknowns. If the damage had been investigated further and found to be extensive, then the fate of the crew would probably have been the same, even if rescue or repair options had been evaluated in light of more accurate information.

Whatever the future, analysis of the Columbia tragedy, along with other fatal setbacks in human space exploration, has proven once again that the provision of an immediate and effective crew escape system is critical to every mission, no matter how advanced and routine access to space becomes and how far we venture into the cosmos. This provision for crew safety and escape has been debated since humans first looked to venture into the higher reaches of the atmosphere and beyond.

REFERENCES

Columbia Accident Investigation Report, Volume I, August 2003.

Columbia Accident Investigation Report, Volume II, October 2003, in particular Appendix D.13 STS-107 In-Flight Options Assessment pp. 391–412.

Kenneth S. Thomas and Harold J. McMann (2006). *US Spacesuits.* Springer/Praxis, Chichester, UK, pp. 36–38.

2

Space: A final frontier

For over a century, the human exploration of space has occupied the minds of visionaries and consumed the working life of a select band of very talented individuals. Between them, they have managed to turn theoretical proposals into scientific fact and created landmarks in the pages of history. Alongside the desire to explore the cosmos came the realization that such a great leap into the unknown would require methods and systems for the protection and rescue of the pioneering explorers. This goal of providing a reliable and effective method of space rescue has been constantly investigated and developed over the last 50 years. It has not just been the preserve of the designers, engineers or even the spaceflight crews either. It has also challenged inventors, medical specialists in rescue systems and even the hallowed halls of politics and law.

Space rescue is still one of the key aspects of human space exploration, one that remains, almost 50 years after the first humans ventured beyond our atmosphere, a challenge that has yet to be fully understood or met with complete guarantee of success. Escape options for the crew have been a consideration for every designer of human spacecraft, but implementing an effective system invariably requires a balance between engineering design, limitations in financial resources and the laws of physics. Many systems that have progressed beyond the drawing board to operational potential have remained unused, either due to the significant work and research required to ensure that the system works first time and every time, or simply due to a lack of time to implement effective rescue options. Several times, the safe recovery of a crew has depended upon contingency planning, sheer good luck and what rescue equipment was available.

In the 2000 title *Accidents and Disasters in Human Spaceflight* published by Springer/Praxis, this author reviewed the incidents and near misses that occurred during the first four decades of the human exploration of space. The book explained that accidents could occur at any stage of a mission, whether in training for a spaceflight, during launch, operating away from Earth or in trying to get back home.

Crew rescue capability has also to be available to cover all these scenarios. From before lift-off to post-landing, the hardware and human crew have to perform as planned, and the rescue systems are vital to the success of the mission and the recovery of the crew—even if (as everyone hopes) they don't have to perform at all.

AMBASSADORS OF HUMANITY

On December 2, 1954, the preparatory committee of the International Geophysical Year proposed that an artificial satellite circling the Earth could be used to study the upper atmosphere of our planet. That notion led to the reality of the "space age", with the launch of the first satellites from the Soviet Union and the United States. It also launched a tangled web of legislation and political debate about who owns what in space and whether Earth-based rules and laws could be applied beyond its boundaries.

The Outer Space Treaty—an International Agreement

According to Article V of the U.N. Treaty, members signing up to that agreement:

> "Shall regard astronauts as envoys of mankind in outer space and shall render them all possible assistance in the event of accident, distress or emergency landing on the territory of another State Party or on the high seas. When an astronaut makes such a landing, they shall be promptly returned to the state of registry of their space vehicle."

The agreement continues to state that, while completing activities in space or on celestial bodies, astronaut members of one state party should render all possible assistance to any astronaut of other state parties should the need arise. It was also required that each party who signed the treaty should disclose any phenomena discovered in space, on the Moon and on celestial bodies that could constitute a danger to the life or health of astronauts.

After a variety of studies and discussions on the legal aspects of spaceflight, the first committee on the peaceful uses of space was created by the General Assembly of the U.N. in 1958, shortly after the world's first satellites were placed in orbit. The following year, the legal sub-committee suggested a wide program of further studies into the topic. Finally, on December 20, 1961, after the first cosmonauts and astronauts had flown into space, the General Assembly of the U.N. issued the first principles of what became known as the "Code of Space", with supplements issued two years later, on December 13, 1963. These proposals were fully discussed and refined over the next few years, with the Outer Space Treaty finally being signed in Washington, D.C., Moscow and London on January 27, 1967. That declaration included the provision for the rescue and return of astronauts from space, and developed into Article V, mentioned above. Somewhat sadly, the treaty on the

Earthrise—Apollo 8.

peaceful exploration of space was signed on the very day that three American astronauts lost their lives in the Apollo 1 pad fire at Cape Canaveral.

Rescue Agreement

The Agreement on the "Rescue of Astronauts, the Return of Astronauts, and the Return of Objects Launched into Outer Space" (commonly known as the *Rescue*

Agreement) was created in 1968 as an extension of the rescue provisions cited in Article V of the 1967 Outer Space Treaty. This entered into force on December 3, 1968, with the first state-signing of the document taking place in Washington, D.C., Moscow and London on December 22, 1968, the day after the Apollo 8 crew left for the Moon. In all, 88 nations have ratified the Rescue Agreement (up to 2006), with 25 of those actually signing the document. In addition, two international and intergovernmental organizations (ESA and the European Organization for the Exploitation on Meteorological Satellites) have both accepted the rights and obligations listed under the 1968 agreement. The agreement includes texts in English, Russian, French, Spanish and Chinese.

More detailed than the original 1967 Treaty, the Rescue Agreement requires that should any state become aware that the personnel of a spacecraft are in distress, they must notify both the launching agency of that spacecraft and the Secretary General of the U.N. In addition, all possible means should be employed to rescue the personnel of that spacecraft should it land within that state's territorial boundaries, whether the landing occurred by accident, distress, emergency or unintended means. Should the distress area fall beyond the territory of any nation, such as international waters, then any state party in the position to do so should, if necessary, extend all assistance in support of the search and rescue operations.

Keeping it updated

Creating an international agreement on the rescue and recovery of international space explorers is one thing. Putting into action an effective method to achieve this is quite something else. Forty years after the creation of this document, there is still no international space crew rescue system available at short notice to respond to life-threatening situations in space.

Each nation's manned spaceflight programs do have their own methods and procedures designed to ensure the safety of the crew. Some work as designed, and have been proven to work in actual flight situations over the years. Others are available for use and have been fully tested, but have never been called upon to save the crew. Some ideas have, over the years, been proposed but rejected in favor of other options.

There have been a number of incidents in the history of manned spaceflight that have led to situations that could have endangered the lives of the crew, but contingency procedures and crew actions have overcome or addressed the immediate problem to enable the mission to continue, albeit often significantly amended from the original plan. But although the international agreements and individual systems exist, and more and more international crews are flying in space, the world still does not have an "international space rescue" organization. This is due, in part at least, to the many countries and agencies involved in operating the global space programs.

The original 1967 Treaty states that "astronauts" were to be rendered all possible aid in rescue and recovery, although the document did not specify the definition of an "astronaut". However, this clearly implied both Soviet cosmonauts and American astronauts, since no other nation was supporting a manned spaceflight program at

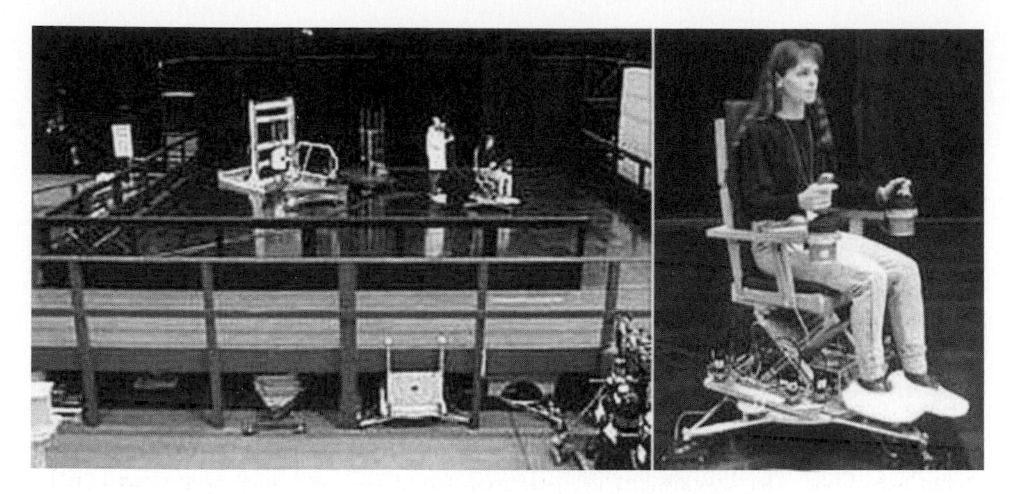

Developing EVA techniques and hardware on Earth.

the time. It also assumed that USAF military astronauts (if flown under the Manned Orbiting Laboratory Program) would, if necessary, fall into this category. The 1968 agreement changed the wording to the "persons of a spacecraft", which sufficiently covered later programs, including payload specialists aboard the U.S. Space Shuttle and guest cosmonaut researchers aboard Soviet missions. The subject of space tourists has also been debated, along with the status of a person who is simply along for a short ride, such as those envisaged on the Virgin Galactic Spaceship II ballistic flights to the fringe of space.

It is clear that, as the world's space programs develop, the agreements covering the legality and responsibility of exploring space have to change. Indeed, one often-debated point of the agreement is who covers the cost of such rescue operations where the spacecraft (or crew) is multi-national. The Agreement states that the launching state must bear that cost, even though it may be launching a vehicle or crew from a different country or countries.

Back to the future

With plans to return to the Moon and send humans out to Mars, effective space rescue systems have to be created to support these long-term ventures. At present we have one crew of between two and seven in space at a time, with perhaps two or three multi-person crew missions overlapping. We do not, as yet, have dozens of space explorers in space, working and living thousands or millions of miles from Earth for months or even years at a time, for whom a space rescue service would be required in a state of readiness at all times—and not necessarily to launch from, or return the stricken crew to, the Earth. Inevitably, the expansion of space activities will have to be balanced by technical limitations, scientific return, economic practicalities and political will, with public support and correct marketing thrown in. There is also the ever-present significant risk.

THE THEORY OF RISK

Developing new technologies for spaceflight is always difficult, not always headline making and usually costly, especially when that technology takes a long time to perfect. Add to that the hazardous (and, in the early days, almost completely unknown) environment in which the technology has to operate, then not only does the cost and testing of that technology increase, but so too does the need to provide appropriate safety systems to protect both the technology and the human crew relying on that technology.

Of course, the development of such safety systems means further costs, testing and quality control, for a product or system that will only be called upon when something fails. The investment of time and resources into such a system can be difficult to justify within a limited budget. Indeed, there is a strong argument for providing primary systems that don't fail, rather than developing a safety system or alternative back-up for an 'acceptable' failure level in the main product.

Recent analysis of space rescue in the current space program provides some interesting observations. This addresses a long-standing argument against space rescue systems, given the lack of aviation rescue systems, especially in the commercial sector. Each passenger who boards a commercial aircraft flight is not provided with a personal parachute. With dozens flying on each civilian aircraft and hundreds of aircraft flying around the globe every hour of the day, each day of the week throughout each year, this is hardly surprising. We accept the risk of flying aboard an aircraft without a personal rescue system every time we get on a plane. Even immediately after hearing of a major accident, people still board planes and fly without a personal escape system. There are procedures and safety systems in place, of course, but they all rely on getting the aircraft safely back on the ground again. So why does a manned spacecraft, with relatively few crewmembers and often on one-shot flights, require the investment in an adequate rescue system which, in all probability, will not actually be needed? Even military aircraft do not always carry personal rescue systems for all crewmembers, so why should a spacecraft?

To address this, a recent NASA study by the Safety and Mission Assurance Branch at JSC in Houston made a comparison between various types of "flight", the closest analogies to space exploration. Their findings revealed that the risk of loss of life in a commercial aircraft is roughly 1 in 1,000,000 flights. This risk increases for a military aircraft to 1 in 100,000 flights. In combat situations, of course, this risk dramatically increases again, to 1 in 10,000 flights. To date, the risk of loss of human life in spaceflight has been estimated at approximately 1 in 100 flights. If data from the Space Shuttle is separated from the NASA study then the risk is about 1 in 60 flights. When looked at this way, the requirement for a rescue system for astronauts, and particularly for the Shuttle becomes more apparent, since it is clearly demonstrated that spaceflight carries a greater risk to human life than any other type of 'flying'.

The question of why then place humans at such risk is answered to a degree by the various situations in which the ability of humans to overcome in-flight situations and perform rescue and repair has salvaged many missions from very costly failures. Equally, the great successes of unmanned missions to Mars such as the rovers Spirit and Opportunity has demonstrated that humans need not always be present. The future of human spaceflight is firmly entwined with that of the unmanned programs if progress to Mars and beyond is to succeed. If this is the case, then rescue and safety systems must still be a major consideration for future programs.

What also is undoubtedly linked to the high risk of the manned space program is the development of technologies and the length of service of that technology. Compared with military and commercial aviation (and even private aviation), spaceflight with human crew aboard is a mode of transportation in its infancy, with relatively few hours of experience gained on the limited number of missions flown in the near 50 years of operations we have logged. In the hundred years or so of aviation across the world, the technology and experience reinvested in the aerospace industry is immense. We are a long way off such a comparable investment in the space industry from the space program. It is true that a significant number of aerospace companies provide space technologies that cross over into the development of materials and systems widely used in commercial aviation, but the reinvestment back into new generations of space vehicles is infinitesimal in comparison. Spaceflight has not had the gestation period of the aircraft industry, with generations of improvements, refinements and investment in new designs.

Aircraft design advanced in a few, very significant steps, from glider flight to powered flight to pressurized compartments, through streamlining to breaking the sound barrier and on to the jet age. Manned spaceflight has developed from single-flight capsules to reusable spaceplanes, but is now returning to (reusable) space capsules. If the Shuttle system has demonstrated anything in comparison of its early design with the configuration actually flown, it is that compromising crew rescue capabilities and relying on one system can seriously limit operational capability, to the detriment of both political and public support, and the fatal consequences for the crews.

Comparisons of the Shuttle with the X-15 research plane have often highlighted the number of flights against operational qualification. The X-15 rocket aircraft flew 199 times during the 1960s and was only ever a *research* aircraft. It was never classed as an operational vehicle, but was used to develop technologies for potential use in future vehicles which could, in turn, lead to operational programs. The Shuttle is only going to fly a total of 133 missions if all the remaining manifested flights are flown. This is far fewer than the entire X-15 program and yet the system was declared operational after only four test flights, clearly an over-optimistic claim. While the design has proven versatile and generally reliable, it still remains that two crews have been lost in just over 100 missions. The X-15 had just one fatality in its near 200 missions, but the system was based on work conducted over decades of high-speed, aeronautical research by other aircraft. This technology had only limited relevance to the Shuttle. The Shuttle launches like a rocket, flies as a spacecraft, and lands like an unpowered aircraft. The X-15 launched under the wing of an aircraft flew to the edge of space and landed unpowered.

Lessons learned from ballistic capsules were also of limited use for the Shuttle, although the old files are being dusted off to examine the Apollo design for the new Constellation spacecraft "Orion". Shuttle technology will be valuable in supporting the developing Constellation Program, but it is in the old files of Apollo that the most applicable lessons learned will be found.

Keeping the public aware

To provide public support for the space program, and through it the political will to fund it, some demonstration of achievement is necessary. Nothing motivates better than success, while nothing hurts more than failure. Trying to "sell" the space program when the immediate returns are not obvious or "attractive" is difficult. Demonstrating the "gee-whiz data" or the spectacular views from space is great for the "wow" factor but not for protracted public interest from the man or woman in the street. For most people, space exploration is not a consideration of everyday life. To get the public to understand what we are doing in space, why we are there and how it benefits all of us on Earth is a vital, but extremely difficult task for any space agency, or space mission. Space spin-offs and Earth-based benefits are major pluses for this argument, but some of these may not become evident for years. With any financial profit likely to take decades, if it comes at all, it is difficult to encourage commercial investment in spaceflight, and government funding is always dependent upon other considerations and public perception. Clearly, the space agencies need to use every method available to generate interest in, and funding for, the space programs.

Hazards of flying in space

It is essential for any spacecraft to utilize the very best technology to deliver performance and safety, while encountering extremes of structural stress, pressure, temperatures and environments. To explain this to the public, former Director of Structures at NASA Marshall Space Flight Center Robert Ryan created a metric that clearly demonstrated the design challenges involved in a launch into space.

There are seven clear categories of human spaceflight in which hazards can affect the safety of the crew and would require an adequate rescue method to protect them.

These are (with examples of such scenarios):

1. *Pre-launch*, from the time the crew accesses the vehicle to the moment it leaves the pad. This does not include the numerous incidents during training for a mission prior to accessing the vehicle on launch day which have also led to delays to the mission, injury or even fatalities. A fire or explosion caused by failing systems, the loss of the structural integrity of the vehicle, a propulsion-related failure or the effects of the natural environment such a electrical storms. (Recorded major incidents: Apollo 1 fire; Gemini 6 pre-launch abort; Soyuz T-10-1 pad abort; five Shuttle RSLS pad aborts.)

Inspection of SSME after a launch abort.

2. *Ascent*, from lift-off to orbital insertion about the Earth. A systems malfunction which leads to the loss of control or structural integrity of the vehicle, a propulsion-related failure or the effects of the natural environment to induce a failure. (Recorded major incidents: Apollo 12 lightning strike; Soyuz 18-1 failure to clear booster separation; Apollo 13 loss of second-stage central engine; STS-51F engine shutdown and abort-to-orbit situation; explosion of STS-51L Challenger, the electrical short during STS-93 ascent, debris impact on ascending Shuttle notably on STS-107 Columbia.)
3. *Flight*, time in space, or orbit. This could include time traveling to and from the Moon, planets or asteroids or other celestial bodies, until the time Earth atmospheric re-entry is performed. A range of systems failures onboard the spacecraft: from explosion, the loss of attitude control, loss of critical functions; release of toxic materials; environmental hazards of space such as solar radiation, micro-meteoroid orbital debris impacts, crew medical health issues. (Recorded major incidents: Gemini 8 thruster failures and loss of control; explosion of oxygen tank on Apollo 13; fuel cell failures on STS-2 and STS-83; fire on Salyut 1 and Mir; Soyuz 33 main engine failure; medical conditions on Salyut and Mir missions.)
4. *Prox ops (proximity operations)*. These include rendezvous, docking, undocking and activities around another vehicle in space. Collision with other spacecraft, sub-systems failures that include explosions, the loss of attitude control or loss of

Launch of a Saturn 1B.

critical function and the release of toxic materials. Hazards from the natural environmental of space such as solar radiation, variations of temperatures, orbital debris impact. Incorrect targeting of trajectory rendering the vehicle(s) off-course as a result of crew health issues. (Recorded major incidents: 1997 collision between Progress and Mir resulting in loss of structural integrity of the Spektr module.)

5. *Descent to and ascent from a non-terrestrial surface* (such as the Moon, a planet or an asteroid). This covers the periods of landing on or taking off from a planetary or lunar surface as a result of systems malfunction, effects of the natural environment, incorrect trajectory or targeting skills, and impact with surface. (Recorded major incidents: loss of control during Apollo 10's simulated landing profile.)
6. *Surface exploration.* EVA on the Moon, planet or asteroid, usually spent outside a spacecraft, but this could also include surface stay time on those celestial bodies in between EVA operations on, for example, a research base or extended-duration surface habitation module. This covers suit systems malfunction, punctures in suit integrity, health of crewmembers on EVA, loss of crew connections to spacecraft where the tether protocol is lost due to a crewmember

Proximity operations.

becoming adrift. (Recorded major EVA incidents: crew helmet fogging (Gemini and Mir); suit ballooning (Voskhod 2); crew exhaustion (Apollo lunar surface); damage to gloves (Shuttle EMU), experiences of extremes of temperature (Shuttle EVAs).)

7. *Re-entry and landing*, the time from initiating an orbital change that results in entry in Earth's atmosphere and landing either on land or water and retrieving the crew safely from that landing mode. This includes the system of structural failure, the natural environment's effect on failures, loss of control and delayed input of entry commands. (Recorded major incidents: re-entry overshoot of Mercury 7, Voskhod and Soyuz TMA-11; Soyuz 1 parachute failures; Soyuz 11 decompression; Apollo 15 single-parachute failure; ASTP Apollo toxic inhalation; Soyuz 23 landing in frozen lake; Soyuz TM5 entry malfunction; STS-51D tyre burst; loss of Columbia.)

This was summarized in the NASA document as: pre-launch and ascent (essentially getting from Earth into space) as 12 serious incidents with 2 resulting in fatalities; flight, prox ops and EVA (essentially working in space—13 major incidents without a single failure). Then entry and landing (getting back on Earth) involving 10 major incidents with 3 resulting in fatalities. The report summarized that the risk of incidents is generally uniform throughout a "typical" flight with the largest risk occurring during ascent and entry, which supports the theory from the aerospace industry that dynamic phases of flight are the greatest hazard. This also supports the adages that the farther away from the launch pad you are the safer you are. Of course, spacecraft are designed to operate in the environment of space, and therefore are less likely to experience a failure once the vehicle has left Earth or before it tries to get back again. Indeed, even the most serious malfunctions actually "in space" to date (the Gemini 8, Apollo 13 and 1997 Mir incidents) were overcome by the crew or

Damaged (Apollo 16) LM panels.

ground controllers resulting in the dangerous situation being stabilized and the crewmembers safely recovered. It is the sheer dynamics of flight that do not normally afford sufficient time to react to or evade a life-threatening situation; however, when a serious situation does exist in spaceflight there has to be sufficient back-up options, contingency plans and more importantly time to overcome what can be life-threatening problems.

Of course, providing a "space rescue capability" for stricken crews in space is not always a viable option, due to time, distance and the level of the problem. Only on the Apollo lunar missions have crews been away from Earth orbit with theoretically an almost immediate or at least short-term ride home. During the Apollo program there was a constant risk of failure throughout the mission profile that could have found the crew stranded away from Earth with no hope of rescue or return. That was a contributing factor to the decision to terminate the program and cancel the final three Apollo missions, especially after the near-tragic events of Apollo 13 and growing financial and social issues both inside and outside the U.S. space program. True life-threatening situations are present in all manned spaceflights but in Earth orbit options for a safe haven or rescue and early return to Earth are somewhat alleviated by the distances involved, though as stated above the dynamic phase of entry and landing is the larger hurdle in concluding a safe landing.

Rescue systems and options

In recognition of these hazards throughout a spaceflight, designers and engineers have adapted a range of novel ideas to help rescue and recover crews in the event of a major incident.

During pre-launch operations, a series of rapid pad egress systems have been developed and incorporated into the pad hardware resulting in rapid evacuation of

Recovery of a Soyuz crew.

Recovery of a Gemini crew, showing the open hatches and seated in their ejection seats.

The first true manned spacecraft—the Apollo 9 LM.

the pad area. These have included slide wires, chutes and protective bunkers on the pads, launch abort towers and rocket systems, and ejector seats. In addition, contingency monitoring systems evaluate the condition of the vehicle in the event of a problem and can terminate the launch at several stages thus saving the crew hopefully for a subsequent launch attempt when the problem has been overcome.

In the ascent to orbit, the launch rocket abort systems that were available on the pad are also adapted to operate up to different altitudes or ejection seats are provided for crew exit. Changes are being made to the launch profile, as the vehicle ascends, to match the changing configuration, power and capabilities of the vehicle, such as aborting to orbit, landing in contingency sites or in some instances bailing out a stricken vehicle.

In Earth orbit the early orbital track of the Soviet Vostok spacecraft meant that in the event of a failed de-orbit burn the vehicle would naturally decay about ten days after launch. Sufficient supplies were aboard to ensure the cosmonaut lived long enough to complete that landing, as long as other essential systems did not fail. In every manned mission there is the primary flight plan, with back-up and contingency options available as alternative missions, so that if the primary objective cannot be achieved for whatever reason, an alternative mission can achieve some or other results as long as it's safe and prudent to continue the flight without endangering the crew. This was a major element in planning Apollo lunar missions in case the vehicle could not get out of Earth orbit or commit to a lunar landing, but were safe enough to allow the crew to fulfill other objectives. Of course, rescue from Earth orbit relies on sufficient information as to the state of the problem, adequate access from a second vehicle capable of rapid launch to a suitable orbital inclination, and time to achieve a rescue (an option not afforded Columbia) . The other method is constantly having a rescue vehicle available to the crew in the event of a major spacecraft failure. In effect, this was the only option for the crew of Apollo 13. When the CM failed due to the explosion on the SM, the still attached and unused Lunar Module took the role of space lifeboat instead of lunar lander. Since the creation of the space station program, crew ferry vehicles (Soyuz for Salyut, Mir and the ISS, and the CSM for Skylab) have been available to return the crew home from the main station. In Skylab a second vehicle was prepared to go to the Orbital Workshop with a core crew of two to return the three Skylab astronauts. In the second manned mission it was almost called upon to do so. There have also been a wealth of concepts to provide adequate escape from orbit since the 1950s, some making it to hardware, others simply theoretical studies, most never reaching final implementation.

Rescue from farther away from Earth is somewhat more difficult and complicated. For Apollo there was simply no Earth-based method of rescue for a crew out at the lunar distance and certainly not for the two astronauts on the lunar surface should anything go wrong. The lone CSM pilot had no provision to reach the surface and rescue his colleagues and it would have been a long and lonely flight home without them. This undesirable scenario is being addressed in Constellation to help prevent a stranded crew at lunar distance having no hope of rescue. Therefore, concepts to recover from failed rendezvous or docking, or to support ascent from and descent to the terrestrial surface from a lower orbit are the only options for rescue

Demonstration of an EVA rescue wearing pressure garments on an air-bearing floor (top). Runway landing of the Space Shuttle (bottom).

support. For EVAs, self-rescue systems such as back-up and secondary life support systems, sharing life support systems (such as the buddy system on Apollo) and the SAFER MMU on Shuttle/ISS help provide spacewalkers with some degree of rescue capability. Also, having crew operate outside in teams with the ability to maneuver the main vehicle to "go and fetch" a drifting crewmember is an option, which was theoretically possible on Shuttle but impossible with the larger mass ISS or whilst the Shuttle is docked to the station, where it would take too much time to separate and chase the drifting crewmember easily.

Entry is usually covered by back-up parachute systems, ejection seats or bailout options, as long as the vehicle is not breaking up and is in almost stable flight. In addition, survival equipment for landing away from primary landing areas is available (adding to the launch mass of the vehicle), such as dinghies for water survival, survival kits for jungle, desert and mountain terrains, signaling equipment and adequate survival clothing.

SAFETY IN MISSION DESIGN

In the planning of any manned flight into space, flight safety should be given the highest priority and around this the trajectory and objectives can be tailored, and constrained within the hardware, software and optional limitations available. To safely achieve this there has also to be constraints on the limits of flight control, the rules of the mission as well as operational procedures. All of this is brought about in conjunction with crewmembers flying in space and the team of controllers and support personnel on the ground. Another factor in ensuring safety is the continual updating from program to program of experiences gained from both success and failure and their carry-over to the new program. Learning by experience is priceless; conversely, not gaining knowledge from those experiences can be costly in time, finance and tragically in human lives. These lessons are fundamental to developing manned space programs that have their roots in the capsule missions of the 1960s, the current Shuttle operation and Space Station programs and planning of the immediate future that will allow a return to the Moon, five decades after the first landings, and place the first humans on Mars. This strategy was decribed in a 1967 NASA internal note reflecting the experiences from Mercury and Gemini in the planning for Apollo, though the model can be applied to any manned space program and is summarized here (Huss *et al.*, 1967).

Mission design

With the priority of flight safety, the primary goal of designing the mission has to be to maximize this objective whilst achieving mission goals without unacceptable risk to the crew and remaining within known constraints. It is therefore obvious that mission objectives must be realistically achievable and have to remain within both flight and crew safety limits. The challenge is to identify the various constraints imposed on the proposed flight plan whilst still achieving the stated objective and ensure safety is not compromised and alternative plans can be employed should the primary objective not be achieved.

Therefore, the mission evolves from the constraints, the planning of the objectives and operations. From these both nominal and alternative missions can be evolved. The constraints can include the launch vehicle, the spacecraft, the operational issues, tracking and communication network, recovery forces and the flight crew.

In planning the desired mission objectives, consideration has to be given to the mission profile, the compatibility of systems and any EVAs planned. In addition, abort and alternative missions have to be studied along with the planning of flight operations. A range of reference trajectories for the mission and the requirements of support also need to be analysed. The program of simulations and training has to be factored into the flight preparations and from this a flight crew plan is evolved. Finally, the scope of recovery operations has to be determined.

Operationally, abort procedures are evaluated, the limits of flight control trajectory defined, alternative missions have to be planned, mission rules have to be set, mission operations trajectories identified, handbooks written (or more recently computer programs written), simulation and training requirements evaluated, requirements for mission support designed and a plan for recovery decided upon.

From the combination of constraints, planning and operations a decision of achievable mission objectives can be defined that also ensures flight safety in either nominal or contingency mission designs. From these a number of aspects are defined; in summary these are

- *nominal mission design:* the orbital altitude, the inclination and techniques of descent
- *contingency mission planning:* abort procedures and the limits of flight control trajectory
- *constraints and considerations:* operational issues, configuration and systems and the software used.

The constraints under consideration for most missions planned include

(i) Operational

- range safety—avoiding land impact, and orbital collisions
- communications and tracking—monitoring of systems, capabilities of ground control
- environmental surroundings—the properties of the atmosphere, winds, lightning, weather considerations, radiation and meteorites
- human factors vs. crew acceleration and deceleration tolerances, crew and ground response times
- procedures—techniques of separating elements of the vehicle, avoiding re-contact, simple and reliable methods for proficiency in training, mission-by-mission carry-over lessons learnt

- landing and recovery—geographical features, lighting conditions, and communications logistics require medical support coverage.

(ii) Configuration and systems

For the launch vehicle

- propulsion—emergency detection and switch-over capabilities
- guidance and control—emergency detection and switch-over, as well as stability
- structural—emergency detection.

For the spacecraft

- propulsion—type, performance and back-up capabilities
- guidance and control—performance, procedures and back-up systems
- structural—landing capabilities and any crew couch/seat supports for impact absorption
- thermal (heat shields)—limitations of the thermal protection, dissipation and space soaking capabilities
- aerodynamics—stability, trim and lift/drag considerations
- window crew geometry—crew visibility of the horizon, manual take-over visibility, visual aids
- consumable—electrical power, environmental systems, propulsion
- sequencing—attitude requirements, time, procedures.

(iii) Software

- launch vehicle—ground control capabilities, up/down data capability
- spacecraft—up/down data capability, systems evaluation
- ground control—up/down data capability, systems evaluation.

Nominal mission design

In defining the design of what would be called a nominal mission, orbital altitude and inclination as well as techniques of descent are the most constraining. In the history of human space exploration to date, of the approximately 260 missions there has only been nine Apollo missions that have operated out of Earth orbit as far as the Moon, therefore most of the considerations for mission design have focused on Earth orbital trajectories.

Above 740 km altitude, radiation is a primary concern. However, another is meteorites, whose consideration is not so much a design constraint though the growing amount of orbital debris is a factor that needs to be taken into account between 417 km and 740 km. Therefore, apogee is the only constraint to dangers posed by radiation, rather than debris impact in orbit and these areas are avoided where possible. The perigee is limited by the requirement of mission duration, atmospheric heating during re-entry and drag for prolonged orbital operations. In addition, landing site locations are also a factor in deciding where the vehicle

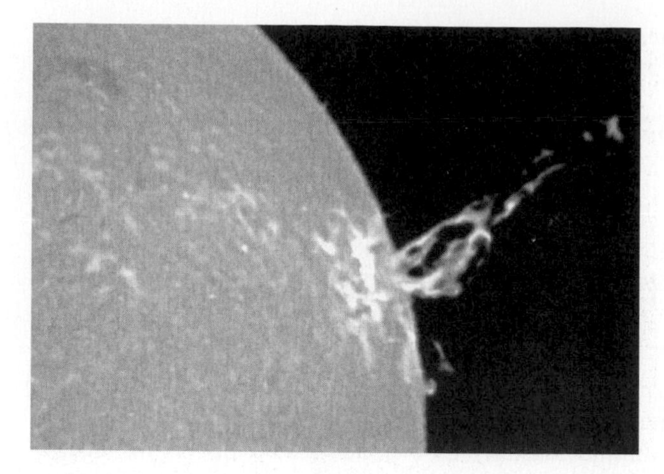

Solar flares radiating into space.

should fly. The techniques used in descent include the type of propulsion system available to create the de-orbit burn, what back-up systems are available, the sequences of the system, vehicle aerodynamic capabilities, the procedures employed for land or water recovery, scope of direct communications with the crew, capabilities of the crew during landing scenarios, etc. Nominal mission altitudes are between 185 km to about 417 km for short-duration flights and 417 km to 740 km for extended-duration orbits at the current time, though odd missions fly higher than this (such as the Hubble service missions) for short periods and ISS operates at an altitude of approximately 340 km.

Contingency mission design

As a planned mission evolves so do the sequence of events and requirements needed to achieve the goals of said mission but remain within the safety guidelines under mission rules. In addition as the "flight plan" evolves so does the range of contingency or alternative mission operations available to planners. This can involve the participation of the flight crew, controllers, management, science and engineering support teams and recovery forces. The largest input in the design of any mission is generally that of developing contingency "what if" planning options.

Abort procedures and the limits to controling the trajectory have a large impact on flight safety and time-critical phases of the mission where rapid crew, ground control or system action is required to respond to in-flight failures and malfunction where deviation from the nominal flight path could cause a danger to the crew and vehicle; this can be broken down into three phases: ascent, orbital and descent.

Ascent

This is defined as the region of the atmosphere through which a space vehicle much pass to get from the ground launch pad to orbital flight. This region is divided into three areas: atmospheric, transition and space.

Leaving Earth.

The atmospheric region is usually considered to extend from the ground/pad up to an 85,344 m altitude where re-entering spacecraft normally "sense" the deceleration forces when coming back to Earth. It's the area of the densest layers of Earth atmosphere. The transition area extends roughly from 85,000 m to 122,000 m where the vehicle moves between atmospheric conditions and space conditions. Space extends above 122,000 m and is where the vehicle operates at its best (i.e., the region in which it was designed to carry out its mission).

In conjunction with these three regions and the constraints and considerations listed above, the ranges of launch abort situations, and the development of hardware systems and procedures to deal with them can be addressed. Consideration has to be given to the operational limitations of the vehicle and systems, the configuration of the aborting vehicle, and the limits of available systems—all of which play a major role in each of the chosen abort modes that, though amended for different vehicles and programs, essentially remain the same.

Defining launch aborts

A Mode I abort is defined as an abort that is almost always in the dense layers of the atmosphere and requires a landing close to the launch area. Any abort system has to cope with exposure to the dense layers of the atmosphere, ground conditions, potential explosion of the launch vehicle, rapid aerodynamic forces and impact loads on the rescue vehicle and crew. The leading considerations for this type of abort are the surrounding environmental conditions; procedures of the abort/spacecraft propulsion system; and sequencing being required to separate from a potentially explosive launch vehicle. High aerodynamic forces will be encountered, and consideration for either land or water recovery has to be factored into the abort profile. Crew capabilities are also a factor in this type of abort mode. Using launch azimuths away from populated areas ensures, to a degree, that land impact is kept to a minimum.

A Mode II abort is usually initiated in the transition region which is at the very edge of the atmosphere, but requires a rather rapid entry back into the atmosphere again. Covering the largest region of ascent is generally categorized as the "the most simple and reliable abort procedure". The explosion and instability characteristics of launch vehicles are generally less at these altitudes, and stress on the descending re-entry vehicle is far less than Mode I. Simple and reliable as this type of mode may be, it is however constrained by the human factor. The crew have to be able to withstand deceleration loads which can still by quite dynamic. The sequence followed

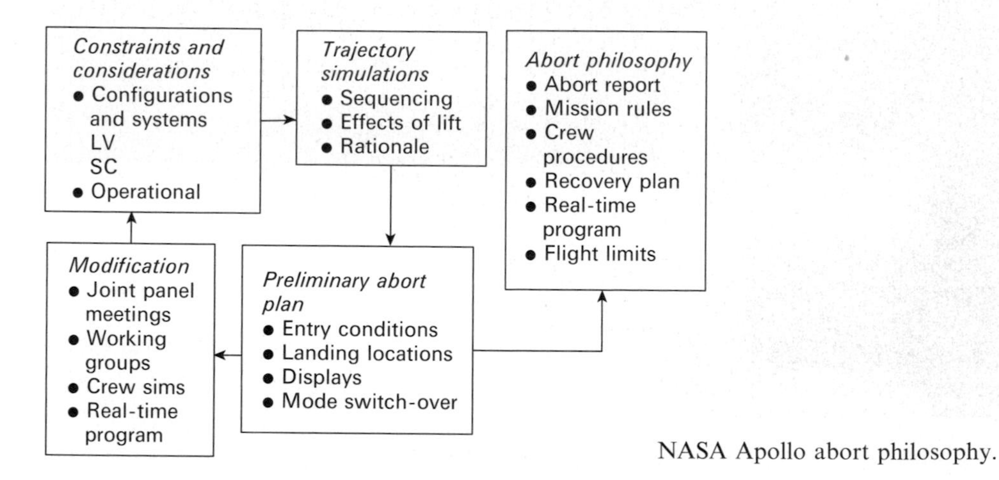

NASA Apollo abort philosophy.

by spacecraft systems requires separating components from the launch vehicle, then distancing itself from explosion damage, jettisoning unwanted hardware, orientating itself to re-entry attitude and compeling a simple entry profile before initiating landing systems. In the Shuttle program this type of recovery would have to be pilot-controlled. Defining minimal population areas is a further restriction at this level of abort.

A Mode III abort remains one of the most critical and usually requires the use of spacecraft propulsion and aerodynamic capabilities for landing, as the ground track of an instant landing shifts rapidly toward an orbital track around the Earth. Here at higher altitudes, the avoidance of populated areas during entry and landing is a factor in planning recovery zones. Therefore, the leading factors are landing and recovery, communications and tracking, and procedures and crew input capabilities. It is also important to include procedures to avoid collision with jettisoned equipment

A Mode IV abort is the contingency orbit mode where orbital insertion maneuvers are normally required for a minimum of one or two orbits to allow ground controllers and the crew to assess the situation and plan a suitable course of action to

Human vulnerability in space.

continue the mission, amend it to an alternative objective or bring the vehicle home as soon as possible. Constraints here focus on communication and tracking, human factors, consumables and procedures.

Orbital operations condensation

Contingency orbital missions have to consider the altitude options limits of the capability of onboard systems. Any maneuvers have to be budgeted for in order to protect de-orbit burns with a back-up capability. Back-up systems allowing manual take-over have to be available and there must be a means for information to be displayed on the failures and levels of various critical systems.

There have to be controls over the length of time spent in orbit under contingency situations, as there are constraints on the heating of components at perigee and environmental constraints at apogee. Communication and tracking has to be maintained as much as possible and any rescue maneuvers have to be carefully considered against available consumables and capabilities.

When the Apollo LM was flown for the first time, astronauts were manning a vehicle without the capability of sustaining a safe return to Earth, and considerable planning was done to provide adequate back-up capabilities of an LM flying solo in Earth orbit and in lunar orbit. Some of the early work on contingency operations during Apollo 9 proved invaluable for the Apollo 13 incident in 1970. As we return to the Moon and extend way from Earth farther than ever before, so orbital consideration are expanded to include deep-space consideration, as well as environmental conditions for extended exploration of the Moon and eventually out at Mars. Details of these plans are being evolved as the program to support those goals develops.

Descent phase consideration

The descent phase is perhaps the most difficult to design for, as it has to consider the de-orbit maneuver, entry control, limits that define the safe entry profile, back-up procedures, landing and recovery techniques, hardware and procedures, and providing for sufficient recovery force logistics and coverage.

The characteristics for entry change with each program and are dependent on the design of the vehicle to be re-entered, its mass, landing capabilities and any limits to input from the crew, as well as the type of landing, be it on water or on land.

When designing contingency de-orbit and landing, consideration had to be taken of the sequence to be followed when separating the elements of the spacecraft, the orientation of the descending spacecraft (which contains the crew), the "free-fall time" from the end of the de-orbit maneuvers to the entry interface at approximately 137,000 m down to 107,000 m, achieving the desired velocity and entry angle. Other considerations concern human crew tolerance to entry loads and thermal conditions. The structure of the spacecraft has to be a leading consideration both in aerodynamic capability, and in increasing or decreasing the loads as required, without affecting onboard instrumentation and controls, or posing a threat of injury to the crew during the descent. Communication during the periods of radio blackout as the spacecraft

Coming home.

cuts through the atmosphere and heats the surrounding ions is difficult during a ballistic capsule–like entry. For the Shuttle, tracking and data relay satellites were used providing a relay link to ground stations during entry and alleviating the time-critical requirement to provide much updated trajectory and landing point information to the crew or ground control as possible before communications were affected.

Landing site selection depends on orbital inclination, the range traveled from the de-orbit maneuver to landing, geographical features, logistics, communications, the weather, lighting conditions, and to some degree political restrictions of the time.

SUMMARY

Flying into space requires countless hours of planning, evaluation, testing, simulation and provision for "what if" situations. International agreements expand the safety parameters of these decisions, and the available technology developed using past experiences, simulations and training assists in making future safety and contingency scenarios much more reliable and hopefully failsafe.

It is clear that the design of the mission, and thus the abilities to abort the flight and/or rescue the crew is driven by decisions in the azimuth of flight, its attitude and profile, and techniques of descent. Contingency mission planning has to be contained within a defined ascent or descent corridor to enable available abort-and-rescue techniques to operate at optimum performance. Flight operations planning also has to address the constraints of the hardware, the crew and the environment. Addressing as many of these potential problems as possible is and will remain a major effort in planning any manned spaceflight.

From the development of the world's first manned spacecraft in the late 1950s to today, operations around the International Space Station, the hope and dreams of returning to the Moon and placing human bootprints on Mars—all require theoretical and practical modeling, simulation and testing, and fabrication of adequate systems to save the crew in the event of something going wrong. Planning for what is thought most likely might fail is combined with a flexible and adaptable range of contingency situations to cope with what was not planned, but is hopefully survivable.

In launching from Earth, either launch escape tower systems, ejection seats or contingency profiles and alternatives have supported the effort to leave our planet. In space, alternative missions, redundant systems and ingenuity in the face of adversity have so far prevailed to bring each crew home to the atmosphere, though not always through it to a safe landing. Land and ocean landing techniques and wilderness survival training have supported the difficult phase of transition from high-speed spaceflight to wheel-stop, a dust-down or splashdown. The challenge in developing these techniques is the story that follows.

REFERENCES

Carl R. Huss, Claiborne R. Hicks Jr. and Charlie C. Allen (1967), *Mission Design for Flight Safety*, Internal Note No. 67-FM-175, 17 November 1967, NASA-TM-X-69696. Mission Planning and Analysis Division, MSC, Houston, TX.

3

Training to survive

No matter how good the equipment provided to protect and rescue a crewmember, none of it would be of any use without the training and familiarity that is an integral part of any crewmember's preparation for a mission into space.

Training for survival, as well as rescue, includes learning about as many systems as possible, including their failure modes and any available back-up or alternative systems. Countless simulations in both nominal and off-nominal scenarios, hands-on practice with survival and emergency egress procedures, and training with each piece of equipment, are all vital to ensure that, should an emergency situation arise, both crew and equipment are able to cope.

Survival and wilderness training has been a major element of spaceflight preparation since the selection of the first American astronauts in April 1959 and the first Soviet cosmonauts in March 1960. As programs changed, so did the emphasis for survival and rescue training for all crews, including guest and specialist crewmembers. Survival training will continue to be an important element in future operations, though perhaps to a lesser degree for future "tourist" suborbital flights to the upper atmosphere. However, extensive awareness of the vehicle's safety and rescue equipment will remain vital and will be more in-depth than a passenger receives for a commercial airline flight, at least for the foreseeable future.

NASA ASTRONAUT SURVIVAL AND WILDERNESS TRAINING

Once the methods and means to place an astronaut into space had been established, then the processes for bringing them home again had to be determined. These had to be developed both for nominal and off-nominal situations. Of course, the ideal is always to achieve a safe, trouble-free landing, but there had to be provision for times when the recovery did not go exactly according to plan. Training had to be adapted to encompass this likelihood. If a spacecraft returned outside the primary or secondary landing areas, the crew might find themselves isolated in remote and harsh conditions. Both the equipment and the astronauts had to be capable of dealing with such real-time situations. Since the creation of the NASA astronaut program, survival and wilderness techniques have been an integral element of training.

Mercury

During the design phase, it was decided to incorporate the capability to achieve a water landing within the Mercury Program. Therefore, the astronaut training program had to include courses to educate and prepare each astronaut for landing on the water, spacecraft egress and helicopter recovery. Because of the location of the launch site, and the direction of launch being aimed away from major habitable areas in the United States, the orbital path of Mercury would take the capsule over the Atlantic Ocean, Africa, the Indian Ocean, Australia, the Pacific Ocean and then the continental United States, with recovery in the Atlantic Ocean. Orbital dynamics would allow Mission Control to predict planned and expected landing sites, based on flight telemetry.

As well as survival training for water landings, it was clear that such training for the astronaut should also address the potential for a landing in remote areas. For Mercury, such remote areas would include the vast deserts of North Africa and central Australia.

It was also soon realized that it would be impossible for each astronaut to absorb all the specialist knowledge from such a complex program. Therefore, in July 1959, each member of the first astronaut group was assigned a speciality area to concentrate on, reporting to the group as a whole once a week, or more often if required. Each individual's past experience and qualifications were taken into account when deciding upon their specialization area. These technical assignments were to become a standard feature of the NASA astronaut groups over the coming years and gave them direct involvement in a variety of support roles in both current and future programs.

Each specialist area for Mercury featured elements of safety and rescue in some way: Scott Carpenter was assigned to communications and navigation (important in emergency situations both in-flight and during recovery). Gordon Cooper looked after the Redstone launch vehicle (and its safety to fly manned suborbital missions). John Glenn was assigned to the layout of the crew compartment of the spacecraft (including visibility, reach of controls and unrestricted method of entry and exit). Gus Grissom focused on manual and automatic control systems (in both nominal and off-nominal situations). Wally Schirra covered the life support system in the spacecraft and the astronaut's personal space suit (integral elements in keeping the occupant alive during the mission and during recovery). Al Shepard covered tracking and recovery (knowing where the spacecraft was at all times), while Deke Slayton handled the Atlas launch vehicles (and associated safety issues).

For the first seven American astronauts (the Mercury 7), training was initially conducted as a group, then individually as a pre-mission refresher course when assigned to a flight or back-up crew position. Periodically, each of the seven astro-

Mercury water egress training.

nauts was given refresher courses, which included revised egress and recovery procedures and participation in egress and recovery exercises.

Allowing the astronaut to practice actually getting out of Mercury soon developed into a three-phase training program. They also completed a scuba-diving course, received a series of lectures and watched a training film before practicing survival operations in a tank and then, a month later, in open water. In all, each of the Mercury astronauts spent about 25 hours in egress training.

Egress Training Phase One was completed in February 1960. Using full-scale boilerplate Spacecraft Number 5 as an egress trainer (located in the Hydrodynamic Basin No. 1 at Langley Research Center), each of the seven astronauts completed several exits using the top hatch. These were conducted both suited in a pressure garment and without pressure suits. Tests were performed in calm water and in simulated waves of up to 0.6 m.

Egress Training Phase Two occurred during March and April 1960. The astronauts went to the Gulf of Mexico near the Pensacola Naval Air Station, Florida to complete their first full-scale open-water egress training over two days. During the first day at sea, both the side and top hatches were used, while on the second day, water survival techniques and drills were practised in a training tank at Pensacola.

Egress Training Phase Three occurred during August 1960. A series of underwater egress exercises was conducted in Hydrodynamic Basin No. 1 at NASA's Langley Research Center, Virginia. Each astronaut completed six egresses from the submerged spacecraft, three of which were completed while wearing a Mercury pressure garment.

When assigned to a flight, the astronauts (both prime and back-up) would participate in a full-scale recovery exercise as part of the mission-training preparation. This included top and side egress, the successful deployment of survival equipment and helicopter recovery operations.

Survival Training Phase Two. During July 1960, all seven astronauts completed a $5\frac{1}{2}$-day desert survival course which was divided into three phases. This included $1\frac{1}{2}$ days of academic training at the USAF Survival School located at Stead Air Force Base, Nevada. Here, the astronauts were instructed on survival operations which focused upon North African or Australian desert conditions. The following day (Phase 2), a field demonstration was completed. This included the care and utilization of available clothing and the use of spacecraft and survival equipment. This was followed by three days (Phase 3) of remote site training. Here, the astronauts applied the knowledge and techniques gained during the earlier phases to "live" desert survival situations.

Gemini

The two-man Gemini Program which followed Mercury would still feature water landing capability, but would have a wider range in which to return from orbit. Therefore, training was expanded to include jungle survival training. In October 1962, the next group of astronauts arrived to begin their training program, primarily for Gemini and early Apollo missions. A program of academic training and a course in geology (to support planned Apollo lunar missions) was included, as well as familiarization with hardware and procedures. In January 1963, new technical assignments were revealed for the nine new astronauts. Each man's specialization featured an element of safety, rescue and survival, but some more than others. Neil Armstrong was assigned to training and simulators and Frank Borman to launch vehicles. Pete Conrad took cockpit layout and systems integration, Jim Lovell handled recovery systems and Jim McDivitt was assigned to guidance and navigation, Elliot See looked after electrical, sequential and mission planning, while Tom Stafford had communication, instrumentation and range integration. Ed White was assigned to flight control systems, while John Young received environmental control system and personal survival equipment assignments.

The new group's training included the same scuba diving, water recovery exercises and wilderness training that the Mercury group had completed. However, unlike Mercury, the Gemini capsule featured ejector seats rather than an escape tower

Gemini water egress training.

for emergency escape on the pad, during ascent or during recovery. Therefore, the new group also underwent parachute training and a session of parasailing as part of their survival training program.

Wilderness training began in August (when the temperatures were at their highest) with the five-day survival course at Stead AFB, modified for the new program. The new astronaut group took part in academic sessions on the characteristics of deserts across the world and in the necessary techniques to survive if they landed in one. A day of field demonstrations included the use of parachutes for clothing, constructing shelters and for signals. This was followed by two days of remote site training in teams of two (reproducing the two-man spacecraft crew) in Carson Sink, Nevada. During the field exercise, the astronauts could carry only their survival kits, parachutes, and 4 L of water for two days. Portions of the parachutes were used to fashion a flowing robe and a burnous to combat the desert heat. The rest of the parachute was then made into a tent, with a life raft oar as a centre pole and the life raft itself as a makeshift bed to provide shelter from the 43°C temperatures.

Water and terrain parachute training began in September 1963 and featured parasailing exercises. The astronauts were towed behind a boat to heights of up to 1,646 m while wearing a 7.3 m ring sail parachute. When the towline was released, the astronaut had to guide himself to a safe water or terrain landing.

Jungle survival training for new astronauts.

Using parachutes for a new wardrobe in the heat.

In January 1964, the third class of astronauts arrived, and they also began the academic, survival and simulator training in preparation for assignment to a Gemini spaceflight. For this group, jungle survival training was completed in Panama in July 1964, and desert training in Nevada in August 1964. As the program expanded, so did the specialist assignments, which were announced in July 1964.

Buzz Aldrin focused on mission planning, including trajectory analysis and flight plans. Bill Anders was assigned to environmental control systems and, in support of Apollo lunar missions, radiation in space and thermal protection. Charlie Bassett worked with training and simulators, while Al Bean handled recovery systems. Gene Cernan was assigned to spacecraft propulsion and the Agena docking target, while Roger Chaffee worked on communications (including the Apollo Deep Space Network). Mike Collins worked on pressure suits and EVA, Walt Cunningham on electrical and sequential systems and Don Eisele on attitude and translational control systems. Ted Freeman was assigned to launch vehicles, Dick Gordon to cockpit integration and Rusty Schweickart to future manned spaceflight programs and experiments. Dave Scott worked on guidance and navigation and C.C. Williams on range operations and crew safety. Most of the members from the first two groups were now heading up Gemini and Apollo Branch Offices in the CB, training for the first Gemini missions, or (as in the case of Glenn and Carpenter) no longer at MSC.

Training for egress from the spacecraft in water totaled approximately 13 hours per crewmember. The training program began with classroom instructions, followed by simulations using an engineering mock-up of the Gemini re-entry module. Egress practice took place both from a mock-up floating in a water tank at the Manned Spacecraft Center and out in the Gulf of Mexico just off the Texas coast near Galveston. The crews practised exiting the spacecraft from the surface and then underwater, although underwater egress was restricted to the water tank at MSC. The final egress training was fully suited.

Apollo

As Gemini gave way to Apollo, more astronauts came onboard the program. In June 1965, the first six scientist-astronauts were selected, although one resigned for personal reasons almost immediately. In the 1960s, it was a requirement for NASA astronauts to have jet piloting experience. Therefore, the three scientist-astronauts without flying experience were sent to flight school for 12 months. The remaining two received technical support assignments pending the return of the other three. It was decided for administration and logistical reasons that the formal academic and survival training of the five scientist-astronauts would be delayed so that it could be combined with that of the next pilot selection (19 in May 1966). This resulted in a survival training group of 24.

Normal launch and abort recovery for the Apollo Program would be in the ocean, so water survival and egress simulations were an integral part of the training program. However, with Apollo flying lunar mission profiles, the mission could be terminated and return to Earth at almost any point around the globe. Landing and recovery emergency situations could see out-of-range landings in almost any part of

Acclimatizing to the jungle.

the world (apart from the polar regions since Apollo's orbital inclination and flight profiles did not encompass these areas). Arctic wilderness training was not included in Apollo, but tropical and desert training again featured, along with water survival courses.

As well as the normal academic and simulator familiarization, the Group 4/ Group 5 wilderness training included $1\frac{1}{2}$ days of water survival (lectures in November and practical training on December 1966). The famous "Dilbert Dunker", at Pensacola Naval Air Station, Pensacola, Florida, provided real experiences of what it was like to exit an inverted aircraft or spacecraft in water.

Tropical (jungle) training was completed at the USAF Tropical Survival School at Albrook AFB in the Panama Canal Zone in June 1967, while desert survival training, again at Stead (3635th Flying Training Wing), was completed during August. Tropical survival training in Panama included two days in a classroom. The highlight of the second day was the "Jungle Buffet", and included such mouth-watering delights as braised boa, iguana thermidor, fried rat, palm hearts, and tara root. The astronauts were then taken into the jungle by helicopter on the morning of the third day. Forming into three-man teams resembling Apollo crews, they had to hike through the jungle until they found a suitable location to construct a lean-to for protection from the afternoon rain. For three days, the astronauts had to "live off the land", with no additional food supplied.

Additional speciality training for this group was in support of Apollo and Apollo Applications Program (extended lunar exploration and space station) plans. The

Desert survival training for Apollo astronauts.

group was split between both branches and specialized in the Apollo CSM, LM and Saturn launch vehicles, AAP and in supporting the relevant safety elements within those branches.

The second group of scientist-astronauts arrived in October 1967 and began an academic and training program to support later Apollo and Apollo Applications program missions. Academic and pilot training took up most of 1968, so their wilderness training took place in 1969. Only eight of the eleven selected completed this phase. Two (O'Leary and Llewellyn) had left the program by then, while Bill Thornton never did survival training due to delays in his graduation from flight school.

Desert training occurred in the summer of 1969 at Fairchild AFB, Spokane, Washington, the actual site being in the Oregon desert located south of Paso in Washington State. In August 1969, jungle/tropical training took place at Albrook AFB in the Panama Canal Zone, near the Chagares River. Water survival was at Perrin AFB in Texas during the summer of 1969.

In August 1969, NASA selected seven former USAF Manned Orbiting Laboratory astronauts to transfer to the NASA Astronaut Group. As all seven had trained for MOL missions using the Gemini spacecraft, they had all qualified for survival recovery operations under the USAF program. This utilized similar training facilities, hardware and procedures to the NASA Gemini Program, so no further wilderness survival training was necessary for this group.

Egress training from the Apollo CSM for all crews assigned to Apollo, Skylab and ASTP was accomplished in a similar format to that of Gemini. The initial phase of training took place in the classroom, followed by simulations using an engineering mock-up of the Apollo Command Module. Initially, these procedures were completed in the MSC water tank before moving on to the Gulf of Mexico near the Texas coastal town of Galveston. The crews practiced exiting the spacecraft from the surface in the Apex up position (Stable I) and then underwater with the Apex down (Stable II). As with Gemini, all underwater egress was restricted to the water tank at MSC. The final egress training was again fully suited.

Generally, about 8 hours training time was allocated for CM ingress, with another 3 hours of training to deal with fires inside the spacecraft. An average of 20 hours was allocated for egress training for Apollo, though this varied based on past experience.

For the Apollo 1 crew, egress training included exiting a boilerplate CM in an open-air swimming pool. This continued until a water tank was incorporated into Building 260 at MSC. Using the first lunar landing crew (Apollo 11) as an example to illustrate the training program, Armstrong and Collins had accumulated 17 hours egress training by 15 July 1969, with Aldrin logging 15 hours. Back-up Lovell had completed 5 hours of such training, Anders 2 hours and Haise 18 hours. Spacecraft fire training had progressed to 2 hours each astronaut, except for Anders (1 hour) and Haise (5 hours). This was typical for all the Apollo flight crews.

The Skylab crews completed 8 (SL2), 16 (SL3) and 24 (SL4) hours of rescue training, following the normal flow of briefings, reviews, and the use of mock-ups and simulators. Each of the ASTP American astronauts on the prime crew received about 13 hours in egress and fire training.

The sea-based water egress capsule was Boilerplate 1102A, used by all Apollo CM crews from 1968 though 1975 for ocean recovery training. It was almost identical in basic construction to the flight spacecraft, except that its outer shell was of painted fiberglass rather than an ablative heat shield. Months prior to a given launch, the NASA motor vessel *Retriever* would carry the astronaut prime and back-up crews, along with BP1102A, into the Gulf of Mexico off Galveston Island to undertake the complete Apollo CM water recovery practice and emergency procedures. The boiler-plate was modified for Skylab to permit a five-person crew simulation, the scenario in the event of a Skylab rescue mission. In December 2000, the *USS Hornet* Museum reported that it had obtained BP1102A on a long-term loan from the National Air and Space Museum in Washington. It was to be located later that month in Hangar Bay 2 aboard the carrier for public display, at its permanent berth at Pier 3, Alameda Point, California.

The majority of the training in recovery and survival for the first seven groups of astronauts had included water-based recovery and emergency egress situations, either by escape rocket (Mercury and Apollo) or by ejection seat (Gemini and MOL). The wilderness training focused mainly on equatorial regions of the world for unplanned landings, specifically focusing on tropical or desert regions. For the next American manned space program, crew rescue and recovery in nominal and emergency situations would change dramatically.

Shuttle

By the time the next group of career NASA astronauts had been selected (in 1978), the Space Shuttle Program was at the forefront of operations. Evaluation of the roles and responsibilities of the crewmembers had been going on for some years and reflected the change from a ballistic, one-flight vehicle to the multi-mission Shuttle, capable of a runway landing in the continental USA. As a consequence, survival and emergency training changed considerably to that of the astronauts selected during the 1959-1969 era.

Survival school for Shuttle-era astronaut candidates included water egress training and, for non-pilot candidates, ejection from T-38 aircraft. With more control over where the Shuttle could land, tropical and desert survival training was no longer considered necessary for astronaut training. However, since emergency ejection from the spacecraft could occur over remote areas of the US, training for forest survival was included in the program.

From 1978, astronauts who were successfully selected would participate in an Astronaut Candidate (Ascan) Program. This included what was termed a training and evaluation period. When they qualified, they would become fully-fledged NASA astronauts (either Pilot or Mission Specialist), eligible for assignment to technical specialities and then hopefully to a flight crew. When the first group was selected, the Ascan Program was expected to take two years, but this was soon cut to one year.

Slide basket training for Space Shuttle.

Leading the way from Orbiter to slide basket.

This continued until basic ISS systems training was included, when the program was extended to 18 months.

Today, Ascan training includes swimming and scuba qualification, and land and water survival training (T-38 related). As the basic training program ended, a group would move into technical assignments related to Advanced Training and to keeping up their aircraft proficiency skills (as pilot or crewmember). This would be followed by flight simulator training, proficiency training, reviews and refresher courses, and more specific training throughout their careers to maintain skill levels, update them with the latest developments and to prepare for mission-specific requirements. Cross-training on safety and emergency procedures would be integral to this progress.

Sixteen members from the first Shuttle era selection (1978) spent three days at Holmstead Water Survival School in Florida in July/August 1978. As many of the 35 members of the 1978 selection had already received water survival training from their military service prior to entering the astronaut program, the 16 who underwent the water survival course included all 6 female astronauts and 10 of the male candidates (McBride, Gardner, Onizuka, Hoffman, McNair, Hawley, Nelson, Hart, Van Hoften and Stewart). Their training included jumping from a tower while wearing a tethered parachute harness as they slid down a wire to a water landing. They were also towed though the water in a parachute harness, to simulate having to release themselves from a parachute that was being dragged across the water. The candidates also endured parasail towing before being released to land in the water

Suited exercise at LC39.

Water egress training for Space Shuttle.

and be recovered by boat. A final test in full survival gear saw them plunge into water from the parasail prior to helicopter recovery from a life raft.

Later in August, 11 candidates visited Vance AFB in Oklahoma to receive proper training in the event of an aircraft ejection. The 11 were Onizuka, Hoffman, McNair, Nelson, Hawley and the six female candidates. The candidates were towed by a pick-up truck while wearing a deployed parachute and harness, then released to land in a wooded area.

This wilderness and survival training has continued since the first Shuttle era selections in the late 1970s to the most recent groups selected for ISS operation. Training for the latter groups has also included additional survival and emergency training for those assigned as part of Russian Soyuz flight crews.

Fire suppression training.

Shuttle flight crews trained for on-the-pad emergency egress in simulators at JSC and at the Cape, as well as for post-landing emergency egress. An example is given in the Training Report for STS-7, where the crew conducted between three and four hours training on Emergency Procedures and four hours on Pad Emergency Egress Procedures. Emergency egress from the overhead flight deck windows, and by harness from the side hatch, was first demonstrated by astronaut Jerry Carr in 1977 during simulations at JSC.

Fire-fighting training was accomplished at JSC, teaching the crew to work together as a team in order to extinguish the fires. After a briefing on equipment and procedures by members of the Houston Fire Department and fire-fighting personnel on site at JSC, the crew would practice with hand-held extinguishers before the training was intensified. After the briefing, the crew would be taken outside to a shallow pool, on which fuel oil was poured and ignited. The crew would then direct a fire hose at the blaze, set to "Fog" to turn the jet of water into a mist which would enable the crew to approach the fire without being burned. The object of this exercise was not necessarily to put out the fire, but to forge a path through it to safety. By aiming at the base of the fire, the stream of water would part the flames. Swishing the stream back and forth would allow the crew to walk through the flames in a tight group to safety on the other side.

When PS Bill Nelson participated in this exercise during training for STS 61-C in 1985, he noted with some concern that although the instructor was wearing an aluminum fire-resistant suit, each of the NASA astronaut crew wore only their blue flight overalls. When he enquired if such a training exercise would be effective at Level 195 (feet) on the launch tower (where the crew access and exit the Shuttle Orbiter), the Fire Chief just rolled his eyes without saying anything. Nelson realized that this type of training was more useful for team building than as emergency training. A fire at the pad would certainly not resemble a small fuel oil fire on a shallow pool.

Approximately every 18 months a simulated pad egress exercise is conducted at KSC to both certify the fire rescue and crew close-out teams and to provided NASA test directors and representatives from both Kennedy and Johnson Space Centers with reality training to help their awareness of just what a Shuttle crew could experience during an actual pad egress. These tests also include representatives from leading contractors, the firing room test teams, pad facility and planning personnel and safety teams.

A three-day program is conducted (the most recent occurring in April 2008) under a flight crew evacuation program termed the Mode II/IV Simulation (Herridge, n.d.). Day 1 consists of a classroom briefing and training in the morning followed that afternoon by rescue techniques at the 59 m level at the pad where a flight crew would normally enter or exit the Shuttle whilst on the pad. The following morning the group practice and demonstrate the techniques of rescuing crewmembers at the base of the slide wire system. The activities here included exiting the basket and using the bunkers located nearby. During the afternoon they return to the 59 m level to practice using the Firex water system as well as other rescue and egress exercises. All of their training and practice is then put into an end-to-end simulation during the third day of the exercise. Here the teams follow procedures from the moment an emergency egress is declared to the simulated dispatch of personnel to local area hospitals.

The scenario featured an emergency during a launch countdown with participants acting as members of the flight (wearing Shuttle launch and entry suits) and ground crews. Fire Rescue and Closeout Crew members assist in the evacuation and rescue of the "flight crew" where simulations included the slide wire basket system demonstrating the loading and unloading operations. Once the members evacuated the basket they moved to the nearby bunker area where they wait until all members are accounted for and the emergency situation is made safe. From there the group were transported to a safe area by using an M113 armored personnel carrier, where simulated minor injuries could be treated and those with more serious injuries could be evacuated to local hospitals for more dedicated attention.

This type of high-fidelity training allowed realistic practice and refinement to the existing systems and operations and provided a wide range of personnel to become acquainted with a situation no one would like to see put into real practice.

During the most recent simulation, representatives from the Constellation Program witnessed the exercise in order to obtain operational information on emergency egress techniques and procedures from LC39 and the use of Shuttle launch and entry suits. This information would be used to develop and implement similar procedures for the Constellation Program once the Shuttle had been retired from flight operations in 2010.

Constellation

It is too early to determine the survival training required for the new Constellation Program, as each NASA astronaut selection from 1978 though to 2004 has essentially completed similar Ascan training. However, with the selection of the next astronaut group planned for 2009 and the retirement of the Shuttle in 2010, Ascan training will have to change for Soyuz or Constellation ballistic training, based on the earlier Apollo Program training.

RUSSIAN COSMONAUT SURVIVAL TRAINING

Russian cosmonaut training has always featured parachute training and a varied survival training program, both for its national team and for guest cosmonauts. Some of this training has changed very little for almost 50 years, while other parts have been specific to a program, or changed to incorporate new operations, procedures and objectives. Combination training is also a feature, in which parachute, water and mountain training can be included in the same program.

Vostok and Voskhod

One of the earliest aspects of a Vostok mission that could be accurately simulated (pending the development of more sophisticated cosmonaut training facilities) was parachute descent from the spacecraft. Though many of the first cosmonauts had only completed the minimum requirement of five mandatory parachute jumps at the start of their Air Force careers, the cosmonaut candidates of the first team selected in March 1960 completed over 40 jumps over a 6-week course, qualifying them as "experts" in the field. Several jumps were from high altitude to simulate the stresses of "weightlessness" followed by accelerated flight. Some of Russia's most experienced parachutists conducted the training.

Ejection seat tests were conducted on an ejection tower, with a Vostok seat bolted to the structure to simulate the short, explosive exit from the spacecraft. The cosmonauts were also strapped into a parachute harness attached to an overhead rig. Though planned for terrain landings, the possibility of water splashdown existed for Vostok, so the cosmonauts trained for the event and practiced water survival training with full egress equipment in the Black Sea. They were also towed through a water tank facility, on their backs, to simulate a parachute being dragged by the wind across the surface of water.

When the first female cosmonauts arrived in March 1962, they also had to complete parachute, ejection seat training and water survival training in order to qualify for Vostok flights. For Voskhod, the removal of ejection seats and the lack of an ejection tower reduced the likely survival scenarios that the crew would encounter. Survival training for Voskhod was thus severely limited. As with Vostok cosmonauts, each Voskhod crew completed wilderness training, in case their spacecraft came down out of the primary recovery site. This was fortunate as Voskhod 2 did just that.

Vostok parachute training (top).
Voskhod winter survival training (bottom).

Soyuz

Since 1967, the only fully operational Soviet/Russian manned spacecraft has been the versatile Soyuz, which has been upgraded over the years to encompass the Soyuz T, TM, and more recently, TMA versions. Though the capabilities of the Soyuz have changed considerably over the ensuing four decades since it was first introduced, the basic flight profile has changed very little. As a result, the ascent and landing procedures for Soyuz have also been essentially the same since the first missions were flown in the late 1960s.

Survival training forms part of the basic cosmonaut training program (OKP) and, as with American astronauts, is repeated throughout each cosmonaut's career to upgrade and refresh their knowledge and skills, to maintain proficiency and to become aware of the latest upgrades and procedures. The survival pack carried on each Soyuz allows the crew to cope with an emergency landing in remote and harsh areas. The training to prepare for such landings falls under the supervision of the Third Directorate of the Yu. Gagarin Cosmonaut Training Center near Moscow, which occupies one building of the training center. The actual training program is a cooperative venture between the training center and the Federal Aerospace Search and Rescue Administration. Experience has proven invaluable in this area and the department is normally headed by either a flown or unflown cosmonaut. Full details of this directorate have been discussed in the companion volume in this series *Russia's Cosmonauts: Inside the Yuri Gagarin Training Centre* (Hall *et al.*, 2005) and are summarized here.

Winter. Early cosmonaut training photos have revealed winter training during the 1960s. The training was put to practical use by the Voskhod 2 cosmonauts in March 1965, when they landed in Siberia outside of their primary landing zone. It was in support of the Soyuz Program that a more in-depth training protocol developed, with cosmonauts travelling to Vorkuta in the Arctic. Here, over two days, they build an igloo and use a parachute and cut tree branches to construct a shelter around their mock-up Soyuz descent capsule. Prior to embarking on this demanding training course, the cosmonauts complete academic and practical training to learn how to deal with a Soyuz landing in deep snow. More recently, the Russians have also identified planned Soyuz emergency landing zones in Canada and Siberia, where similar winter conditions could be encountered. As well as providing the cosmonauts with experience in handling equipment in such harsh conditions, the training has also provided psychologists and doctors with medical evidence on each cosmonaut's state of health, physical strength and ability to work under stress conditions. International Space Station resident crews also complete Soyuz winter training and have used remote areas within the grounds of TsPK as part of this preparation.

Mountain. As the Soyuz could land in mountainous regions, survival training to cope with this eventually became part of the training program. Once again, this has been put into practice during a real space mission, when the Soyuz 18 crew intended to man the Salyut 4 Space Station had their ascent aborted and landed on the side of

Swamp training for Soyuz. Courtesy Bert Vis.

Soyuz desert training. Courtesy Bert Vis.

Winter training for Soyuz.

a mountain in April 1975. This became known as the April 5 Anomaly and was not given a formal Soyuz mission designation (though it has been called Soyuz 18-1 in the West). Mountain training can also include elements of winter survival training.

Desert. Heat chamber endurance testing is conducted at the cosmonaut training center and is known for being one of the more arduous elements of early cosmonaut training. Occupants have to endure up to 40°C, cope with changes in the composition of gases inside the chamber, and also wear fur-lined flight-training suits in temperatures of up to 80°C. Again, this type of training also provides useful information on the physical and psychological condition of the cosmonaut for the medical team. There is also a 24-hour wilderness exercise in desert conditions. When this type of training started in the late 1970s, the area used was the Kara Kum desert near the town of Mary in Turkmenistan. Since the collapse of the Soviet Union in the early 1990s, this training has been conducted at the Baikonur Cosmodrome in Kazakhstan. As with their American counterparts, the cosmonauts have to learn to use equipment from their spacecraft to provide shelter, supply additional water and signal rescue teams while enduring the harsh heat of the day and the bitter cold of night.

Swamp. Occasional training photos have been released showing cosmonauts undergoing survival training in swamp-like conditions. It is not clear if this is undertaken by every crew or limited to certain missions. Nor is it known precisely where this type of training is conducted.

Sea recovery. Soyuz is normally targeted for a terrain landing, but the possibility of a sea or water landing is always present if a mission is terminated early or under emergency conditions. Therefore, all crews train for water recovery. This type of training was again put to practical use, in the 1976 landing of Soyuz 23 in Lake Tengiz in Kazakhstan, under severe weather conditions and at night. Before embarking in sea trials in the Black Sea near Sochi, crewmembers practise their techniques and skills on the deck of a ship, using the same Soyuz mock-up capsule as they do in the water. They can also use a small hydro-tank on the ship or at Star City to practice the sequence of events before trying them in open water, where motion sickness can add to the challenge of the training. Two capsules have been identified for use during this type of training: a blue-painted capsule called *Ocean*, and a bright orange one with an image of a dolphin on its side.

Dropped from the side of *Sevan* (a ship), the cosmonaut crew inside the mock-up Soyuz practice putting on their survival or extreme weather gear. They can then either exit the spacecraft into an inflatable dinghy wearing their orange *Forel* survival suits, or be recovered by sling from a helicopter. This is another strenuous exercise which can last for over two hours. Bobbing around in the swell in the small confines of the Soyuz, the training crew frequently has to endure heat exhaustion and motion sickness on this exercise.

Parachute training. All cosmonauts are given an extensive course in parachute training to gain experience of stressful and potentially dangerous situations. Many cosmonauts have continued their parachute jumping beyond mission training to achieve Instructor Parachutist awards. This often includes free-fall descents, and some cosmonauts have logged over 500 jumps. During the Vostok Program (1960–1963), cosmonaut ejection from the capsule and descent by individual parachute was the main method of recovery, so parachute training was a necessary requirement for assignment to a Vostok flight crew.

Water egress training for Soyuz in a water tank. Courtesy ESA.

Zond

During preparations for the manned Zond lunar program, ocean landing was an option in the flight profile, so cosmonaut training for ocean recovery was a major part of their short training program before the project was canceled. Other survival training would probably have been similar to that of a returning Soyuz, as Zond was a variant of the basic Soyuz design.

Buran

For the Buran Shuttle, ejection from the spacecraft was a possibility in emergency situations, so mountainous regions were included as part of the Buran cosmonaut group survival training program. A week-long trek on foot was completed in the

Open sea egress training for Soyuz.

Pamir mountain range, departing from the city of Frunze and heading to Issyk-Kul. They also completed a mountain ski trip as part of their survival training, this time in the Dombai region. Prior to assignment to Buran missions, pilots conducted cosmonaut OKP training in addition to their test pilot and Buran flight training program. Several were also assigned to Soyuz flight crews and completed the required Soyuz capsule survival training program.

CHINESE SURVIVAL TRAINING

As Shenzhou is similar to the Russian Soyuz spacecraft, the Chinese training program is probably similar, including survival training in extreme and hazardous conditions. Photographs of Chinese taikonauts conducting water, desert, winter and mountain survival from a Shenzhou capsule have been released, indicating that preparations for emergency situations and their support equipment are similar to that of the Russian Soyuz Program.

PASSENGER AND SPECIALIST SURVIVAL TRAINING

Between 1959 and 1976, American astronauts and Soviet cosmonauts were selected for training on domestic spacecraft and missions. The exception to this was the joint US/USSR international docking mission, the Apollo-Soyuz Test Project, in which an

American Apollo docked to a Russian Soyuz. Though joint training sessions were completed, along with familiarity briefings and training sessions on each other's spacecraft, emergency and survival equipment, no joint survival training was completed by either crew outside of their own training protocol.

In the early 1970s, NASA announced that the forthcoming Shuttle Program would allow non-career astronauts and foreign guest scientists to fly short, one-off missions. Training for such "passengers" was abbreviated to that of the NASA career astronauts, but nevertheless featured elements of survival training from the Space Shuttle Program. In 1976, the Soviets announced that it, too, would train cosmonaut guests from Eastern Bloc and foreign countries to fly scientific, week-long research missions on their Space Station programs. Again, full cosmonaut training was not a requirement, but an abbreviated period of preparation would be completed, which of course included the required survival training as part of a Soyuz crew. Since 1992, the American Shuttle and Russian Soyuz Programs have become almost combined, initially to support the Shuttle-Mir Program, but more recently to support the International Space Station Program. Survival training for foreign astronauts and cosmonauts has become a regular feature of spaceflight preparations for both the U.S. and Russian national programs.

American payload specialists

In 1978, the first Shuttle payload specialists were identified from ESA and the United States. The training protocol for these non-career and foreign astronauts was being developed at NASA and was constantly revised over the course of the Shuttle Program. The loss of Challenger in 1986 temporarily suspended Payload Specialist activities. When they resumed in 1990, it was primarily for major science payloads, rather than as "guest positions" to support a satellite deployment, or as an "observer". As the program moved more towards ISS operations, the flights of non-career astronauts declined. Training focused more on inviting partner space agencies to send national representatives to NASA's Astronaut Candidate training program in preparation for Mission Specialist roles on Shuttle ISS assembly missions or for docking to the Russian Mir Space Station.

The program to prepare Payload Specialists for flights on Shuttle fell under the PS Flight Preparation Plan. This listed the requirements necessary to prepare assigned PSs for STS missions. This was also supported by the PS Operations and Integration Plan, which detailed the integration process for a PS as a member of an assigned flight crew. Specific requirements changed from flight to flight, depending on the PS objectives, but the emergency and safety training was normally standardized across the program. The Plans stated that survival training should "provide sufficient emergency and safety training [so that] those hazardous situations can be handled with speed and skill."

Though detailed in a document related to Payload Specialist training, this overview is applicable to all Shuttle emergency and survival training.

The training program began with a series of orientation and familiarization sessions, including a series of videos and tours related to safety and survival training.

These included fire protection and toxic propellant safety at KSC, LC 39 Pad A and Pad B safety familiarization, as well as safety information on the Shuttle Orbiter and Spacelab facility. There were video presentations on the Emergency Egress System (the slide wire) at LC Pad A and Pad B, and the hypergolic fire suppression system. Courses completed as part of integration into a specific crew included three and a half hours of briefings and demonstrations of fire-fighting techniques, which included hands-on practice.

There were 14 hours of extraction training (mainly bailout-related) and 24 hours of water survival training, scheduled over three days. The *Bailout Workbook* (Bailout 2102) was covered over one hour and contained information on the Crew Escape System and related procedures. This was followed by a series of courses. Bailout Intro 2101 was a one-hour activity supporting this system and its operation. A six-hour (two 3-hour sessions) course (Bailout 3120) covered all aspects of the bailout activity, including the simulated depressurization of the cabin, simulated side hatch ejection and simulated bailout. Deployment of flight–type escape equipment was also practiced.

Water survival certification was a prerequisite for the Bailout 3127 water survival training in the WETF and addressed the deployment and use of all Shuttle unique water survival equipment.

Bailout 3127 included a poolside briefing. Each crewmember, wearing full flight gear (launch and entry suits), participated in simulated water landings, including water drag simulations, flotation device activation, ingress and configuration of the life raft, and a simulated water rescue.

One week before the actual training session, a walkthrough (Bailout 3220) was conducted for equipment familiarization. Five weeks prior to the planned launch (L-5), the crewmember undertook a refresher course for equipment and procedures, covered in Bailout 3120.

Russian cosmonaut researchers

From 1976, cosmonaut guest researchers (initially from the Eastern Bloc, then international partners) were included in the cosmonaut training program. As the third crewmember on Soyuz, they all had to complete the Soyuz emergency and survival training course to qualify for a flight seat.

Parachute training continued, although only a couple of jumps were required. In her 1993 book, UK cosmonaut Helen Sharman recalled her strong feelings of nervousness prior to her first parachute jump. Once she approached the open door to make her jump, then all her nerves had gone. The first descent was enjoyable, the second even more so. But she was not allowed to complete any more jumps: "My well being and safety were no longer my exclusive concerns. Two jumps were all I was to be allowed" (Sharman and Priest, 1993). She also recalled her experiences of water survival training during June 1990, in temperatures of 30°C. Taking off pressure suits and putting on survival gear, gathering emergency rations and equipment, and then exiting the spacecraft as it swayed in the sea seemed simple enough. But in practice, it was all physically tiring. In reality, it took almost three hours for the TM-12 crew to exit the spacecraft: "My skin temperature went up 2°C during this time, and because of the wave motion there was a strong smell of vomit by the time we'd finished. This is an experience in which your relationship with the other people in the crew is tested to the limit. However, we made it" (Sharman and Priest, 1993, p. 165). Once outside the spacecraft, the training is far from over, however. Each crewmember had to hold on to the next in line, or form a triangle in case of heavy seas or a storm, and be prepared to remain this way for up to three days. Sharman reasoned that although the training gave you the confidence to survive, everyone hoped that the Automated Landing Systems worked and a nominal landing was achieved.

International crewing

As multi-national crews became common, survival training had to be incorporated into their training profile. For those flying on the American Shuttle, the bailout system and pad egress training on LC39 was completed. For Soyuz, survival and pad abort situations were covered.

REFERENCES

Rex D. Hall, David J. Shayler and Bert Vis (2005). *Russia's Cosmonauts: Inside the Yuri Gagarin Training Centre*. Springer-Verlag, pp. 91–104.

Linda Herridge (NASA JFK Space Center writer). *Mode II/IV Training at Kennedy*. Available at *http://www.nasa.gov/mission_pages/shuttle/behindscenes/modell-lv_prt.htm* [last accessed 1 May 2008].

NASA (n.d.). *Bailout Workbook*. NASA (Washington, D.C.).

Helen Sharman with Christopher Priest (1993). *Seize the Moment*. Victor Gollancz, pp. 160–161.

4

Pad escape

The first phase of a flight into space is potentially the most dangerous, from the point of view of the crew onboard. Sitting atop a fully fuelled launch vehicle on the pad or in the early stages of a flight, the chances of anything going wrong (and the magnitude of the disaster, should that happen) are probably at their greatest here, given the complexity of the vehicle, the dynamics of the flight and variations in atmospheric conditions. Getting off the Earth may be one of the most thrilling rides of your life, but it carries the greatest risk of being your last. Protecting and recovering the crew from catastrophic and fatal incidents at launch has been a challenge for spacecraft designers since humans first ventured into space.

IF A LAUNCH GOES WRONG

Generally, two types of emergency recovery systems have been used on manned spaceflights to date. Ejection seats were employed for Vostok cosmonauts and Gemini astronauts and ejection seat technology was available for the first four two-man crews on the American Shuttle. It was also planned as a primary rescue system for the canceled Soviet Buran space shuttle. The second mode of rescue featured escape towers, pulling the crew compartment clear of impending disaster. These were used on the American Mercury and Apollo spacecraft, the Russian Soyuz spacecraft and, more recently, on the Chinese Shenzhou missions.

With the development of a multi-person vehicle such as the American Shuttle, where astronauts were seated on two levels, the use of ejection seats became impractical, and escape towers and separation of the crew module too complicated and expensive. Instead, a series of procedures and flight profiles were developed which would allow Mission Control to abort the ascent at various stages of launch. In theory, these aborts could still save the crew and, hopefully, the Orbiter.

Over the past five decades, only three crews have actually used escape systems or abort modes during a mission. In 1975, a Soyuz crew intended to reoccupy the Salyut 4 Space Station had to abort their ascent due to the faulty separation of a rocket stage. In 1983, a second Soyuz crew performed the first off-the-pad manned abort just seconds before their launch vehicle exploded beneath them. In 1985, the 19th Shuttle ascent suffered a main engine failure, resulting in the first use of a Shuttle launch abort mode to safely deliver the vehicle to orbit.

There have also been one Gemini and five Shuttle missions aborted on the pad. In addition, there have been several problems during ascent which have not required a full abort situation. Many of these have been resolved by alternative systems recovering the situation, such as Apollo 12 or Apollo 13, or have failed simply due to the lack of time to initiate any type of rescue, such as the loss of Challenger. Of course, being aware of damage caused to the vehicle during launch would perhaps have given those few vital extra seconds to evaluate whether repair or rescue was possible or practical.

Learning from past accidents and mistakes can only enhance the development of new and safer protection systems for the crew. This is clearly the thinking behind the development of the launch protection system for the Constellation Program on spacecraft Orion. Plans for a launch escape tower have been central to the design for some time and are currently undergoing testing prior to employment in operational flights. As long as we use current rocket technology to reach space, then we must include a reliable and effective rescue option for the crew onboard.

AT THE PAD

The first options for the crew to escape from a launch problem occur at the pad prior to lift-off. These can include equipment or procedures incorporated into the launch support equipment, or from the vehicle itself utilising escape towers or ejection systems. The latter two are discussed in the next chapter. Here, we focus on facilities available at the various launch pads in the United States, Russia and China.

Baikonur Cosmodrome

The first launch facility to support a human flight into space was constructed at Tyuratam, in the Kzyl Odra Raion, in the Republic of Kazakhstan. This is more commonly known as the Baikonur Cosmodrome. It has supported the launch of all Soviet and Russian manned spaceflights since 1961 and despite numerous funding and political difficulties, remains an important launch complex for the sustained exploration of space. When the Shuttle was grounded following the loss of Columbia in 2003, the launch facilities at Baikonur allowed space agencies to continue to support International Space Station operations. They will continue to do so for some years after the Shuttle is retired.

Vostok and Voskhod. With the R-7 on the launch pad, service structures and access support platforms surrounded the crew compartment in the hours leading up to the launch. When all was ready, the crew were sealed into their spacecraft and the access arms retracted. The launch support team moved to nearby bunkers. Any problem prior to launch would require replacing the access arms and support gantry, removing the side hatch and extracting the cosmonauts. The whole team would then descend from the launch tower and retreat to a safe distance—should there be sufficient time. Problems during the final stage of launch during Vostok would require the lone cosmonaut to use the ejection system. But this had been removed for Voskhod, so crew evacuation on the pad would have been more difficult for the three unsuited cosmonauts on Voskhod 1 and the two suited cosmonauts on Voskhod 2. Fortunately, pad egress was never called upon for Vostok nor Voskhod, as it would probably not have been without fatalities.

Soyuz. Soyuz uses the same pad as Vostok, and its three-module design hinders quick access to and extraction of crewmembers. With the crew strapped in the Descent Module (beneath the Orbital Module and inside the launch shroud) extracting the crew close to launch and under pressure would be difficult in an emergency. Therefore, for emergencies leading to launch and during ascent, the Soyuz is fitted with an escape tower, and on one occasion in September 1983 this was called upon to rescue a crew from impending disaster (Shayler, 2000).

Cosmonauts for the Moon. Larger launch vehicles were to be used in the attempt to send cosmonauts around the Moon (the UR-500K Proton) and support a manned landing mission (using the huge N-1). Both the manned circumlunar (L-1 Zond) and lunar orbiting spacecraft (the Luniy Orbitalny Korabl or LOK) were fitted with an escape tower which could be used in the event of an explosion of the launch vehicle.

Buran. Buran was the Russian space shuttle system, which had a long gestation from rocket research aircraft and military spaceplane projects in the 1960s (Hendrickx and Vis, 2007). After an enormous, and ultimately disappointing, effort to send cosmonauts to the Moon in the late 1960s and early 1970s, Vladimir Barmin, the chief designer of KBOM (Design Bureau of General Machine Building) launch pad design bureau, had insisted on using at least some of the investment at Baikonur to support launch operations for Buran. The launch pads of the N1 (Pad 37 and Pad 38) were modified to support the proposed launches of the Buran/Energia system. Other, more cost-effective and less complex plans were put forward to develop launch facilities for Buran, but these were not pursued.

The rotating service towers on both former N-1 pads were retained, though they were shortened by about 60 m to correspond to the two fixed service structures. Apparently, this was to minimise the possibility of exhaust flames impinging on the structures during launch, though both were 'parked' at relatively safe distances from the pads. In position on the launch pad, the Buran faced the rotating service structure, which may have prevented the use of the crew ejection seats in the event of a pad emergency, had the structures remained at their original height.

In a departure from the slide-wire-and-basket escape system available to American Shuttle crews, the Buran launch support facilities featured two tubes leading

Buran on pad showing escape tubes.

down to a pair of separate underground rooms from the access arm. The upper tube featured 12 person trolleys used to hoist launch and flight crew personnel to the access level on Buran. The lower tube featured the pad emergency egress system. In the event of an emergency on the launch pad, workers or the crew could slide down the lower tube. The slope they had to negotiate to reach one of the underground rooms was relatively gentle but sufficient to provide a quick slide down an "escape chute". At the bottom, a giant mattress cushioned the landing. Once underground, the crew or work force could hermetically seal themselves inside an adjacent bunker to protect them from an impending explosion, or the leaks of poisonous gases. Once the area was deemed safe to enter, they could be rescued by other ground personnel. In evaluating such a design in 1986 the Scientific Research Institute of Chemical and Building Machines (NIIKhSM) located in Zagorsk, to the north of Moscow, constructed a test stand reproducing the pad escape slide. The device at the pad was also tested several times by engineers and, it appears, by off-duty soldiers at Baikonur who reportedly enjoyed the unusual attraction to occupy off-duty hours.

Cape Canaveral

Early American space missions took place at adapted facilities at the Eastern Test Range in Florida, more commonly called Cape Canaveral. With the creation of the Apollo lunar program, new facilities were constructed north of the ETR. This was designated LC39, and later became known as the Kennedy Space Center. It was from these facilities that the Space Shuttle first launched in 1981. Planned launches of Ares will be from modified facilities at LC39.

Launch Complex 5/Launch Complex 6 (Mercury-Redstone). In the final 90 minutes prior to lift-off, with the astronaut aboard the spacecraft on the pad, an elevating boom known as the "cherry picker" was available to remove the astronaut during any emergency situation in which the escape rocket system could not be used.

Launch Complex 14 (Mercury-Atlas). For the Atlas orbital missions, an emergency egress tower ramp was available to extract the astronaut. This facility was evaluated during the launch simulation test, which was designed to evaluate both the spacecraft and launch vehicle systems in a launch configuration, as well as evaluating procedures to be used on the day of launch.

When the platform was extended, it aligned adjacent to the side hatch of the spacecraft. When not required it was stowed vertically, locked in position until needed so that the platform did not strike the launch vehicle as it ascended. It could be lowered in about 30 seconds by remote command from the blockhouse and gave the astronaut a means of escape unaided by other pad workers. If the astronaut was incapacitated, the external egress crew could use the tower to remove the astronaut to a safer location for recovery or medical attention. The "cherry picker" mobile access tower device used during the Redstone flights was available, but tests had revealed that the structure could potentially interfere with radio communications. In addition, the greater pressure generated by the Atlas engines had the potential to significantly damage the tower. The static egress tower was therefore chosen for Mercury–Atlas support. The "cherry picker" was located behind the blockhouse as a substitute if the primary egress method failed.

To deliver an egress crew to the pad, or to evacuate the team from the area, special rescue and fire-fighting vehicles were positioned just outside the complex

Cherry picker for Mercury-Redstone.

perimeter. These vehicles were protected by special thermal insulation and were able to extract an astronaut to a safe zone outside LC14 in approximately 150 seconds (2.5 minutes). In addition to the mobile vehicles, there were four nozzles of a fire-fighting system installed on the pad. These were remotely controlled from the

Mercury-Atlas launch.

blockhouse. The nozzles could be directed to deliver either water or fire-smothering foam to any area within the complex. The radio command subsystem allowed controllers within the blockhouse to fire the escape rocket and abort the mission prior to launch, and to rearm the primary system of abort for Mercury-Atlas launches in the first 10 seconds of flight (through "tower-clear").

Launch Complex 19 (Gemini-Titan II). At the top of the new extended launch tower developed for Gemini missions was a "white room", which was used for access to the Titan upper stage and the spacecraft. Escape from the spacecraft prior to launch would, ideally, utilise the twin ejection seats. This system was almost put to the test on December 12, 1965 by the crew of Gemini 6.

After the loss of their Agena target vehicle on October 25, plans were made to send Gemini 6 to rendezvous with Gemini 7, which was already in orbit and nearing the end of its 14-day space marathon. On December 12, though ignition of the Titan was achieved, the engines automatically shut down 1.2 seconds later, prior to lift-off. Walter Schirra correctly assessed the situation and elected to remain in the vehicle rather than eject, thereby preserving the option to try to launch again on December 15. Had he ejected, then it would not have been possible to refurbish the vehicle on the pad in time and the chance to achieve a milestone in space history and gain important experience in rendezvous and proximity operations in orbit would have been lost.

Gemini 6 crew leave the pad after a launch abort.

Launch Complex 34 (Apollo Block I–Saturn 1B). At the 67 m level, the Apollo access arm provided the prime crew and support team access to the CM atop a Saturn 1B via the White Room. The umbilical tower elevator, which could travel at 137 m per minute, was the primary emergency escape method. This was set to run non-stop from the 67 m level of the umbilical tower down to the ground. A slide-wire-and-basket system was available as an alternative means of quick pad exit. This would have been used for immediate pad evacuation. The 365 m slide wire was attached at the 67 m level and it would have taken only 30 seconds to reach the pad perimeter and safety bunkers.

Following the Apollo 204 pad fire, a number of additional changes were incorporated into pad facilities at LC34 and into the new facilities at LC39. These included structural improvements and additional fire-fighting equipment, emergency egress routes, and emergency access to the spacecraft during tests and launch operations where possible.

Electrical equipment in the White Room was purged with nitrogen. Large exhaust fans drew smoke and fumes from the area, and the room was decorated with a fire-resistant paint. To provide easier access to the spacecraft, certain structural fixtures were removed. In addition, a hand-held water hose was available for fire fighting. A water spray system was installed to cool the launch escape system (located above the CM) in the event of a pad fire and additional water spray systems were incorporated throughout the egress route from spacecraft to ground level.

Launch Complex 37 (Apollo Block II–Saturn 1B). Facilities similar to those at LC34 described above.

Kennedy Space Center

To support the launch of the huge Saturn V rockets designed to take Apollo spacecraft to the Moon by 1970, NASA constructed a new complex north of the Cape Canaveral facilities. They eventually chose an area around Merritt Island which became known as Launch Complex 39, or more commonly 'Moonport', during the Apollo Program. These facilities were later converted to support the launch of Saturn 1B vehicles for Skylab and Apollo Soyuz, the Space Shuttle, and more recently, the Constellation Program.

Launch Complex 39. Part of the contract for the fabrication of Pad A was to provide an Emergency Egress System for pad workers and the crew in the minutes leading up to a launch. This was in conjunction with the Apollo Launch Escape Tower, designed for off-the-pad aborts as well as during the early stages of ascent.

Apollo

In the event of a hazardous condition occurring during countdown, but without sufficient time to leave the launch area, the Apollo crew were provided with two methods of evacuating the pad, as long as it was still safe to do so.

Apollo Saturn V on the launch pad.

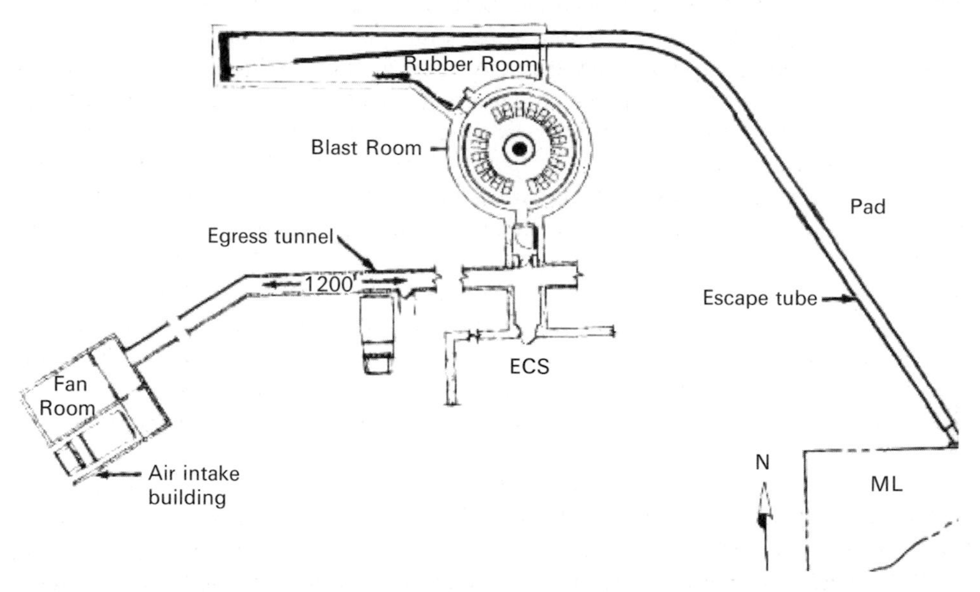

Layout of LC39 (typical).

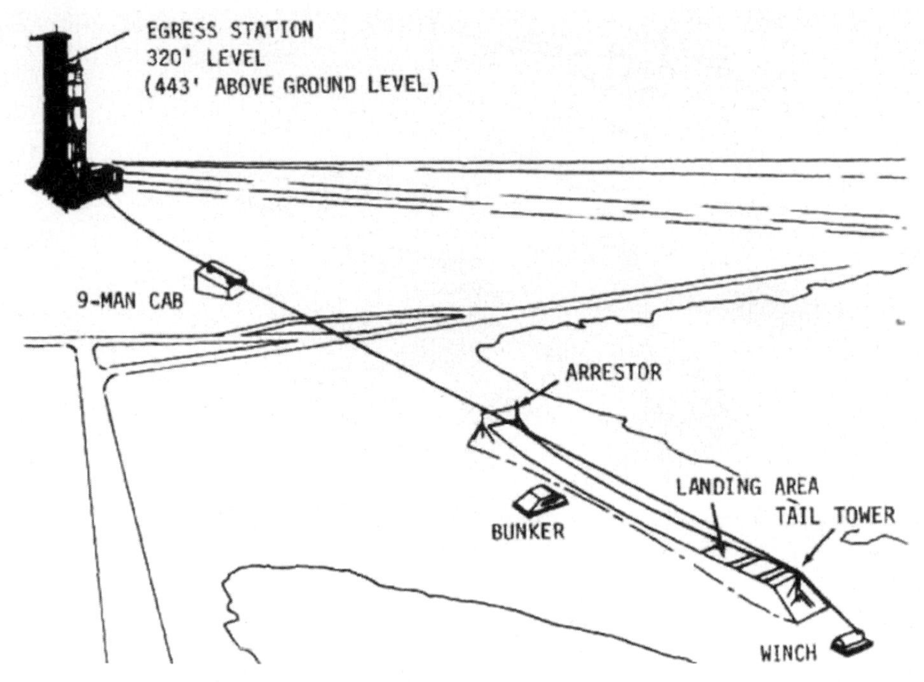

Diagram of LC39 pad slide wire facilities.

Known as the Apollo Emergency Ingress/Egress and Escape System, it provided access to and from the CM, as well as an escape route if pad evacuation was impossible. The astronauts (and support crew) could exit the spacecraft and walk over to the mobile launcher via a swing arm. From there, they could board a high-speed elevator from the 104 metre level and drop 30 stories, at 183 m per minute, to Level A in approximately 30 seconds. They would then exit the elevator and slide down an elevated curve inside a 61 m long flexible slide tube to a thickly padded deceleration room (called the termination room). They would then enter a blast room and close huge steel doors resembling those of a bank vault. Seated in contour chairs and wearing safety harnesses, up to 20 people could theoretically have survived the explosion of a Saturn V and used the facilities inside to stay alive for the next 24 hours. Rescue crews could probably have dug them out once the pad fires had been extinguished. However, it is worth pointing out that a smaller Soyuz R-7 exploded on the pad at Baikonur in September 1983 and the debris burned for 24 hours. A much larger Saturn V explosion would probably have burned for much longer than one day.

The dome-shaped facility lies 12 m below the pad structure. It is 12 m in diameter, with 0.8 m thick steel and concrete walls, and huge steel doors designed to withstand blast pressures of 3.5 bar and accelerations of 75*g*. The 3.3 m high domed room was built on a spring suspension system, which reduced the potential 75*g* that could have been exerted on the room to a mere 4*g*. The whole facility was covered with an Earth

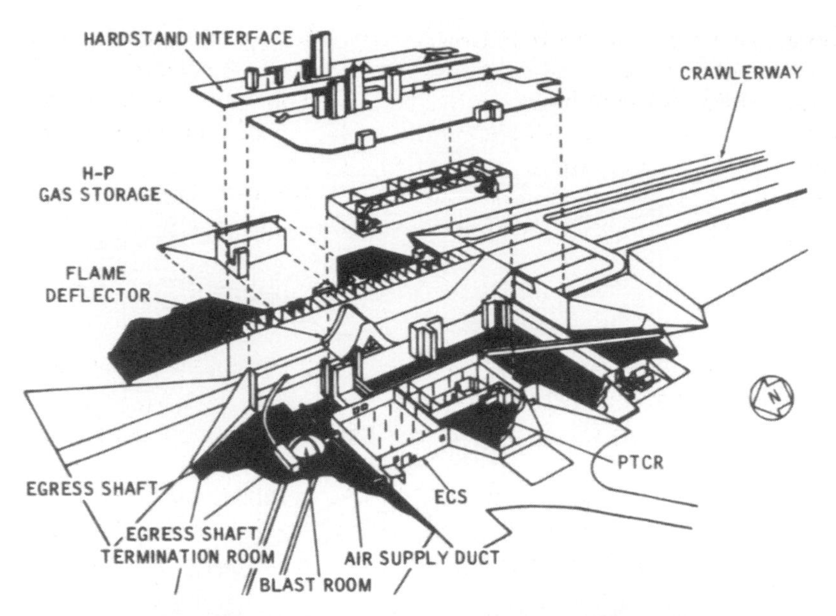

Cutaway of Blast Room Pad 39 (top). Blast room under construction at Pad 39 (bottom).

revetment for added protection. Fortunately, this system was never called upon during Apollo operations, although it was evaluated by astronaut Charles Duke in 1968, who became the first astronaut to slide (partially suited) down the tube into the termination room.

An alternative method of escape from the crew assess arm level was via a cable cage. The cable cage used a slide-wire-and-basket system leading to a bunker approximately one-half mile from the launch site. Astronaut Stuart Roosa evaluated the slide wire system during Apollo.

A fire suppression system called FIREX was incorporated into the pad facilities during construction. This supplied fire-fighting water to the pad propellant storage facilities, the high-pressure gaseous hydrogen facility, the Mobile Service Structure, the perimeter fire hydrants and the Mobile Launcher fire hose connections, from separate pumps located at the industrial water–pumping station.

Specially designed and protected vans were on standby for such emergency situations, until the area was cleared and the countdown neared its end.

Shuttle

After Apollo era operations came to an end following the Skylab and Apollo Soyuz Test Project launches in 1973 through 1975, attention turned to converting the facilities at Launch Complex 39 to support Space Shuttle operations from Pad A and Pad B.

Apart from the six fixed pedestals which supported the Mobile Launch Platform, all of the structures developed and installed for servicing the Saturn vehicles were removed, rerouted or relocated. Reconstruction of Pad A was completed in mid-1978 and Pad B in 1985.

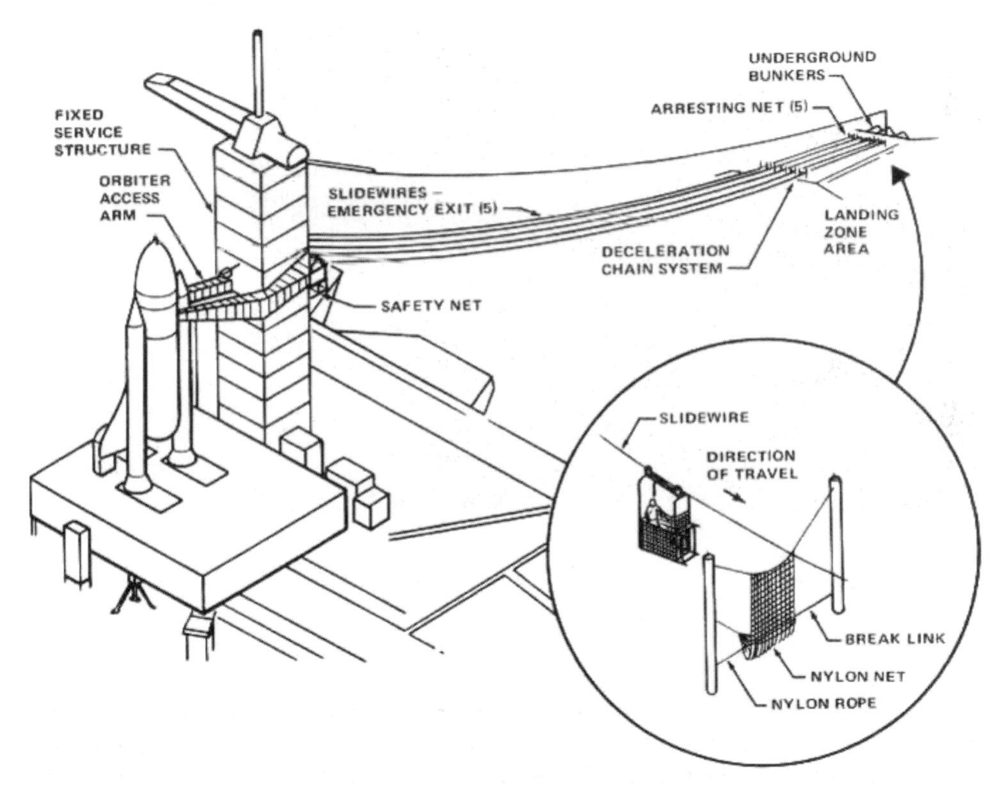

Shuttle slide wire facilities at Pad 39.

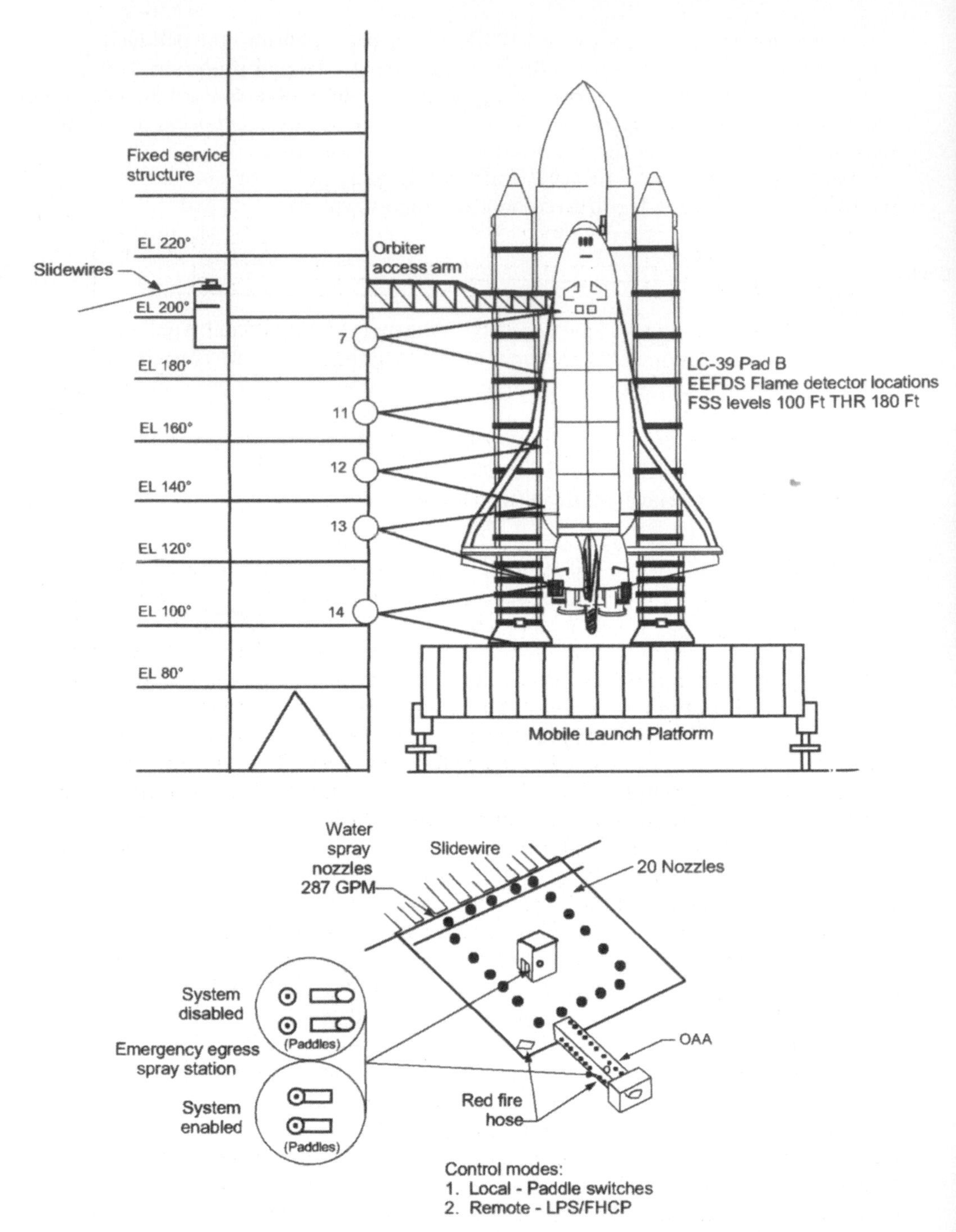

Shuttle on Mobile Launch Platform with tower. (Top) Flame detection (EEFDS). (Bottom) Emergency water spray control panels.

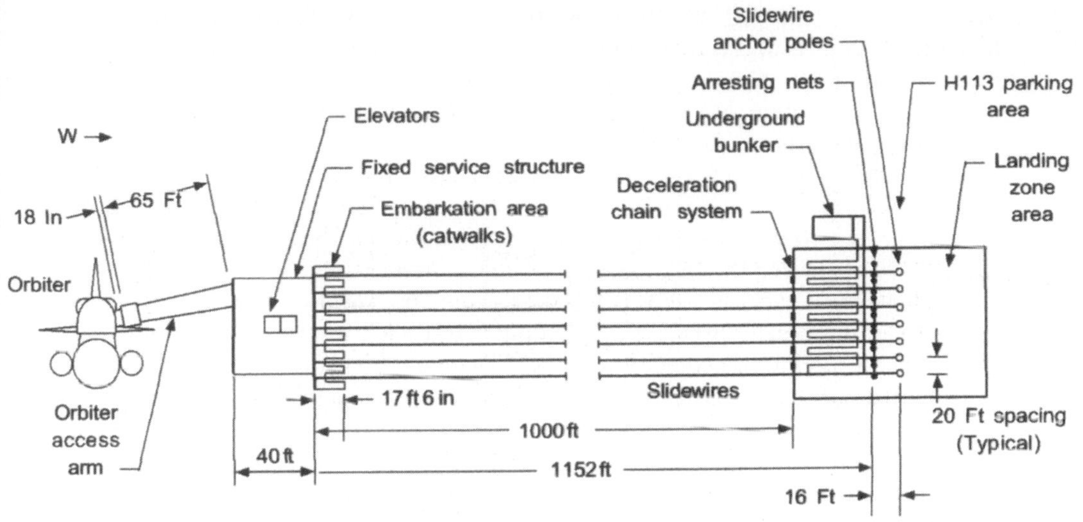

Shuttle slide wire detail of the Launch Pad Emergency Egress System.

The upper portion of the umbilical towers were removed from the Mobile Launcher Platforms and relocated at each pad to serve as a Fixed Service Structure (FSS). The Shuttle FSS also includes an upgraded Emergency Exit System ('the slide wire'). These are accessible to personnel on the Shuttle or the Orbiter Access Arm until the final 30 seconds of countdown. Initially, there were five two-man slide wire baskets. These featured flat bottoms and netting surround the structure, and were positioned on the FSS in a state of readiness. A maximum of four persons could be carried in each basket (maximum 20 persons for the system). The number of baskets was later increased to seven, 1.5 m wide and 1.1 m deep and capable of carrying three persons. Slide wires extend from the Orbiter Access Arm level on the FSS down to the ground on the edge of the pads, where bunkers are provided for added protection. In operation, the baskets slide down the 365 m wire to the arresting nets at the base of the system. This consists of a catch net and drag chain to slow and arrest the buckets at the landing zone. A lightning shield wire is part of the lightning mast protection system to prevent additional electrical discharges during storms. Water delivery systems and escape vehicles located near to the pads provide additional support during off-the-pad exit in the event of a vehicle pad abort.

Crew (and pad team) training includes various emergency scenarios, which are practiced regularly both in preparation for a mission and in between launch operations. During the down time created by the loss of Challenger in 1986, emergency egress procedures were evaluated by an Emergency Egress Test Crew. Not classed as an official "flight crew", they undertook these evaluations as a program support role to develop their own experiences in pad operations and emergency training and to evaluate systems and procedures at the pad and for the launch team in a simulated situation. Tests such as these provide an opportunity to maintain proficiency and to develop personal and teams skills during a period of flight inactivity. On November 2,

1986, and again on April 28, 1987, the Emergency Egress Test Crew comprised astronauts Frank Culbertson (taking the role as 'Commander'), Steve Oswald ("Pilot") and Carl Meade, Kathryn Thornton and David Low ("Mission Specialists"). Two other NASA MS astronauts, Pierre Thuot and Jay Apt, acted as Payload Specialists for the simulations.

In his 1988 book, STS 61-C Payload Specialist and U.S. Congressman Bill Nelson recalled his participation in the pad escape systems tests at the Cape, which were conducted two weeks prior to their scheduled launch (Nelson, 1988). Nelson tried his hand at driving the tracked escape vehicle, in his words "very carefully", so as not to make newspaper headlines the next morning by crashing the vehicle. Escape training for pad evacuation began at Building 9A at JSC, where a Crew Compartment Trainer is used for emergency training. The simulation could be moved from a horizontal (landing) to vertical (launch) configuration to practice ingress and egress in both nominal and emergency situations. In 58 seconds following the "go" from Commander Robert "Hoot" Gibson, all seven astronauts unstrapped from their seats, pulled down their visors on the Launch and Entry Helmets and switched from the Shuttle Oxygen System to a portable air supply located on their belts. The side hatch was opened and each man crawled out and completed the exercise. The sequence from the hatch to the escape baskets was practised at the Cape, together with fire-fighting techniques and driving the armored personnel carrier (used for a sheltered, quick escape) to the edge of the pad area. Nelson knew that they would have to move much faster in a real situation to prevent being fried or blown apart in any ensuing explosion.

Redundant Sequence Launch Sequencer (RSLS) Aborts

One of the Shuttle Abort Modes is available for instances on the pad where a serious problem prevents the launch after onboard computers have taken over from the ground launch sequencer but just prior to ignition of the Solid Rocket Boosters. In the history of Shuttle flight operations since April 1981 there had been five instances of a Redundant Sequence Launch Sequencer (RSLS) Abort.

June 26, 1984. During a second launch attempt for the maiden launch of OV-103 (Discovery) the new Orbiter's onboard computers detected a "sluggish" valve in main engine number 3 initiating the RSLS at $T-4$ seconds. The faulty engine was subsequently replaced before Discovery finally left the pad on 30 August 1984 (Shayler, 2000). After warning his crew at $T-31$ seconds that they were soon to start their mission and feeling the initial SSME ignition Commander Hank Hartsfield realized that the master alarm and two red lights indicating engine shutdown meant they were not going anywhere. After confirming that all was safe and a small fire at the base of the Orbiter was only residual hydrogen burning off, the crew prepared to leave the Orbiter. Mission Specialist Steve Hawley expressed the disappointment in stating he expected to have been a little higher at Main Engine Cut-Off than he now found himself. Thoughts of using the slide wire system were soon dispelled as the crew were not enthusiastic about the prospect of being the first to use them in a real emergency. As it was they all got a soaking from the water used to spray down the tower and pad as a precautionary measure against fire.

July 12, 1985. During the countdown for STS-51F, the onboard computers of Challenger (OV-099) detected a problem with a coolant valve located on the SSME #2 unit. This was replaced and Challenger finally left the ground on 29 July 1985, but encountered another abort situation on the way to orbit. Mission Specialist Tony England was sitting on the middeck of Challenger reviewing his role in making sure the two Payload Specialists seated next to him would be safe in the event of a problem. As a career astronaut England's training was far more involved than that on the one-flight Payload Specialists. When the pad abort was called, England was far more concerned in successfully performing his assigned tasks under the emergency situations should they be called upon than the disappointment of not flying into space. Once safely outside the Orbiter the disappointment set in but a crew visit to Disneyworld helped relax the crew in preparation for their next launch attempt, which unbeknown to them would be far more stressful that the first attempt.

March 22, 1993. It was almost eight years later after the recovery from the loss of Challenger in 1986, and several pad and Orbiter fuel leak problems in 1990 that the next RSLS abort was to affect the program. This time Columbia (OV-102) was about to fly the STS-55 mission when at $T - 3$ seconds a problem with the purge pressure readings in the oxidizer preburner on SSME #2 halted the launch. All three engines were subsequently replaced on the pad but the Columbia mission was rescheduled after STS-56. STS-55 finally left the pad on 26 April. Commander Steve Nagel stated that the prominent reaction of the crew was one of disappointment. The procedures in saving the Orbiter and pad area after this the program's third RSLS abort were by now as well trained and performed as they could be. KSC Launch Director Robert Sieck later commented: "As far as pad aborts go, this is probably the best we've executed" (Gaston, 1993).

Inside the Orbiter the crew certainly felt the vibration of the SSME ignition then the master alarm indicated the shutdown. From then on, training and routine took over their actions. They had during months of training practiced this situation many times before and though a certain level of stress was natural in such a situation they were assured all was fine outside and so they just waited until it was safe to leave the Orbiter. It was probably more stressful for their families who watched from three miles away as the engines ignited and then shut down with Columbia silently holding their loved ones firmly attached to a potential bomb that had already displayed a reluctance to leave Earth.

August 12, 1993. Another aborted launch also at $T - 3$ seconds occurred just five months later during the attempt to get Discovery (OV-103) off the ground. This was the third attempt at launching STS-51. This time the abort was called when the onboard computers detected a problem in SSME #2. One of four sensors that are used to monitor the flow of hydrogen fuel from the External Tank to the engine detected a problem. Again all three engines were replaced on the pad delaying the launch until 12 September.

18 August 1994. This RSLS abort, 12 months later, was the closest to SRB ignition in the program occurring at just 1.9 seconds prior to lift-off of STS-68. This time the problem lay in SSME #3 where the onboard computers recorded higher-than-acceptable readings in one of the channels of the sensor that were used in monitoring the discharge temperature of the high-pressure oxidizer turbopump. Later test firings at Stennis Space Center in Mississippi on September 2 helped confirm the cause of the problem of an increase in the turbopump temperature was the result of a small drift in a fuel flow meter. Endeavour (OV-105) was rolled back to the pad and three replacement engines fitted, resulting in a new launch date of October 2.

In 2003 Mission Specialist Thomas Jones wrote about his experiences onboard Endeavour that day (Jones, 2003). He felt the SSME ignite and Endeavour sway on the pad waiting for the 'kick-in-the-pants' ignition of the twin SRBs. However, they never ignited, being replaced by the loud blare of the master alarm and silence on the three main engines. Endeavour still swayed as the crew set the controls to safe the orbiter and prepared to quickly evacuate via the escape slide wire system. But there was no pad fire or explosion, and so they sat waiting for the ground crews to open the hatch and extract them. Thirty minutes after he should have been pinned into his seat by the acceleration thrust of ascent to orbit, Jones found himself sitting up on his seat, parachute harness off, and calmly eating a peanut butter sandwich.

Constellation

With the Shuttle scheduled to be retired in 2010 and the first manned flight of Ares I scheduled for 2015, plans for the conversion of facilities at LC39 have again begun to take shape. As with Apollo-to-Shuttle modifications, restructuring, refurbishing or amending facilities to incorporate the new Constellation Program are underway to support the first unmanned test flight of Ares I, hopefully in the spring of 2009.

Part of these modifications is the Orion Emergency Egress System. This features a group of multi-passenger "rollercoaster" cars on a set of rails to move the astronauts and ground crew as quickly as possible from the vehicle entry position some 116 m above ground level to protective concrete bunkers. Though not what might be called a "white knuckle thrill ride", some of the world's leading roller coaster developers were used as consultants for the plans at the Cape. It is intended to be more efficient than the slide wire system of the Shuttle, with rapid relocation from the spacecraft to the bunker within three minutes (180 seconds). This would be especially helpful, and possibly life-saving, for an incapacitated crewmember, who could be lifted into the "cars" and carried out the other end. This would be much easier than trying to carry a person from the spacecraft to the basket slide wire and then into a bunker at the other end. This new technology would enable prompt escape from a

Artist impression of Pad Escape System for Project Constellation (top). Roller coaster to safety (bottom).

Artist's impression of the Ares launch vehicle on the pad with the Orion manned spacecraft and Launch Escape System in place.

pending huge pad explosion. As with Apollo and Shuttle, everyone hopes it works as advertized but that they will never actually need to use it.

Vandenberg Air Force Base

A second manned launching facility was considered in both the 1960s and 1980s, to support military missions and polar orbital trajectories. To support manned launches, the launch complex at the Vandenberg AFB needed to incorporate facilities for pad escape into the ground facilities for both the planned Gemini Titan IIIM, and later the Space Shuttle.

Space Launch Complex 6 (Gemini-Titan II (MOL)/Shuttle DOD)

For MOL/Gemini launches, the two USAF MOL astronauts were provided with ejection seats, as with the NASA Gemini astronauts. Construction of pad support facilities, including umbilical and mobile service towers, began on January 27, 1967, the day that three Apollo astronauts were killed in the Apollo 204 (Apollo 1) pad fire at the Cape in Florida. The work continued throughout most of the next two years, until the end of November 1968. The facilities included access platform and fire suppression systems similar to those found at LC19 at the Cape in support of the Gemini missions launched by NASA.

The MOL program was canceled in June 1969 without a manned launch being completed and only one unmanned launch occurring. In April 1974, Vandenberg AFB was selected as a second launch site for Shuttle missions. Construction to convert the facility from the MOL to the Shuttle program was a challenge in itself, but included pad escape facilities compatible with the Shuttle Orbiter. New pad construction began in late 1979. One facility change at Vandenberg focused on the crew access arm on the access tower which, unlike the swing arm at LC39, would roll back from the Orbiter on tracks. Slide baskets were installed at the facility, running from the access tower westwards to a set of bunkers. In unmanned tests, it was found that the fully loaded basket would strike the south corner of the structure designed to protect the Ready Building from the effects of launching a Titan IIIM. In the change to Shuttle facilities, it was decided not to demolish this structure made of sand-filled concrete, as it would be too expensive to take down. The problem was solved by removing about 3 m from the structure (known as "the flower box") to allow the slide basket to pass cleanly by.

China

In October 2003, China became the third nation to place its own citizen into Earth orbit and thus the third nation to achieve manned spaceflight capability. Interest in placing Chinese citizens into space had originated in the 1960s and resulted in a program called Project 911, or *Shuguang (*meaning "Dawn"). However, funding for the project was not forthcoming and the project was terminated in May 1972. Despite several rumors, photos of space simulations, and the offer to fly a Chinese citizen on the Space Shuttle as a Payload Specialist, it would be a further 20 years before renewed interest would develop in a manned spacecraft project. This became Project 921 and the development of a new manned spacecraft began. From 1992 until the end of the decade, a whole new infrastructure had to be created to support such an ambitious program, though the Chinese clearly benefited from the experiences of the Americans and Soviets.

Jiuquan Complex (Shenzhou-Long March 2F)

Reports of the first manned Chinese spaceflight revealed that, with the launch vehicle standing alone atop the pad at a height of nine storeys, the "White Room" access facility on top of the launch gantry would be pulled back some two hours prior to launch. Should the crew need to be evacuated, a team of 14 support personnel was on standby (presumably in a nearby bunker) to move in and replace the access arm to extract the crew as quickly as possible. Though full details are still to become clear, the pad at Jiuquan features an explosion-proof elevator or an escape slide and explosion-proof bunker close to the pad, similar to the American facilities at LC39. In the event of an explosion prior to launch, the escape tower system would jettison the crew compartment and carry it to a safe distance for parachute landing. As the Shenzhou program develops, more details of the structure of the pad and its launch abort support facilities will become clear.

REFERENCES

Nancy Gaston (1993). 5 ... 4 ... 3 ... 3 ... Launch is shut down. *Houston Citizen*, 23 March, p. 1.

Bart Hendrickx and Bert Vis (2007). *Energiya–Buran: The Soviet Space Shuttle*. Springer/Praxis, Chichester, U.K. [an excellent in-depth overview of the Soviet shuttle program and system].

Thomas D. Jones (2003). Escape velocity. *Aerospace America*, September. A publication of the American Institute of Aeronautics & Astronautics, Reston, VA. Available at *http://aiaa.org/aerospace/article.cfm?issuetocid=403&archiveissueID=42*

Nelson, W. (1988). *Mission: An American Congressman's Voyage to Space*, U.S. Congressman Bill Nelson with Jamie Buckingham. Harcourt, Brace, Jovanovich, Orlando, FL.

David J. Shayler (2000). STS41-D incident entry. *Disasters and Accidents in Manned Spaceflight*. Springer/Praxis, Chichester, U.K.

5

Launch escape, 1: Escape towers

Having achieved lift-off, the next hazardous phase of the mission is getting from the pad and safely into orbit while riding a potential explosive rocket. In addressing this challenge the designers of manned spacecraft have evolved two main systems of crew evacuations during ascent: by means either of an escape tower or ejection seats. In addition, there are a number of flight profiles developed in the event of an erroneous launch profile that does not require the crew to be evacuated from the vehicle but to "fly" a trajectory that *hopefully* will ensure either their safe return to Earth or to orbit.

To date (summer 2008) the launch escape tower has only been employed for one incident: successfully rescuing a Soyuz crew from a launch vehicle explosion on the pad in September 1983. Contingency or abort flight profiles have also been employed on two American missions: Apollo 13 in April 1970 and Shuttle 51F in July 1985.

For the first manned spacecraft, the Soviets adopted an ejector seat approach while the Americans followed the escape tower system. These towers continued on Apollo and Soyuz and will be reintroduced in the new Constellation Program that will replace the Shuttle. Ejection seat systems were incorporated on the American Gemini Program and Earth-based Space Shuttle flight tests. Ejection seats were also incorporated in the subsequently canceled Buran Soviet space shuttle program. These are covered in Chapter 6; here we focus on the escape tower concept.

ESCAPE TOWERS

An "escape tower" can be defined as essentially a rocket motor mounted in a support framework attached to the crew compartment, which can explosively separate and propel the crew away from an errant or exploding booster in situations of emergency or accident. The towers are mounted at the top of the spacecraft and jettisoned prior to entry into orbit when they are no longer required. They have been used on America's first manned spacecraft (Mercury), the Apollo Command and Service

Modules for lunar and Earth orbital missions, the venerable Russian Soyuz and recently the Chinese Shenzhou vehicle. They were also to be used for the Soviet manned lunar spacecraft, considered for the American Gemini Program and are to be reintroduced for the US Orion spacecraft in the Constellation Program.

MERCURY LAUNCH ESCAPE TOWER

The objectives of the Mercury Project, the initial US manned space program, were stated to be to place a manned spacecraft in orbital flight around the Earth and after investigating the astronaut capabilities and ability to function, recover both man and spacecraft safely. Using existing technology and off-the-shelf hardware wherever possible, it was decided to develop a staged program of flight operations that included a demonstration flight of production hardware leading to manned sub-orbital missions using the Redstone Launch Vehicle and finally manned orbital flights utilizing the Atlas Launch Vehicle. Early in the initial design phase of the Mercury spacecraft a number of mandatory design requirements were imposed on McDonnell Douglas, the primary contractor. One of these being:

> "The spacecraft must be fitted with a reliable launch escape system which would rapidly separate the spacecraft with its crew from the launch vehicle in case of immediate disaster" (NASA, 1963a, b, 1966a; Catchpole, 2001).

Reliability and safety in flight

In choosing the Redstone and Atlas launch vehicles the program management and development teams recognized the need for a fine balance between system reliability and safety whilst activating the program's objectives in the shortest possible time. Both missiles had a long development program and required minimal adjustments to allow a human payload to be carried safely into orbit. Reliability and flight safety features regarding the launch vehicles were organized into groups: some under the missiles themselves, a safety program for the pilot and an adequate abort-sensing system. Previous experience in using the missiles along with the retention of proven components where possible offered an advantage over developing new systems, which would have lengthened the development processes, and provided proven flight data for developing an adequate safety program for the pilot.

It was clear that significant changes to the reliability program in the design of the launch vehicles could not be achieved within the time frame or scope of Project Mercury. As a result a three-stage program was developed for assessing the quality of the workmanship, an inspection at the factory during rollout prior to shipment to the Cape and a review of safety elements by key personnel prior to committing the vehicle to flight in a series of "go" or "no-go" recommendations.

An Abort Sensing and Implementation System (ASIS) was developed to enhance the safety of the astronaut during the period of powered ascent. By evaluating past performance records of Atlas (and Redstone) launches it could be determined which parameters could be monitored that would indicate an approaching infringement to abort parameters in enough time to allow the abort system to work and at the same time ensure a false reading was not interpreted as a pending abort situation. Qualification of the system subject to a variety of extreme environmental conditions was completed under an extensive program of ground tests as well as flight demonstrations. From this evaluation, discrepancies that were revealed were able to be corrected prior to the manned Atlas missions. A successful abort was initiated during the unmanned MA-3 mission and thus saved the spacecraft, enabling it to be launched again on MA-4.

The function of the Mercury ASIS was to sense an impending failure of the launch vehicle, then automatically generate the command to abort the flight and activate the escape tower in enough time to ensure the safety of the astronaut. The system was supplemented by additional abort capabilities by means of human intervention. While on the pad up to the time the vehicle had lifted 5 cm off the pad "off-the-pad aborts" would have been initiated by the Test Conductor via direct electrical circuits. From the 5 cm point to the end of powered flight, Mission Control would have used a radio frequency link to initiate an abort. Indirect abort capability was the responsibility of the Range Safety Officer. Here a manual command from the ground would cut off the engines, which would automatically initiate the abort command to the escape rocket to pull clear of the spacecraft. A 3-second delay was incorporated into the system to enable the tower to separate from the spacecraft at a safe distance from the launch vehicle should a command destruct signal be sent to destroy the vehicle.

This system also included a step-by-step approach to qualifying systems, including the launch escape system, for operational manned spaceflight. Ground tests were followed by a program of beach aborts for qualifying both the launch escape and Earth landing systems. A program of test flights using the Little Joe Launch Vehicle was followed by unmanned ballistic flights of the Mercury Redstone and Mercury Atlas concepts, initially instrumented at that time with primates and finally human crewmembers. The orbital flights progressed in a logical 3-orbit to 6-orbit to 22-orbit mode through to the conclusion of the program. It is as a result of stringent testing, evaluation and incorporating changes in the system as this program developed that, during the manned Mercury launches, no instances requiring any abort action by the ASIS were encountered and no false ASIS signals were recorded which could have led to unnecessary abort situations.

Mercury Escape Tower

Evaluating the type of abort system, engineers found that by far the most effective would include a design incorporating a solid rocket motor to propel the spacecraft containing the astronaut away from a possibly exploding or breaking up launch vehicle, whether on the pad or during the ascent. In addition, combination of a reliable initiation system, use of a rapid thrust but short-lived flight time and effective separation devices would be required. Servicing and maintenance should also be kept

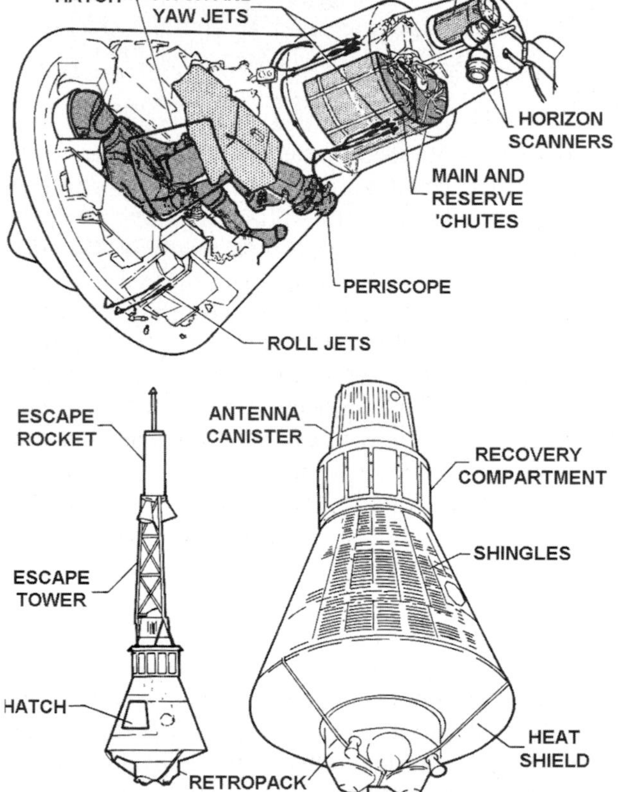

Mercury astronauts inspect a model of the launch vehicle, spacecraft and escape tower (top). Mercury spacecraft in detail (bottom).

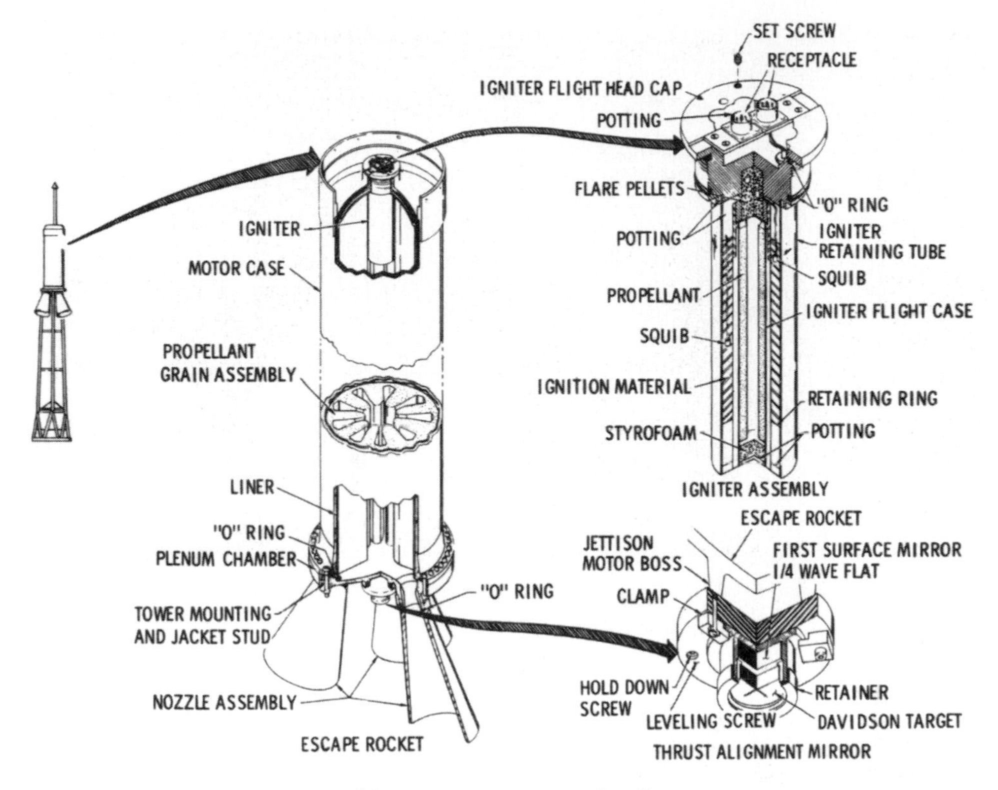

Mercury escape tower detail.

to the minimum and levels of safety should be maintained at all times whether the system was used in a real-time scenario or discarded during ascent.

The final configuration of the Mercury Escape Tower comprised a tripod tubular steel support structure strengthened by cross-members onto which was mounted a single solid motor casing, with three exhaust nozzles spaced 120° apart. The requirement for a short burn was achieved by burning propellant in just one second, producing a thrust of 231,296 N and thus meeting the design requirement. On top of the motor was a single solid propellant tower jettison motor which produced 3,558 N of thrust in a 1.5-second burn time to move the tower away from the spacecraft. The overall length of the system from the base of the tower to the apex of the aerodynamic spike was 5.1 m with a mass of 580 kg.

System operation

An abort situation during the Mercury Program that required the use of the escape tower was classified as either an "off-the-pad abort" or an "in-flight abort". If a problem arose during the final moments of the countdown with the launch vehicle still on the pad, or immediately after lift-off (up to Ground Elapsed Time of 2

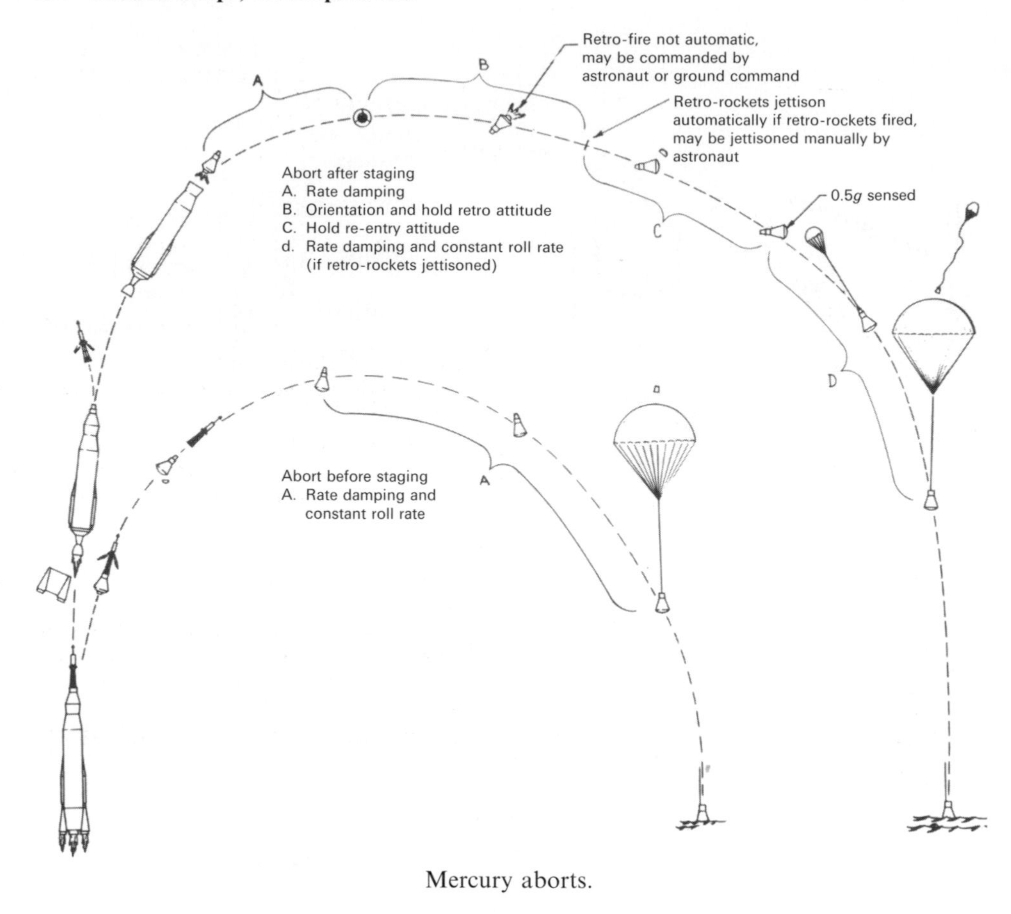

Mercury aborts.

seconds), onboard sensors of the automatic failure detection systems would detect the problem. As soon as explosive bolts attaching the spacecraft to the launch vehicle were severed the retro-rocket package was separated and the Launch Escape System (LES) fired. The LES would have pulled the Mercury spacecraft and its astronaut occupant on a rapid ballistic trajectory, with a peak altitude of about 610 m, to the ocean side of the launch pad. At peak altitude the tower was separated from the spacecraft by explosive bolts, releasing a securing clamp ring. The momentum generated by the initial thrust was enough to clear the pad and a launch vehicle about to explode. The tower jettison rocket then fired to propel the used escape tower away from the spacecraft and enable recovery by parachute descent into the Atlantic Ocean off Cape Canaveral. Pilot recovery would be by support helicopters from the DESFLOT-Four unit.

In-flight aborts took place between the 2 seconds of flight and GET of 2 minutes 23 seconds. After that time the launch vehicle would have produced enough thrust to enable the spacecraft to separate without the use of the tower and perform a nominal ballistic parachute descent. During the first 30 seconds of a flight, should a problem

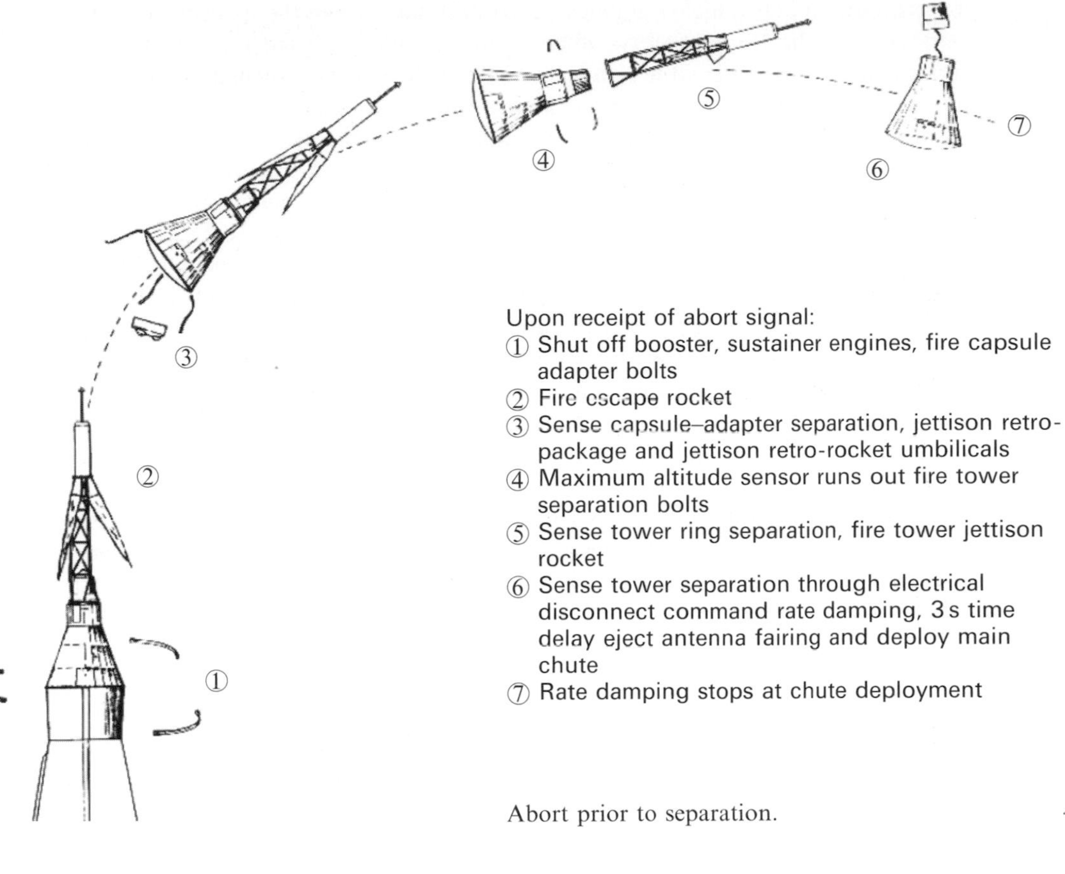

Abort prior to separation.

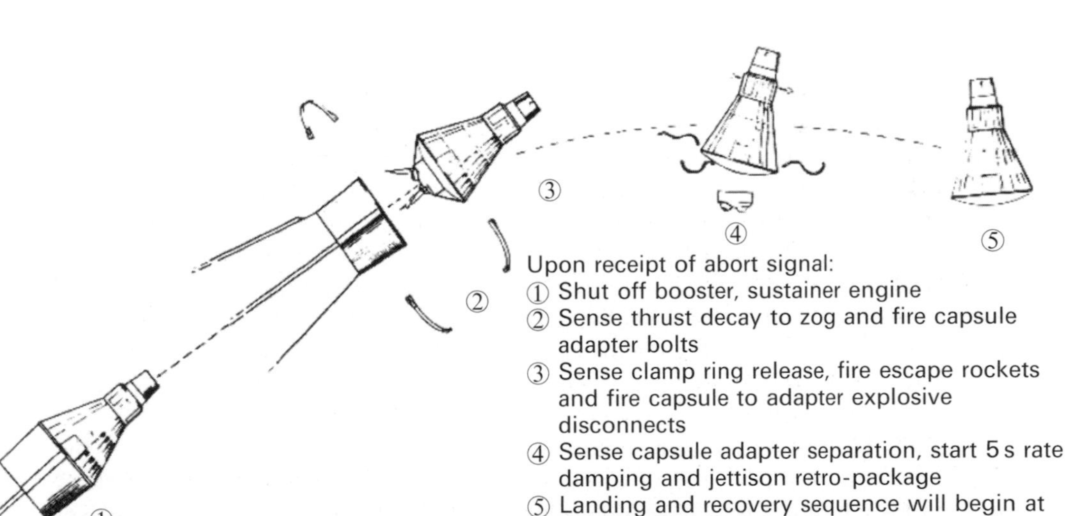

Abort following separation.

occur and the launch vehicle's engines fail to shut down, then the spacecraft would be separated, and the Range Safety Officer would initiate a destruct signal detonating onboard charges to destroy the errant launch vehicle before it crashed into the launch pad or, more importantly, inhabited areas. Should the engines be shut down, the system would be the same as a pad abort but this time the apogee would be higher.

In a nominal operation where an abort situation is not encountered, the tower would be jettisoned at a GET of 2 minutes 23 seconds when the clamps holding the tower to the spacecraft would be jettisoned. During the sub-orbital Redstone launches this occurred on shutdown of the main propulsion system. On the Atlas missions the tower was ejected following the cut-off of the booster and separation of the booster skirt. To clear it from the ascending trajectory of the Atlas, the launch vehicle sustainer's engine was slightly pitched down prior to separation and then pitched back up again following the departure of the unwanted tower. In all occasions once propellant had run out in the system, the tower and motor were allowed to fall into the Atlantic Ocean. No attempt was made at recovery during the operational flights of the program.

The electrical system on the spacecraft provided the capability of abort at any time after the gantry had been removed. If an abort had been initiated then the capsule adapter clamp ring would have been released, the escape rocket ignited and the capsule carried to about 760 m altitude. Signals from the altitude sensor would have released the tower clamp ring, ignited the tower jettison rocket and the unwanted tower would have been released. Two seconds later the drogue chute would have been deployed followed two seconds after that by the release of the antenna fairing. Twelve seconds later the heat shield would have been released and landing bag deployed.

The astronaut "chicken switch"

Inside the crew compartment of Mercury was the abort handle. Located forward of the support couch left arm rest, its function was to initiate the abort sequence from within the spacecraft. During nominal launches it was also used as a restraint handle. To operate the handle the astronaut had to depress a release button on the top of the handle and this allowed it to be rotated outboard activating the "abort" position. Then an electrical switch was activated that signaled the detonation of the capsule-to-adapter clamp ring bolts. The escape system was then initiated, as long as the main umbilical to the ground had been disconnected. Until umbilical release the abort handle was inoperative.

As all the astronauts were former test pilots, their flying skills had been the leading criteria for their selection, and not wishing to harm their image (later termed by American author Tom Wolfe as "the right stuff") astronauts termed this control lever the "chicken switch". Fortunately, none of them needed to use it during any of their missions.

Evolving the system

Using ballistic missiles to launch manned spacecraft was, from the very beginning, always a risk. In the 1950s this new technology had a nasty tendency of blowing up as theory was put into practice. Therefore, a reliable and effective means of escape would have to be included to allow the occupant at least the chance of escaping a potential fireball. Earlier designs for the USAF Man In Space program were found to be too complicated and complex for what the engineers at NACA's Pilotless Aircraft Research Division (PARD) at the Langley Aeronautical Laboratory in Virginia were developing. The USAF designs featured a pusher rocket escape system where rockets mounted at the base of the "manned capsule" would in effect push the crew compartment upwards clear of danger. Mercury's prime contractor McDonnell Douglas had originally proposed installing rocket motors on fins around the base of the spacecraft in order to provide emergency separation from a failing launch vehicle. The PARD team were considering an alternative design. Led by Max A. Faget, in July 1958 the team suggested a tractor rocket above the capsule powered by solid propellant that would pull the crew compartment clear (NASA, 1966a, pp. 96–97).

By August a design emerged that featured a slender rocket casing and nozzle which was attached to the capsule by three slim struts. By using a solid rocket tractor system fewer components would be required, simplifying the design and hopefully its operation should the need arise. Ease of operation and dependability were the driving force behind the PARD team. Plans were made to test these escape systems down at Wallops Island Pilotless Aircraft Research Station in Virginia using "boilerplate" (mock-up) capsules or full-scale metal models before moving on to testing them on ballistic launch vehicles.

Little rocket for a big role

The vehicle used for these tests was called Little Joe. Based on an idea first suggested by Max Faget and Paul E. Purser in January 1958, Little Joe was used to support research and development in the manned satellite vehicle project. Originally it was intended to test full-size and mass versions of the developmental manned satellite to provide test results on the conditions and environment the operational spacecraft might encounter. It was envisaged to use the launcher to develop escape maneuvers during an aborted launch.

Design work commenced in October 1958, shortly after the creation of NASA. On November 26 the manned satellite project officially became known as Project Mercury. The contract for Little Joe was awarded to North American Aviation on December 29, 1959, with the remit of designing and fabricating the air frame. It was planned to deliver the rocket motors early and the air frames every three weeks. At the test stand, spacecraft assigned to these tests were being designed and constructed.

By January 29, 1959 the flight test program had been devised. In addition to tests of the abort system a program of flight dynamics and spacecraft aerodynamics would be investigated. Taking advantage of flying small primates and in some cases pigs on

Preparing for an abort test.

Mercury capsule on a Little Joe.

Little Joe flights would provide baseline data on which to refine the design of the proposed manned capsule, especially during the increased aerodynamic and physical strains encountered during aborted flight or hard landings. Tests were also planned on parachute operation and the physical effects on biological specimens.

Little Joe stood 14.6 m high with a maximum mass of 18,747 kg and had a 2.3 m diameter. Powered by four Pollex and four Recruit cluster solid rockets a thrust of 1,110,000 N could be reached, with a total of 1,788 kg being lifted.

Form-fitting couch

Using a ballistic capsule design the occupant could expect loads of between 5*g* and 6*g* during a nominal ascent and 8*g* to 9*g* during a shallow entry and water landing. Should an abort situation occur, however, as much as 20*g* could be experienced. Once again Air Force ideas to support crewmembers during such accelerations and decelerations were overly complicated. Max Faget again had a more simple solution. What Faget and his colleagues proposed was, instead of a crewmember lying on a couch, why not have it formed around him with the crewmember in effect inside it? It could be lightweight and stationary but strong enough to support the body form, fabricated as it was of fiberglass. By May 1958 test couches had be molded and by the end of July tests at the large USN centrifuge at Johnsonville, Pennsylvania indicated an occupant could withstand up to 20*g* safely.

Wind tunnels and ground tests

During 1959, to evaluate the most appropriate escape system for inclusion on Mercury, engineers completed a program of wind tunnel tests at the Arnold Engineering Development Center in Tullahoma, Tennessee. Here both the tractor rocket design of Max Faget's STG and the original idea of a posigrade rocket with fins from McDonnell were evaluated in wind tunnels. In addition, the rocket motors were evaluated in the wind tunnels and in ground tests to determine the effects of engine ignition up to 3,048 m simulated altitude and trajectory characteristics during planned burn times.

Tractor rocket definition

On March 8, 1959 a test of the abort system was performed at Wallops Island. These types of simulations were termed "beach aborts". Using a full-scale model of the spacecraft and escape tower with a Recruit escape rocket, the configuration encountered erratic motion in flight and as a result Langley Research Center was tasked to test smaller scaled flight models in order to evaluate abort system motions in flight. This design of the escape rocket featured three exhaust outlets each spaced 120° apart and angled 15° to the vertical. Three tubular steel struts were used as a "tower" allowing the exhaust jets to clear the spacecraft beneath when fired, arcing the spacecraft towards an ocean recovery away from the launch area. Analyses of the flight data indicated that all was well for the first 60 seconds of powered flight, then the spacecraft tumbled three times end over end before completing a nominal water landing about 304 m offshore.

The tumbling was not planned and probably would have been an uncomfortable ride for the occupant had the spacecraft been manned. Two theories resulted from the post-flight examination of data. It was determined that one of the graphite throats of one of the exhaust outlets had burned through, affecting the thrust vector and thus causing the unplanned tumbling. To prevent this occurring again the throat area of the nozzles was redesigned. In addition the angle of the nozzles was thought to be a contributing factor to the tumbling. As a result a series of test flights evaluating alternative angles of the outlet nozzles were performed by the Langley workshops between April 13 and 15 using five 0.33-scale models. To simulate the Recruit rocket in the full-sized version a smaller 8.25 cm rocket was used.

On April 13 the first two tests were conducted. The first, designated U-1, featured an angle outlet of 10° and clearly was very unstable in flight. So was the second test (U-1) at 15°. The following day a similar test (U-3) recorded unstable characteristics at 20°. Finally, on the 15th two further tests were completed. U-4 was set at 25° while U-5 was set at 30°; both were found to be stable during operation. From these tests it was decided that the exhaust outlets on the Grand Central Escape Rocket motor would be set at 19° to the vertical.

Beach abort programme

Further beach aborts were scheduled after the March 8, 1959 demonstration. On April 14 a boilerplate spacecraft was launched from Wallops. This time the thrust of the tractor rocket motor was deliberately offset. This simulated an off-the-pad abort reproducing similar pitch and translation rates to those of a real abort situation. Further simulation of an operational abort by the Grand Central rocket motor was performed using a Recruit rocket rated at 154,078 N. The burn time was 1.5 seconds, and following separation of the spent tractor rocket the spacecraft completed one tumble before falling into the water. This was deemed a successful demonstration of the system.

The next beach abort occurred on July 22, 1959, this time using a production spacecraft. Maximum altitude attained was 609 m. One tumble was noted prior to the ejection of the escape system, but following separation the spacecraft continued to tumble until the drogue parachute deployed. It was only after deployment of the main parachute that the vehicle fully established. Six days later a second beach abort recorded total success. Two further beach aborts simulating off-the-pad aborts were flown, the first resulted in a failure of the spacecraft to separate from the spacecraft correctly. The results from this experience resulted in a redesign of the single exhaust outlet to a three-point outlet spaced 120° apart from the other two and angled 30° to the vertical.

The beach abort program qualified the LES for off-the-pad aborts for future manned launches.

Little Joe

The first launch of the Little Joe Vehicle (LJ-1) was planned for July 1959 but was delayed until August 21. This was a research and development mission but inadvertently evaluated the abort system. As the vehicle's battery power supply was being

Ready for testing.

charged late in the count at Wallops Station the escape system was inadvertently initiated and the tower ignited sending the spacecraft on what was a pad abort sequence though triggered by a fault. However, though the tower was successfully jettisoned and the drogue parachute released, the main parachute deployment circuitry suffered from the lack of enough electrical power for it to operate properly and as a result the spacecraft was destroyed upon impacting the water.

Little Joe 6 was successfully launched on October 4, 1959 from Wallops. This was a test of the Little Joe system itself after the experience of LJ-1. It was decided to fly an empty Grand Central rocket casing filled with ballast to demonstrate that the launch vehicle could fly without completing an ejection sequence.

Returning to the task of qualifying the abort system, the next launch on November 4, 1959 was designated Little Joe 1A, repeating the flight planned for LJ-1 two months previously. Though an in-flight abort was completed the primary objective was not accomplished due to slow ignition of the escape rocket motor. This resulted in a delay to spacecraft launch vehicle separation until the launch vehicle had passed through the designated test region. All secondary test objectives were met and the recovered spacecraft was returned to the launch site. As a result of another test, LJ-1B was added to the program to complete the test objectives planned for this flight.

Before this, however, on December 4, 1959 LJ-2 was launched from Wallops. This was planned as a demonstration of a high-altitude abort at an altitude of 30,480 m. The abort sequence was initiated reproducing the possible scenario for a Mercury Atlas launch. A successful demonstration was recorded as was the evaluation of re-entry characteristics without the benefit of a control system aboard. Dynamic stability through the increasingly dense layers of the atmosphere was as predicted. Recovery and floating devices were also evaluated satisfactorily. The rhesus monkey passenger (Miss Sam) was also recovered successfully with no side-effects from the experience. This gave the designers a welcome boost to confidence in the system.

Escape rocket test operation.

Little Joe 1B was successfully launched on January 21, 1960 from Wallops. This time all went according to plan and the spacecraft and its rhesus monkey passenger were recovered successfully.

On 29 July 1960 Mercury Atlas 1 was launched from Cape Canaveral to demonstrate the structural integrity of the Mercury spacecraft and its re-entry elements, therefore no escape system

Sequence of an abort test.

was installed. This was a pity, as just 60 seconds into the flight the mission failed and the launch vehicle and spacecraft plummeted into the ocean, thus delaying the Mercury Atlas Program for seven months whilst the structural integrity of the combination was beefed up.

The next Little Joe launch, LJ-5, occurred on November 8, 1960. This was to be a production spacecraft qualification flight, though things did not go according to plan. During powered flight the mission failed as a result of the escape rocket igniting before the spacecraft separated from the launch vehicle. The spacecraft remained attached to the launch vehicle and impacted the ground destroyed. Elements of the hardware were retrieved from the ocean floor but the exact cause could not be determined. In addition a lack of flight measurements added to the problem. It was thought the sequential system failed, though this could not be precisely determined. As a result the sequence systems aboard future spacecraft were changed to preclude possible premature firing of the escape rocket motor.

Little Joe 5A was added to the test program as a result of the loss of LJ-5. Again a production spacecraft was used in the test, but was configured with just those systems required for the tests planned. Once again early during the ascent the escape rocket motor ignited prematurely, making the mission unsuccessful. However, a back-up system initiated by ground control enabled the spacecraft to be separated from the launch vehicle, released from the tower and successfully recovered. The early ignition of the escape rocket was determined to be due to structural deformation in the interface area of the spacecraft adapter. The system falsely recorded movement and triggered the escape rocket. By stiffening the switches, recycling electrical signals and reducing air loads on the area such corrective action prevented a repeat of the problem.

Little Joe 5B was launched on April 28, 1961, just a few days prior to the launch of Al Shepard aboard Mercury Redstone 3 and 16 days after Soviet cosmonaut Yuri Gagarin had ridden Vostok into space and the history books as the first manned spaceflight. LJ-5B used the refurbished Mercury Spacecraft No. 14A employed previously in LJ-5A, while incorporating improvements to the spacecraft adapter mountain, fairings and electrical circuitry to prevent repeats of previous failures. A more severe test profile was flown as a result of delayed ignition of one of the two main launch vehicle motors. The flight plan had called for an abort at a dynamic pressure of 449 kg/9.29 dm^2 instead the level had risen to 871 kg/9.29 dm^2 at initiation. The abort sequence and recovery of the spacecraft performed flawlessly with the Mercury capsule splashing down in the ocean five minutes after launch. It was safely retrieved and returned to the launch site within 30 minutes of leaving the pad. It was now time to put a man in the capsule.

The series of Little Joe flights were considered a great success in qualifying the abort system and providing flight data useful in the planning of forthcoming manned flights. Human-rating of the system was possible by flying Little Joe launches and simulating Atlas aborts at the point of maximum dynamic pressure.

Mercury-Redstone

This vehicle was to be used in the suborbital flights of the Mercury spacecraft. Constructed by the Chrysler Corporation it was 29 m long and had a 1.7 m fuselage diameter. The single North American A-7 engine produced a thrust of 346,944 N using a propellant of liquid oxygen (75% alcohol and 25% water). The launch site was Pad 5 at the North Atlantic Missile Range (Cape Canaveral, Florida). Initially in January 1959 there were two spacecraft and launch vehicle qualification flights followed by a series of six manned suborbital "training" missions for the astronaut prior to assignment to an orbital mission. This was later amended to result in two unmanned launches (one carrying a chimpanzee), a developmental launch and two manned missions. The other manned flights were cancelled as unnecessary.

The day they launched the escape tower

The first Mercury-Redstone flight of November 21, 1960 was planned as a qualification flight for the production Mercury (Spacecraft No. 2) with a Redstone launch vehicle. At least, that was the plan. There was a "launch" but not the one expected. At lift-off the engines of the Redstone shut down and after rising just a few inches settled down back on its launcher. The spacecraft also shut down and deactivated its systems as it was supposed to in such a situation. The Launch Escape System motor fired but

The day they launched the escape tower—Mercury MR-1.

the clamp securing the system to the neck of the spacecraft released and as a result the tower climbed away from the pad, its exhaust covering the pad area in a thick smoke cloud. At 1 second the escape motor shut down and after peaking at 1.2 km the escape system continued its trajectory, arcing down towards observers on the beach who dived for cover as the tower slammed into the sand some 365 m from the pad.

Despite media claims of a malfunction, the system had worked as designed given the information presented to it.

A refurbished spacecraft was launched atop a Mercury-Redstone 1A successfully on December 19, 1960. Separation of the escape tower was the nominal ten seconds after shutdown of the booster. The next flight MR-2 carried chimpanzee Ham into history. An abort signal was received in the spacecraft when the propellant aboard the Redstone depleted before the velocity cut-off system was armed, and prior to the thrust chamber abort switch becoming disarmed. The abort system worked as planned and added to the velocity which increased the recovery point about 186 km farther downrange than anticipated. Following a booster development flight on March 24, 1961 the Redstone vehicle was cleared for manned flight.

Mercury-Atlas

The launch vehicle assigned to place the first American astronauts in Earth orbit was fabricated by General Dynamics Aerospace Division and measured 28.7 m in length. Its booster was 4.8 m in diameter and its sustainer was 3 m in diameter. Two Rocketdyne LR89-5 engines (734,057.9 N) were assigned to the booster stage while one Rocketdyne LR105-5 (253,581.7 N) was installed as the sustainer in addition to two sustainer vernier engines (4,452.7 N). The propellant was liquid oxygen and kerosene (RP-1). Following a prototype launch, four unmanned and one chimpanzee flight was followed by four manned launches (a fifth manned launch was canceled as unnecessary).

Mercury-Atlas 1 was not provided with an abort escape tower as its purpose was to test the structural integrity of the spacecraft launch vehicle combination. Unfortunately, the vehicle failed at 60 seconds into the mission, and provision of an escape tower may have added further information into the operation of the system on an Atlas vehicle. MA-2 in February 1961 repeated the objective originally planned for MA-1 the previous July.

Mercury-Atlas 3 was planned to be the first production spacecraft to enter orbit but was aborted 40 seconds after launch by the Range Safety Officer on April 25, 1961 due to electrical faults on the Atlas that affected roll and pitch-over into the planned direction of flight (its azimuth). Despite this setback the abort system worked as designed and the spacecraft was recovered successfully shortly after splashdown, with only minor damage recorded.

Mercury Atlas 4, launched September 13, 1961, became the first Mercury spacecraft to complete its primary mission profile of manned Earth orbital spaceflight and recovery. From here on all Mercury-Atlas missions performed as programmed with nominal separation of the escape tower and achievement of orbital flight. On February 20, 1962 MA-6 carried John Glenn into history as the first American to orbit the Earth; he completed three orbits. This feat was repeated by Scott Carpenter on MA-7 in May of that year. Then in October MA-8's Wally Schirra extended the orbital experience to six months before Gordon Cooper completed a 22-orbit flight on MA-9 in May 1963 closing out the one-man Mercury Program. Though a three-day MA-10 mission was discussed for a while, it was decided to move onto the two-man Gemini to gain valuable experience for the Apollo lunar landing program. For Gemini, however, the designers had evolved a different method of extracting the crew in the event of a problem on the way to orbit.

Summary

One of the problem areas in the Mercury system had been in the sequencing system. A reason for separating different systems in the Gemini spacecraft, which was a descendant of Mercury, was the added complexity of building sequential systems around even some of the more simple operations. The pilots had control of most of the systems onboard Mercury, but safety also required an automated back-up.

This required a veritable spaghetti-like tangle of wiring and electrical connections. For Gemini this would be significantly reduced.

GEMINI LAUNCH ESCAPE TOWER

In the early design studies for the follow-on spacecraft to Mercury (initially called Mercury II, but from December 1961 termed Gemini) illustrations featured both an escape tower and deployable landing bag similar to the original Mercury spacecraft. The tower system had been developed for America's first manned spacecraft and was being evaluated for the three-man Apollo, also under development. Confidence in placing ejector seats in a manned spacecraft was affected by suggestions that developments in this area might be troublesome. Once the launch profiles of the Titan II rocket carrying a Gemini spacecraft had been established then the idea of using ejection seats instead of an escape tower for Gemini grew. In addition the need to open the hatch for spacewalking (extravehicular activity or EVA) required the opening of hatches in flight. The idea of using a tractor rocket similar to Mercury to clear the crew compartment away from the troubled launch vehicle was soon abandoned in favor of twin ejector seats.

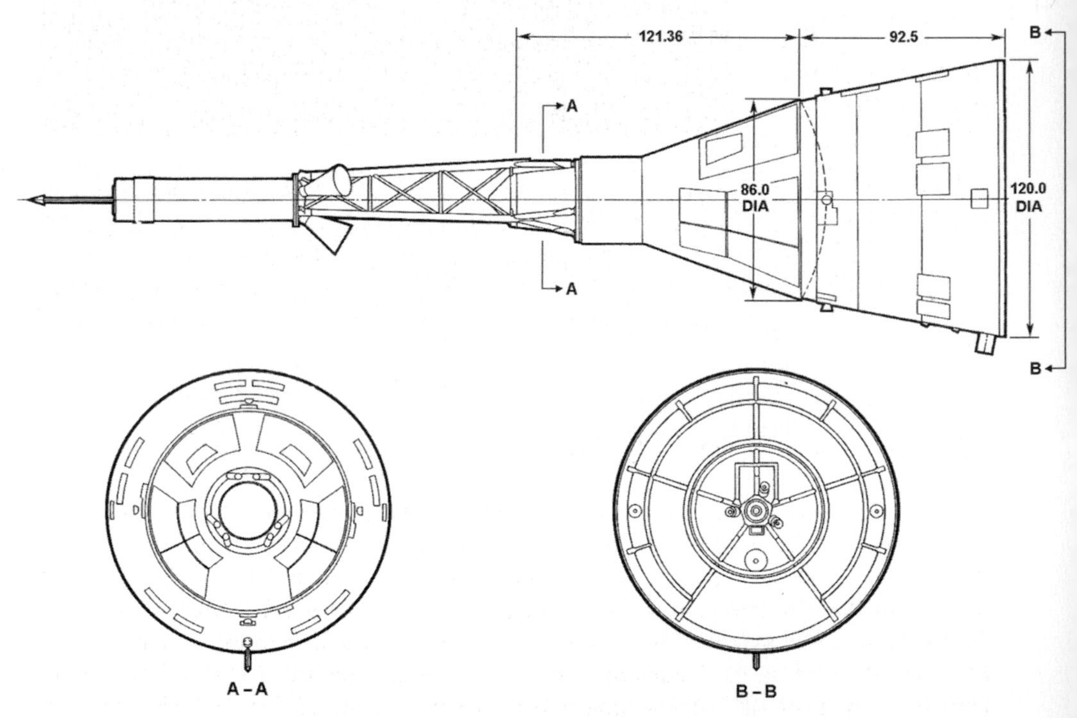

Proposed escape tower for Mercury Mark II (Gemini).

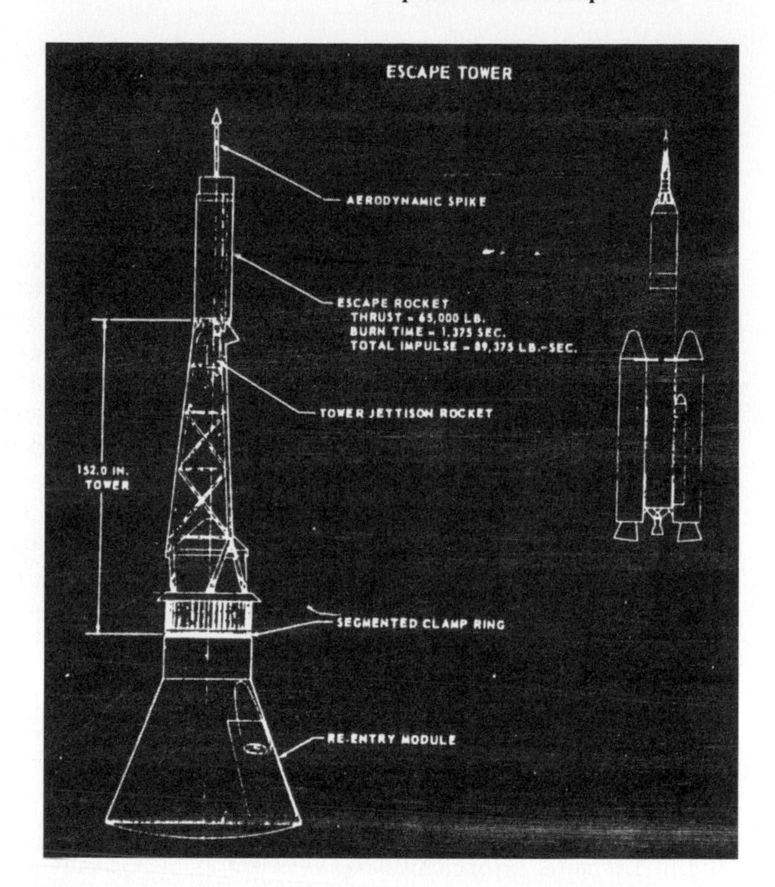

A short-lived escape tower for the USAF Manned Orbiting Laboratory.

One of the most complicated sequencing systems in Mercury had indeed been the automated abort modes. Apart from the electrical circuitry, much of which was left behind when the tower was ejected, unwanted, adding to the launch mass, was the overall mass of the tower itself. The redesigned Gemini evolved from Mercury II, with ejector seats eliminating much of the mass of the launch tower and associated wiring.

APOLLO LAUNCH ESCAPE TOWER

Between 1967 and 1975 the Apollo/Saturn combination was the primary U.S. manned spaceflight program. The system utilized a launch escape propulsion system that would in the case of an emergency on the pad or during ascent to orbit carry the three-man Apollo crew away to a safe distance for parachute descent using the Earth landing sub-system. Its operation continued work developed under the U.S. Mercury Program and is currently being used as a reference for the forthcoming Constellation Program, and a similar system will be utilized for the Orion manned spacecraft.

Overview

The system features three solid propellant motors, the launch escape motor, a tower jettison motor, and the pitch control motor. Resembling a large rocket connected to a command module, which was protected by a Boost Protective Cover, it was attached to the conical CM by a latticework tower. Measuring 10 m long and with a mass of 3,628 kg the maximum diameter was 1.2 m.

The main launch escape motor had a thrust of 70,308 kg, capable of moving the CM rapidly away from danger to a safe altitude for recovery. Once clear of the tower, the jettison motor would separate the system from the CM allowing deployment of the parachute system. The tower jettison motor was also the primary separation system under nominal flight conditions. To achieve the desired safe trajectory away from the launch vehicle a pitch control motor was essential in the design. Between 1961 and 1966 the Apollo escape system was designed, qualified and tested; it became operational between 1967 and 1975. Throughout this period of 14 years the system remained free of failures, emphasizing the quality and effectiveness of the design and that of the Apollo "system" in its reliability, despite some in-flight difficulties during launches. The Apollo LES was not called upon to retrieve a CM during nominal mission operations.

Hardware

The Apollo Launch Escape System comprised several subsystems (NAA, 1968).

Launch escape motor (Contractor Lockheed Propulsion Company). This was a solid rocket motor (propellant being a composite of polysulphide) with a steel casing measuring approximately 4.7 m long and 66 cm in diameter. On the pad thrust was another 652,680 N for 3.2 seconds which increased with altitude. It was designed to provide the required thrust to rescue a crew from a dangerous situation early in the launch phase. This was the largest motor in the subsystem. Its mass was 21,319 kg, two-third of which was the propellant. The nominal thrust vector angle was 2.75° from the center of gravity line. The two nozzles in the pitch plane had a different throat area to those in the yaw plane: one being 5% larger, the second 5% smaller than the two in the yaw plane. The structural skin to which the tower was attached was made of titanium.

Pitch control motor (Lockheed Propulsion Company). This was another solid rocket motor measuring 0.6 m long and 22.8 cm in diameter, again within a steel casing. The propellant was also a composite of polysulfide and the motor generated 10,656 N of thrust for just 0.5 seconds. This was designed to provide an initial pitch maneuver of the CM towards the Atlantic Ocean in the case of a pad abort or an abort at very low altitudes. This would also prevent the descending CM from encountering the debris from a damaged or destroyed launch vehicle. Having just escaped from an errant Saturn, the astronauts would not want to return to a possible

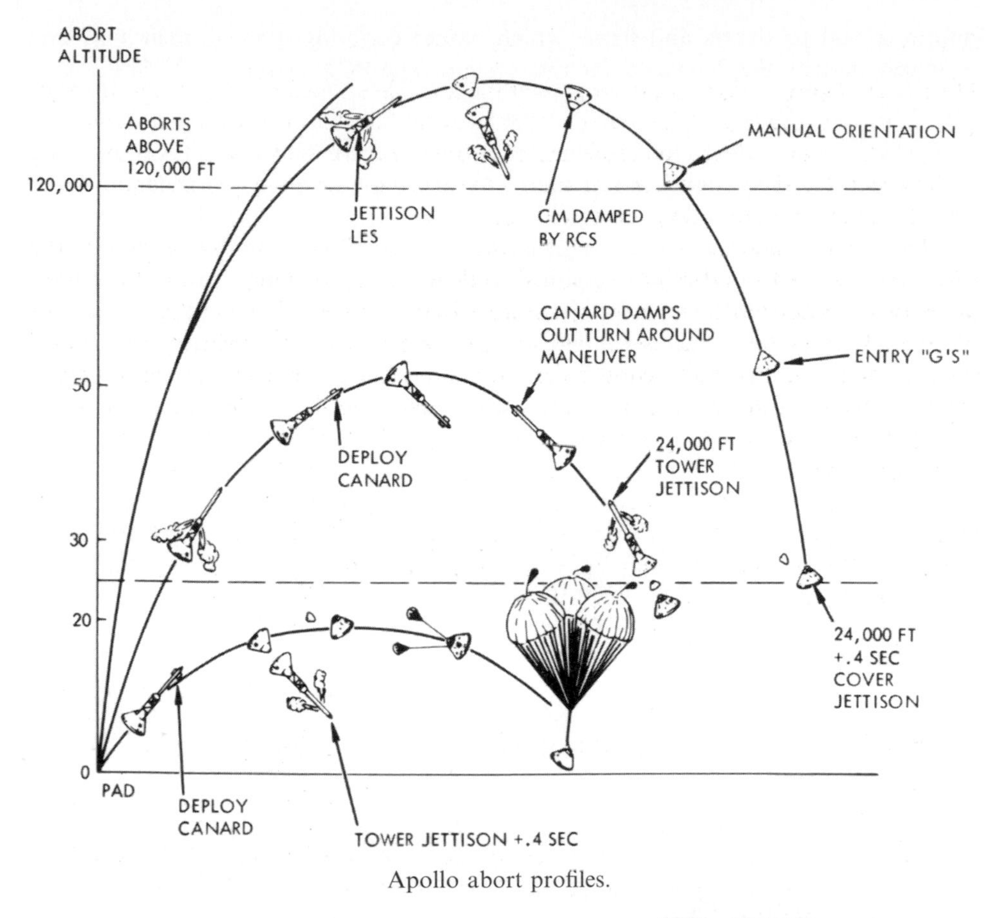

Apollo abort profiles.

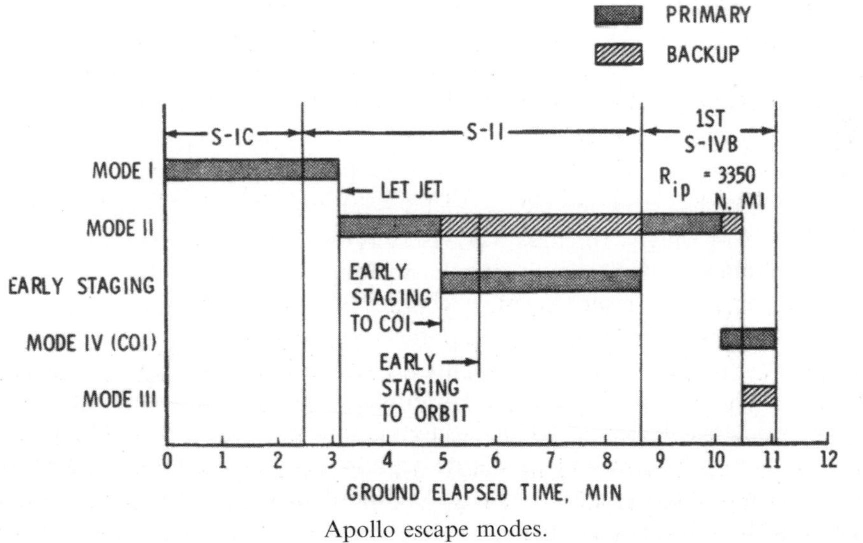

Apollo escape modes.

inferno cloud of debris and flame which would certainly damage their CM and probably destroy the descent parachute system. The PCM mass was 22.7 kg.

Tower jettison motor (Thiokol Chemical Corporation). This was a 1.4 m long and 66 cm diameter solid rocket motor with a high-carbon chrome–molybdenum steel-forced casing. The polysulfide composite propellant provided 139,860 N of thrust for 1 second. This motor was used to jettison the LES after the first-stage separation when the system was no longer required or during a launch abort prior to recovery parachute deployment. The tower jettison motor mass was 238 kg. Located at the top of the launch escape motor it had two fixed nozzles 180° apart, and it projected from the steel casing at a downward and outward angle. This skewed nozzle design provided a thrust vector angle of about 4°, which effectively tipped the tower and pulled it free of the CM.

Launch Escape Tower. This was a 3 m long welded 6.3 cm and 8.8 cm diameter titanium tubing truss structure. It was covered with Buna-N rubber insulation that protected it against the heat of the rocket motor exhausts. At the top it was 0.9 m square tapering down at the base to 1.2 m where four legs fitted into wells on the Command Module through the Boost Protective Cover. The mass was 227 kg which included all the attachment fittings, associated wiring and insulation. The tower supported the three motors described above. It was fastened to the CM by studs and brittle nuts containing a small explosive charge which broke them to ensure CM separation when the tower and LES were jettisoned.

Boost Protective Cover. Completly covering the Apollo Command Module was a cone-shaped covering, designed to prevent the charring of the CM's external surfaces during ascent through the lower atmosphere and during aborted flight profiles until tower jettison when it was discarded. Constructed of layers of resin-impregnated fiberglass, honeycomb core–laminated fiberglass and cork. The design featured 12 "blow-out" ports for reaction control motors, vents, and an 20.3 cm diameter window in front of the Commander's (center seat) viewing window. There was also a hinged opening to gain access to the CM's sidehatch during operations in the White Room. The structure measured 3.3 m tall and 3.9 m in diameter at its base and fitted the CM snugly. Its mass was 317 kg. During launch a passive tension tie connected the apex of the boost cover to the docking probe of the CM. On a nominal ascent the tower jettison motor is jettisoned after first-stage separation and its thrust pulled the cover away snapping the tension tie from the docking probe. Had an abort been necessary during one of the Apollo missions, then the Master Events Controller and Lunar Docking Events Sequencer Controller would have sent electrical signals to fire the ordinance devices thus separating the docking ring on the CM. When the LES was jettisoned for recovery parachute deployment then the tower, still attached to the probe, would have separated the probe and docking ring from the CM along with the Boost Protective Cover.

Canards. These were only deployed during abort profiles, 11 seconds after the initiation of an abort, and were used to orientate the CM to point the heat shield forward and parachute compartment to the aft of the direction of flight. They feature two metallic clamshell-shaped aerodynamic control surfaces approximately 1.2 m long.

Q-ball. At the apex of the tower structure was the nosecone, 34 cm long and 40 cm at its base, made of aluminum. Called the Q-ball assembly it comprised different pressure transducers and electronic modules, which measured the differential of dynamic pressures about the pitch and yaw axis in order to modify the angle of attack of the spacecraft as it followed the abort profile. The Q-ball itself sent electrical signals to a display on the main console in the CM and to the ground. There were eight static openings (ports) on the ball that measured changes in pressure as the angle of attack changed. Pitch and yaw pressure change signals were then electrically presented on visual displays indicated in the spacecraft and on the ground. This information acted as baseline data for crew abort decisions, should the Saturn be underperforming. A tapered ballast compartment completed the assembly.

Master Events Sequence Controllers. These consisted of a pair of box-like structures measuring 35.5 × 25.4 × 20.3 cm, each of which contained time delays, relays, fuses and fusitors. They were located in the Forward Right Hand Equipment Bay of the CM and controlled a variety of launch abort functions as well as nominal mission functions.

Emergency Detection System. This was the monitoring system that analyzed critical conditions aboard the launch vehicle during powered ascent. Should an emergency situation develop then the necessity to abort was indicated to the crew by means of the main control console in the CM. The EDS also made available to a crew-initiated abort use of the Launch Escape System or the Service Propulsion System of the Service Module, following the separation of the tower assembly. An automated abort capability was also included in the system to support time-critical conditions such as the loss of thrust on two or more engines on the first stage (S-1C) of the launch vehicle or excessive angular rates of pitch, yaw or roll on the Saturn.

Operation

During normal flight the jettison of the LES was initiated manually. The tower leg explosive bolts and tower jettison motor were ignited simultaneously, pulling the system out of the way of the ascending launch vehicle. The lateral separation maneuver created a minimum separation of about 46 m (called a miss distance). In the worst case scenario the LES could separate and still avoid recontact with the launch vehicle. If the tower jettison motor had malfunctioned then the launch escape motor could have been fired for LES jettison without affecting the safety of the crew or progress of the launch vehicle towards orbit.

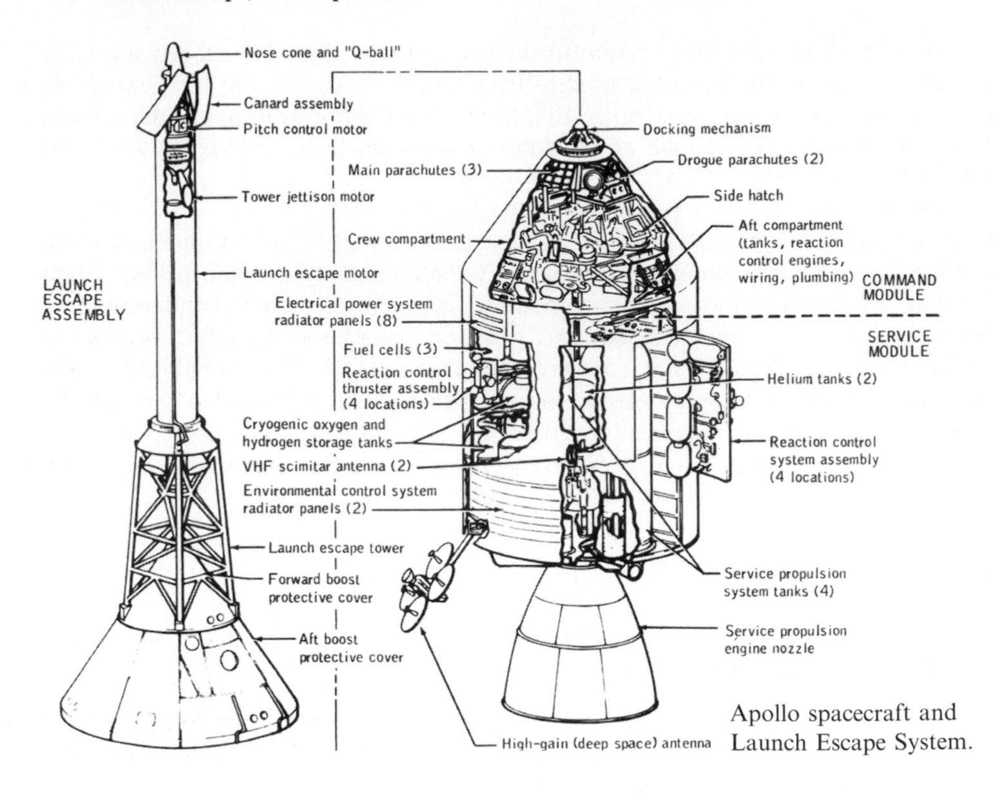

Apollo spacecraft and Launch Escape System.

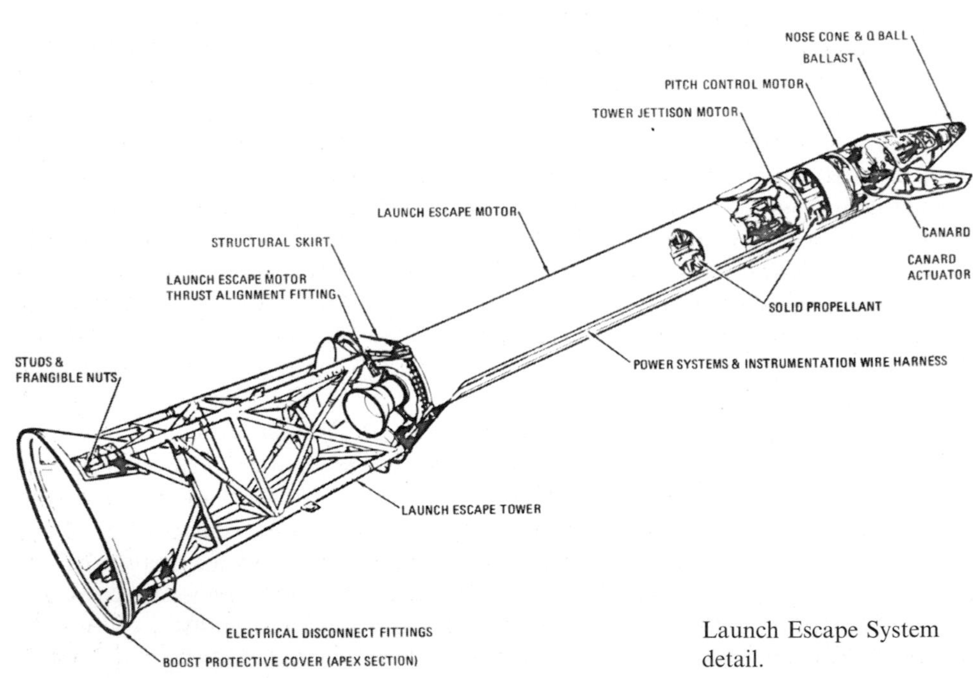

Launch Escape System detail.

Had an abort been initiated either by manual or automated means the EDS would have cut off the booster engines after the time restriction had been exceeded. Due to range safety restrictions a time delay was imported on both Saturn 1B and Saturn V manned launches: 40 seconds into the flight for the Saturn 1B or 30 seconds for the Saturn V. This did not, however, impair the LES operation.

The EDS circuits aboard the CM were activated automatically at lift-off and deactivated 100 seconds into the mission. There were three switches on the main display console which were placed in the automatic position at lift-off. They could deactivate the complete automatic abort capability or just the "two engines out" or "excessive rate" portions independently. They were switched off prior to separation of the first stage. As a back-up, two automatic abort circuits were located in the Saturn Instrument Unit which were also deactivated just prior to separation of the first stage.

Inside the CM three indicator lights warned of a problem in the associated field. They were

- *LV rate*—launch vehicle roll, pitch or yaw rates exceeding predetermined limits.
- *LV guidance*—a red light indicating loss-of-attitude references in the guidance unit.
- *LV engines*—a yellow light would come on when an engine was developing less than the required thrust.

There were visual clues to the operational status of engines on the Saturn, these were ignition, cut-off, below thrust and physical stage separation. As the vehicle climbed, so the abort capabilities changed accordingly.

The red abort light provided high luminance in the crew compartment and could have been initiated by the Launch Control Center for a pad abort, the Range Safety Officer after lift-off or via uplinked data from the Manned Spacecraft Flight Network.

After ignition of the second stage, the LES was jettisoned. There were two tower jettison switches available for this operation, and though both would normally be used either one could have initiated the function. The crew used the digital events timer in conjunction with the visual lights clue to jettison the tower at the appropriate time. For the Saturn 1B launch the No. 1 Engine status and Stage 2 (S-II) separation light for the Saturn V were used to aid the jettison of the tower on a nominal mission.

To initiate an abort from the crew position the Mission Commander had the responsibility from launch pad to orbit. Constantly scanning the controls and displays for signs of a problem when more than one cue from the EDS was seen, to avoid an automatic signal triggering an erroneous abort the Commander would manually twist the T-handle in his left hand counterclockwise to initiate the LES. To ensure automated systems did not trigger an unnecessary abort, the different systems were triple-redundant and "voted" on the decision to abort or continue. The abort handle (which was in effect a "chicken switch") was a cause of concern for Apollo 11 CM Pilot Mike Collins who during the final moments of countdown on the historic first manned lunar landing mission saw Armstrong's pressure suit rubbing against the handle. He pictured next-day newspaper headlines had the system been inadvertently initiated, added to which there would be the embarrassment of owning up to the mistake—in what could have been a quite expensive "oops".

It was from Apollo 11 that the Commander had the option of "flying" the Saturn to orbit if he determined from onboard information that the huge rocket was unable to guide itself. The option was to let control of the Saturn pass to the Command Module or the Commander himself. He would also have information presented to him as to whether the Range Safety Officer or Launch Control at KSC Florida or Mission Control in JSC in Houston thought an abort should be initiated. To aid in the crew's decisions a computer program (Program P11) was run during ascent in the CM updating displays on their velocity, altitude and rate of change; in addition the attitude of the ascending vehicle was also presented which, should they need to use the P11 program as clues to "fly" the vehicle, the commander had an updated reference of how fast they were flying, at what altitude they were and which direction they were heading and pointing.

Apollo abort capabilities

For Apollo several categories were available during ascent, termed modes. Those under Mode 1 were associated with the LES. There were six types of abort capabilities concerned necessary for LES operation. Modes 2–4 came into play after separation of the LES and required the uses of the Service Propulsion System. Like all aborts each mode was subdivided according to the altitude of the vehicle at the time of the abort call. The crew would be informed of the next mode available to them as the previous one was surpassed in the air-to-ground commentary from Capcom in Houston.

The events controller was used in all cases, though some sequences of events changed from nominal operations depending on the type of abort (pad, altitude, high altitude) and dumping of RSC propellant, which enabled control of the CM attitude automatically. The SM initiated aborts allowing nominal entry and landing techniques and procedures.

Mode I

This covered the period of time from sitting on the pad to separation of the Launch Escape System on a nominal mission at a GET of 2 min 44 s. The LES automatically activated (it could also be activated by a crew command input) when up to two engines in the first stage failed or excessive rates built up on the ascending vehicle. After separation of the CM, drogue and main parachutes were deployed and the CM would splash down up to 741 km downrange.

- Pad abort—for use just prior or shortly after lift-off prior to tower clear. Tension-tied pyrotechnics would sever connections between the CM and SM. The LES would have separated the CM from the SM and Saturn. It would then be propelled to the appropriate height to enable proper operation of the CM Earth landing system. Wind drift problems at the time would be factored into the range

Hoping it all works as designed.

of the abort. The abort trajectory plane was fixed in a downrange direction and for pad aborts this was 914 m at apogee.

- Low-altitude abort —the LES had the capability to operate even though range safety limits prohibited the cut-off of boosters for the first 30–40 seconds of powered flight.
- Abort at high dynamic pressure—the LES had to operate during the time when the maximum dynamic pressure was exerted on the launch vehicle. The abort would of course be initiated prior to the structural break-up of the launch vehicle.
- Abort below 30,480 m altitude—a manual or automatic abort could be initiated from signals generated from the EDS. The launch escape motors would have ignited for 0.6 s and the pitch control motor for 4.0 s. The pitch control motor could deliver a large pitching motion of relatively short duration increasing the range capability of pad abort and increasing lateral separation from the launch vehicle at higher altitude aborts. The motor would have been deactivated at 42 s after lift-off either automatically by a time-controlled relay or a crewman-operated switch. After 11 seconds canards would have been deployed to orientate the combination with the CM heat shield forward. After a further three seconds or descent to 7,620 m the tower would have been jettisoned allowing the CM to perform parachute descent.
- Abort over 30,480 m—the crew would use the CM RCS to provide positive pitching motion for the LES and CM, canards would then be deployed at 11 seconds into the abort attitude, and the sequence would then follow that in the preceding bullet.
- Maximum-altitude abort—the use of the LES could not exceed the altitude of second-stage ignition. There were limits on dynamic pressure speed and separated components hindering LES operation. Parameters for LES operations for Apollo were defined as a powered altitude of 97,536 m, Mach 8.0 and dynamic pressure of 0.2 kg m^2 to 0.4 kg m^2.

Within Mode 1 were three divisions depending on the trajectory and velocity of the vehicle during the first-stage operation.

During the first 42 seconds of powered flight up to about 3 km in altitude the mission was flown under *Abort Mode One—Alpha*. In this status the crew were ready to react to any launch deviation as the huge launch vehicle left the pad, cleared the tower and slowly ascended away from the launch complex area. Once the 42-second mark had been surpassed then *Abort Mode One—Bravo* took over. By now the Saturn Apollo "stack" had gained substantially more horizontal speed, tipping over the Atlantic as it raced skywards following the program azimuth (direction) towards orbit. By now the CM would not be in a position to fall back into the debris of the Saturn launch vehicle and as a result the Pitch Control Motor would not be used in this mode. However, something was needed to ensure the CM turned around in flight to ensure the parachute Earth landing system would operate nominally. It was determined in tests that if the Boost Protective Cover/LES jettisoned, then the cover might impact the CM structure, damaging the Earth landing system and possibly preventing clean separation. Therefore, the canards could be used to reorientate the combination, using drag to turn around the vehicle allowing the CM heat shield to face the correct direction in flight. *Abort Mode One—Charlie* operated from the 30 km altitude until the second stage of Saturn had taken over. At these altitudes the canards on the escape tower would be ineffective in the rarefied atmosphere. Therefore, the orientation of the CM would have been completed by the small reaction control thrusters around the base of the CM normally used at the end of the mission following separation of the SM prior to re-entering the atmosphere.

Mode II

This covered the period of time from 2 min 44 s into the flight up to 9 min 33 s into the mission. The CSM would separate from the S-IVB followed by a 20-second SM Reaction Control System posigrade burn (with direction of flight used to increase velocity). The CM would then separate from the SM, and perform a full lift re-entry landing between 741 km and 5,930 km downrange

Abort Mode 2 came into effect following ejection of the Escape Tower. As aerodynamic forces were no longer a threat to vehicle break-up, the spacecraft could separate on its own to effect a recovery.

Mode III

This covered the period from the 9 min 33 s GET point to orbital insertion (9 min 53 s GET). Here the separation sequence was the same as for Mode II but the Service Module Propulsion System (SPS) burn would have been retrograde (against the direction-of-flight slowdown). The CM would fly an open-loop (half-lift) re-entry to the recovery area at 5,930 km for an immediate return to Earth, if an abort to orbit was not a viable option. The Mode III abort to orbit was also known as *contingency orbit insertion* where a failure on S-II meant that a combination burn of the S-IVB then the SM would place the spacecraft in Earth orbit, but by separating the S-IVB third stage there would be no provision for a translunar injection burn, and so one of the contingency Earth orbital alternative mission scenarios could safely be flown (see Chapter 7).

Mode IV

This covered the period from 9 min 27 s into the mission and up to orbital insertion. The SM propulsion system would be utilised in a posigrade burn to push the spacecraft into orbit with at least a 105 km perigee but leaving enough to complete a de-orbit burn. Essentially an abort-to-orbit profile, if an alternative mission could not be accomplished then recovery would be attempted in the West Atlantic or Central Pacific Ocean after just one orbit. This is a more preferable mode than Mode II and would have been chosen unless an immediate return to Earth was called for under Mode III.

Apogee Kick (AK) Mode

This was a variation of Mode IV in which the SPS was fired to raise perigee to 105 km. This would be when the velocity cut-off at S-IVB was greater than 30.5 m/s. This is again the preferred option unless an immediate return to Earth was required.

Launch Escape System development

On February 13, 1962, Lockheed Propulsion Company was awarded the contract by Apollo prime contractor North American Aviation for the Apollo Launch Escape System. This came just six weeks after NAA had won the lucrative contract to design and develop the Apollo Command and Service Modules for NASA's manned lunar programme. Originally the plans were for an 888,000 N thrust solid-propelled rocket motor featuring an active thrust vector control subsystem. However, following more extensive studies the control subsystem was removed and by June the design was changed to a pitch control motor subsystem in conjunction with a pitch control motor. As a result the thrust of the escape motor was reduced to 688,200 N.

Three months later on 6 April, Thiokol Chemical Corporation was chosen by NAA to fabricate the solid-fuel rocket motor to be used to jettison the escape tower whether during a nominal mission or following a launch abort situation. In January 1963 a pair of aerodynamic stakes was added to the CM to prevent the danger of a crew module re-entering apex-forward after an abort. If the abort was at low altitude then there was the problem of ensuring the apex cover cleared the CM once jettisoned. Adding the stakes increased the mass considerably and resulted in a change in the CM center of gravity; additional ablative covering added to the problems. If the stakes were removed a major redesign would be required to allow successful apex jettison (and clearing of the Earth landing system) in the apex-forward orientation.

By 1963 it was decided by NAA that the stakes could be replaced by flaps that would be jettisoned with the tower during normal missions or retained on the CM during aborts. Tests by NAA suggested it might be more advantageous to use "flaps" on the upper end of the tower which could be exposed to the air stream, thus turning the spacecraft around for landing. Further tests were completed at NAA whilst an independent investigation of deployable aerodynamic surfaces, called canards, located on the forward end of the LES could act as a lifting surface sufficient to orientate the CM with its heat shield forward following an abort situation.

On April 18, 1963, Lockheed Propulsion Company received the Solid Propellant Motors Contract from NAA. The design of the system by July included a redundant tower separation device which included explosive bolts and umbilical cutters.

As part of the development of new spacecraft, the NASA Astronaut Office (CB) was closely involved by giving astronauts various technical assignments such as representing the CB in meetings, reviews and visits to contractors. Astronauts would then report back to the group in regular meetings to keep the rest of the group abreast of developments in key areas that were beyond the ability of each individual to keep track of.

Saturn V Apollo stack on the pad. The Launch Escape System is at the top of the vehicle.

During a January 1964 meeting between astronauts and the System Engineering Division at the Manned Spacecraft Center (MSC, the precursor to the Johnson Space Center) in Houston the engineers wanted to know if the astronauts thought a manual reorientation maneuver was feasible if canards were added to the Apollo abort system. The astronauts felt that it was feasible and manual control was a valuable

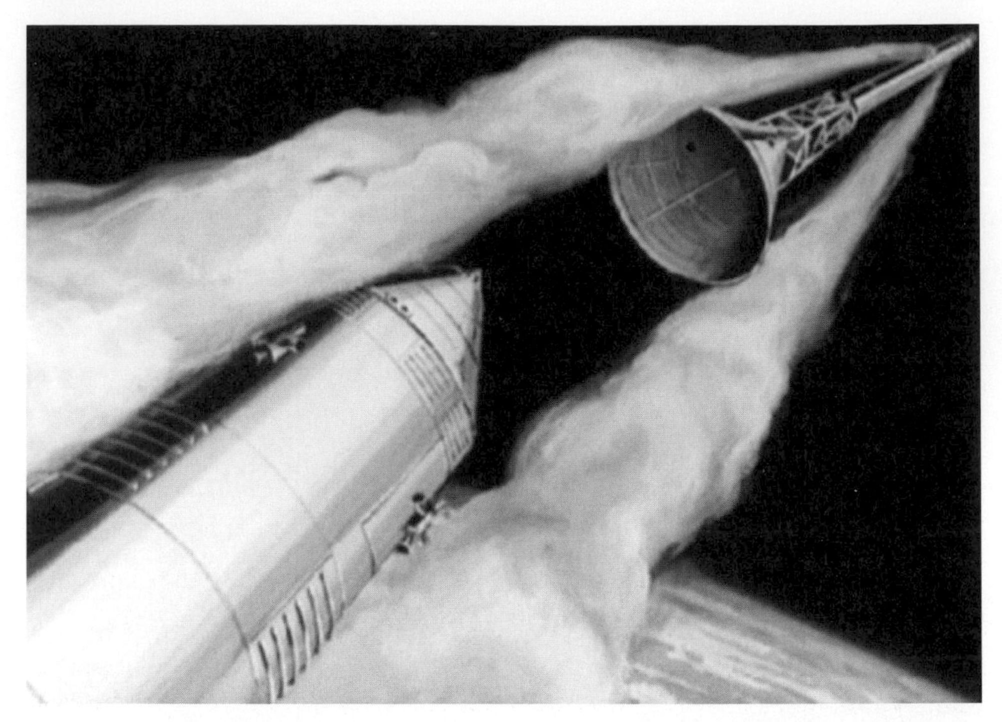

Boost protective cover separation as it would appear in flight.

option at high altitude; the problem they envisaged was the potential for sooting of the CM windows, thus restricting visibility during both normal and aborted flight profiles. The suggestion was made to install a protective cover over the CM to help eliminate this problem. This could then be ejected when the LES was separated in either the normal or abort modes. This discussion continued and in 1966 a technical note was issued detailing the studies of high-altitude manual aborts for Apollo launch profiles (NASA, 1966b).

Meanwhile, discussions as to the advantages of either canards or flaps continued, with a strong case being put forward in favor of canards. It was suggested that canards be installed on the Block I (Earth orbital, no docking system) LES, and further studies continued until the time was reached to choose the system for Block II (lunar distance) LES. By February 1964 studies indicated that with the use of flaps there was a high probability of exceeding crew acceleration limits during high-altitude aborts. Therefore, canards were added to Block I with further studies needed before they were assigned to Block II. The following month the concept of a hard Boost Protective Cover was selected which would be removed at the same time as jettisoning the tower.

In November 1964, Bellcomm Inc. had evaluated the addition of a Q-ball device in the emergency detection system. This was to be enclosed in the nose of the escape tower and was designed to record dynamic pressure on the vehicle during the angle of attack and inform the crew of an impending aerodynamic break-up of the vehicle, allowing activation of the abort system. This was critical to crew safety and the device was ideal for the task and incorporated into the escape system design.

During 1965, added mass revealed that the LES would be incapable of lifting the CM a safe distance from a booster. Studies continued but it appeared that ballistic effects and intense heat from an exploding launch vehicle could have serious effects on the parachutes used for crew recovery.

Studies on improving the system continued throughout the flight operations period. In March 1969, Manager of the Apollo Spacecraft Program Office George Low responded to a briefing he had at Downey the previous October into the suggestion of deleting the Boost Protective Cover in order to save weight, which could be used to incorporate new systems and upgrades into proposed later generation Apollo CSM vehicles to support expanded lunar and Earth orbital scientific activities. The decision was taken to retain the cover.

White Sands Launch Facility

To qualify the Apollo Launch Escape System, NASA utilized the facilities at White Sands Missile Range in New Mexico between 1963 and 1965. The plan to test the Apollo LES was originally to launch from the USAF Eastern Test Range at Cape Canaveral in Florida, but a heavy schedule of high-priority launches precluded that plan. By using WSMR the requirement for water recovery was also eliminated, which would have been the case from the Cape or NASA Wallops Island, thus keeping costs down. In addition to supporting the development of the escape system, facilities at the WSMR were also used to perform certification testing of CSM structures, onboard batteries and some of the flight instrumentation, saving time at the Cape. The opportunity was also taken to perform some fit-and-function tests on the CSMs and checkout of ground support equipment. Once the work at the facility supporting Apollo had been completed then the property and equipment assets of the Flight Test Office there were disposed of (within 60 days of the final Apollo Little Joe II flight). Since the 1970s nearby facilities at the White Sands Space Harbor have been used as a contingency landing site for the Space Shuttle, though only one mission (STS-3 in 1982) has actually landed there to date (see Chapter 8) (NASA, 1975, pp. 10-1 to 10-4).

Launch Complex 36. The original basic complex consisted of facilities to fire Redstone boosters. These were a launch pad, blockhouse and service tower. When that program ended the site was upgraded to support the Apollo Little Joe II series. The service tower and half the blockhouse were utilized, the other half of the blockhouse was used for another program. A new pad had to be constructed, and permanent tracks to move the service tower to the new pad and power supply faculties and connections had to be laid. To accept the Apollo/Little Joe II configuration the service tower required extensive modifications, and a clean room was added to enclose the spacecraft for the installation and checkout of equipment and instrumentation. The pad also included flood lights, shower and water-flushing systems for cleaning the pad and added safety. Other support equipment included a hydrogen peroxide–servicing tower and portable recording and monitoring equipment.

Vehicle Assembly Building. One mile from the launch pad was the newly constructed Vehicle Assembly Building. This new facility featured a high-bay, laboratories, storage facilities and checkout areas. Upon arrival at the facility the Little Joe II vehicles were stored inside the VAB, sometimes accompanied by the Agol rocket motor. Subsequently, a clean room was added to the facility. As the program was of short duration, permanent office space was rejected in favor of 15 mobile trailer offices for vehicle assembly and checkout support. Those in the launch pad area were moved prior to installation of final system checks and then returned after the next launch.

Control System Test Facility. This was the location of all the equipment necessary for the testing and servicing of the Reaction Control System and Hydraulic Powered Aerodynamic Control System. A concrete test pad and environmentally controlled prefabricated steel building were used for systems testing.

Launcher. This was the mechanical structure for the final assembly of the launch vehicle and checkout of systems. It performed as designed in all launches with minimal modifications but some improvements were required in light of flight test experiences and system upgrades.

Ground Support Equipment. There were 248 items of support equipment required for Little Joe II. Some were commercially obtained by the contractor or NASA, while others were specially manufactured for the program. The CSM needed 45 units of handling equipment, 22 units of checkout equipment and 9 units for servicing (a further 28 items were obtained).

The program at White Sands in support of Little Joe II reflects the often overlooked additional material, equipment, operational requirements and personnel support required for developing a system that is suitable for operational use but is not highlighted unless called upon to be used in an emergency situation.

Ground-based test program

A program of ground-based tests qualified each element of the Apollo Launch Escape System for both flight tests and operational use. This program included static test-firing of the Launch Escape System pitch control, the jettisoning of the tower and operation of the launch escape motors. In addition a program of environmental tests was completed to subject the hardware and system operation to a variety of conditions.

Once the propellant had stabilized at selected pre-fire temperatures, test-firing was completed at a pressure of 1 bar. These temperatures reproduced the minimum (−6°C), nominal (21°C) or maximum (49°C) expected operational temperatures.

Pitch control motor. A total of 14 pitch control motors were environmentally tested in five groups. These were temperature cycling and vibration; temperature cycling and drop testing; accelerated aging; accelerated testing; and temperature cycle testing. This was followed by a program of static test-firing of 17 pitch control motors, 10 of which had already been used in environmental testing. Here three groups were tested based on the pre-fire temperatures of −6°C (six motors), 21°C (four motors) and 49°C (seven motors). Lockheed Propulsion Company successfully static test-fired four pitch control motors in December 1962. Development testing on the motor was completed by August 1963.

Tower jettison motor. In this phase of testing, static test-firing used 21 tower jettison motors which included 15 that were tested environmentally. This time four test groups were identified: temperature recycling; accelerated aging; temperature recycling/impact testing; and vibration testing—again using the three predetermined temperature levels of −6°C (nine motors), 21°C (five motors) and 49°C (seven motors). The first static test-firing of this motor occurred on December 1, 1962 by the Thiokol Chemical Corporation. They completed qualification testing in February 1965 though a modified seal had to be developed and tested due to an ignition delay on a February test, resulting in a redesign of the ignition cartridge. This was tested successfully during August and September 1965.

Launch escape motor. A series of 20 static test-firings of the LES were completed. There were four separate tests which included test-firings (seven motors used), accelerated aging (two motors), temperature cycling (four motors) and sequential testing (seven motors). From these tests seven launch escape motors were subjected to a specific sequence of environment testing. Six motors were each subjected initially to a temperature cycle test, then a drop test, followed by a firing test. A further motor was used in a vibration test, a temperature cycle test, then a firing test. The program encompassed all launch escapes that specifically related to ballistic performance, environmental testing and thrift alignment (apart from roll moment testing). Analysis of test data revealed that five motors exceeded the maximum roll limit of 40 m/0.45 kg. A subsequent optical check of the nozzle's throat indicated that maximum misalignments would result in a roll of no more that 1.2 m/0.45 kg. The excessive high-roll rates were attributed more to errors in the test stand load cell than a deficiency in the design. December 1962 saw NAA complete three static test-firings of the launch escape motor.

Qualifying and production problems

A number of problems were encountered in the qualification and production of the three engines used in the LES, as was expected in the development of such a system. Further evaluation, changes in procedures and refining the process answered or brought more serious problems to the fore, allowing the system to be qualified for use in manned launches without excessive delay to the manifest.

Apollo LES flight test program

There were eight qualification flights for the Apollo Launch Escape System. These included two pad aborts and four Little Joe II flights. All six, conducted at White Sands Missile Range, were designed to investigate the performance of the Apollo system under abort flight scenarios. The remaining two were under normal flight conditions on Saturn launch vehicles out of Kennedy Space Center.

Pad abort tests

For the pad aborts, two boilerplates were used to simulate an LES-propelled spacecraft from a launch vehicle to a height great enough to utilize the Earth landing system.

Pad Abort 1 (PA-1), November 7, 1963

The flight test program commenced with a pad abort. All primary objectives were met. The plan was to determine the aerodynamic stability characteristics of the Apollo LES on a pad abort profile. The capability of the LES, designed to pull away a CM to a safe enough distance from the launch vehicle, was demonstrated in this test. The stability of the LES/CM configuration was also demonstrated successfully.

Table 5.1. Apollo Launch Escape System qualification flights.

Designation	*Spacecraft*	*Launch date*	*Launch site*	*Notes*
PA-1	BP-6	1963 Nov 07	WSMR	First pad abort
A-001	BP-12	1964 May 13	WSMR	Transonic abort
AS-101	BP-13	1964 May 28	KSC	Nominal launch and exit environment
AS-102	BP-15	1964 Sep 18	KSC	Nominal launch and exit environment
A-002	BP-23	1964 Dec 08	WSMR	Maximum dynamic pressure abort
A-003	BP-22	1965 May 19	WSMR	Low-altitude abort (high-altitude abort planned)
PA-2	BP-23A	1965 Jun 29	WSMR	Second pad abort
A-004	SC-002	1966 Jan 20	WSMR	Power-on tumbling boundary abort

Apollo Little Joe II.

However, during the powered phase of the flight the pitch yaw and roll were not as predicted. It was during post-flight examination that exhaust particles from the escape motor had impinged on the CM causing soot deposits. This was a significant factor that led to the provision of a Boost Protective Cover on the CM for all subsequent vehicles.

A production LES was used with Boiler Plate CM6. This represented the first boilerplate CM to be flown under the Apollo Program. However, there was no instrumentation onboard to determine structural loads during the test flight as it

Apollo abort test launches.

was not a representation of a flight spacecraft. What was measured were the accelerations on the ascending vehicle, its angle of attack, the Mach number and the dynamic pressure. From this data engineers could determine the in-flight loads resulting from the external loads of the dynamics on the vehicle in flight.

Mounted on three bearing points of a supporting structure which were attached to a concrete pad, the CM was installed in a vertical position with the LES atop. A ground signal at the appropriate time initiated the launch escape sequences that activated the launch escape and pitch control motors almost simultaneously. This resulted in the command module being lifted along a pre-planned flight path with the escape tower separating 15 seconds later and following its own ballistic trajectory. The CM made a nominal parachute descent at a velocity of 7.3 m/s, 165.1 seconds after ignition.

Though the stability of the vehicle was less than predicted, all leading objectives were met. The vehicle actually exceeded the minimum altitude and range requirements for an Apollo pad abort (by 296 m and 465 m, respectively).

Pad Abort 2 (PA-2), June 29, 1965

Due to the changes in mass concentration of the CM, the implementation of a canard system on the escape tower and a Boost Protective Cover, a second pad abort test was scheduled, some 19 months after the first. This was the penultimate abort test in the manifest and all objectives were achieved. The escape motor and pitch control motor were, as planned, ignited at the same time, with the canards deployed successfully, which enabled the combination to turn the vehicle around to heat-shield-forward attitude. The LES and the apex cover were both jettisoned as planned. One discrepancy was that a moderate roll rate developed at lift-off, but this did not affect the outcome of the test.

In this pad abort a qualified LES was used with the BP-23A spacecraft, which had previously been used on A-002 (see p. 149) and was refurbished to represent as near as possible a Block I CM in mass and with other characteristics to give a more representative demonstration of an operational spacecraft. Pad 36 was used for the launch. It was also determined that the Boost Protective Cover which was attached to the tower should remain intact through the canard-induced pitch maneuver, but at tower separation the cover collapsed due to the different pressure levels experienced in separating from the CM. There was no evidence of re-contact or any interference between the components during the rest of the flight.

The parachute system worked as designed and the maximum altitude achieved was 2,822 m above mean sea level (LC 36 at White Sands is about 1,219 m above mean sea level), or 1,600 m above the launch pad. This was about 200 m higher than predicted and the CM landed 610 m farther away than planned, some 2,316 m from the pad. An additional test on this flight was the mounting of glass samples on the CM boilerplate in the general area of where the rendezvous and crew windows were planned to be located on the later spacecraft. Though no soot from the abort motors was found on the glass three of four samples did have oily film deposits. It was determined that there would be no degradation to the horizon scan or crew observation ability with regard to the ground with this amount of debris had it been a real manned abort.

With this flight there was no requirement for further pad abort tests prior to qualifying the system for operational use.

Little Joe test program

An early decision in the Apollo Program was the human-rating of the Launch Escape System at the minimum level and as early as possible. There was not a suitable (and reasonably priced) launch vehicle available to NASA to support the tests and be capable of handling the payload mass and a range of thrust levels intended for the series. A contract was agreed for the design and fabrication of a special vehicle capable of meeting the goals of the program.

Launch vehicle development

There were two configurations of Little Joe II as the vehicle was designated. The original featured fixed fins, whereas the latter model used flight controls. The dimensions were chosen to match those of the Apollo Service Module and to match the length of the rockets chosen for propulsion. Stability in flight was assured by the use of four aerodynamic fins at the base of the vehicle. The contract for the vehicle was awarded to General Dynamics/Convair who had a joint program agreement with NAA (for the spacecraft) and NASA MSC in Houston. The fabrication of the first vehicle commenced in August 1962 and factory check-out was completed by July 1963. As far as possible simplified tools and off-the-shelf hardware were utilized in addition to a less complicated manufacturing program, all of which kept costs down. This also had the effect of reducing the number of components and thus keeping the construction time to a minimum.

The mass was based on 67,056 kg which included a 24,384 kg payload. The vehicle was designed for sequential firing of four first-stage and second-stage Algol (465 kN thrust each) solid-propellant motors. A wide range of performances were achieved by firing a varied number of rockets, up to seven, clustered to meet the mission objectives. When extra boost was required to supplement the overall thrust at lift-off, Recruit (Thiokol XM19) rockets with 167 kN thrust each were added to the vehicle (up to six were capable of being attached). The overall length of Little Joe II was 10.1 m without the CM/SM/LES attached. When added to the vehicle the overall length extended to 26.2 m with a main body diameter of 3.9 m; burn time was about 50 seconds.

Qualification test vehicle

A Little Joe II qualification test vehicle without a spacecraft attached was launched on August 28, 1963. This test was designed to demonstrate the capability of the vehicle to perform the required launch trajectory for the A-001 mission, to demonstrate that Little Joe II could indeed clear the launch pad and to demonstrate the Algol thrust termination system. The flight also demonstrated that there was no flutter on the fins, its structure was sound, it could compensate for wind deviations in both elevation (height) and azimuth (direction), and evaluate the operation of ground support equipment. All of which were achieved.

This test flight was followed by a series of four launches under the Little Joe II program. These demonstrated the ability of the LES to safely abort a CM under the various critical conditions a manned mission might encounter. In addition the Earth landing system was tested under abort-induced stress conditions confirming not only its structural integrity but also the reliability of the system to perform as designed.

A-001, May 13, 1964. This was, in chronological order, the second launch under the program and was designed to evaluate the escape system capability to propel away the CM from the launch vehicle while in the high-dynamic-pressure (transonic) region of a simulated Saturn trajectory. One Algol sustainer that burned for about 42 seconds and six Recruit motors (duration 1.5 seconds) launched BP-12 to an altitude of 9,075 m above sea level. Launched from WSMR there had been a 24-hour postponement due to unacceptable wind conditions, but all went according to plan on the second attempt until a ground signal terminated the thrust of the launch vehicle by rupturing the casing of the Algol motor. The LES worked as designed though some structural damage was sustained by the CM aft heat shield which re-contacted the booster at the end of the thrusting maneuver. The tower was jettisoned at 44 seconds into the test. Though landing on two instead of three parachutes due to damage sustained in rubbing against the upper CM structure, the CM successfully landed 6,827.5 m downrange after a 350.3-second duration.

A-002, December 8, 1964. This time the objective was to demonstrate the abort capability under the maximum dynamic pressures of a simulated Saturn trajectory at approximately the same altitude at which a Saturn emergency detection system would trigger an abort situation. Improvements to the launch vehicle included flight controls and instrumentation; it was powered by two Algol and four Recruit rockets. Canards were added to the escape tower and the CM (BP-23) was the first to use a Boost Protective Cover. The abort and pitch-up maneuver was implemented using a real-time plot of dynamic pressures vs. the Mach number. However, incorrect data was input from meteorological data resulting in a 2.4-second early pitch-up maneuver. Though this meant the planned test point based on Saturn trajectories and selecting a nominal point for the Max *Q* region was not achieved, the early pitch-up created a higher dynamic pressure than the design value.

Canard deployment was initiated 11.1 seconds after the signal to abort, and the CM tumbled four times prior to stabilization with the aft heat shield forward as planned. It was found that during the first tumble the Boost Protective Cover was not structurally sound for the environment encountered and its soft portion was torn away. This demonstration flight was more than adequate for verification of the system at Max *Q*. Maximum altitude attained was 1,535 m above mean sea level, with the CM landing 10,000 m downrange 443.4 seconds after launch. The CM BP-23 was reused as BP-23 on the second pad abort test.

Launch Escape System mounted on top of the Apollo 8 stack.

A-003, May 19, 1965. The objective of this test was to demonstrate the system's capability at an altitude approximating the upper limits of the canard subsystem to reorient the CM for safe landing. The propulsion system for this test was changed to a cluster of six Algol motors, and BP-22 was used. A planned altitude of 36,575 m was scheduled; however, 2.5 seconds after launch a malfunction in Little Joe II caused the vehicle to roll excessively out of control and break up prior to ignition of the second stage. As a result a low-altitude abort was initiated instead of the planned high-altitude objective. An unplanned but successful low-altitude abort was initiated after 26.3 seconds with a successful land landing about 4.5 minutes later. Though the high-altitude test point was not reached, an unplanned low-altitude abort (11,149 m) from a rapidly rolling (approximately 335°/s) launch vehicle was demonstrated. At the time of the abort signal it was determined that the Mach number, the dynamic pressure and altitude were all close to nominal Saturn 1B or Saturn V launch trajectories. Post-flight evaluation revealed that part of the soft portion of the Boost Protective Cover remained with the CM for a short time, though most of it moved away with the tower upon jettison. This malfunction also revealed the ineffectiveness of the canards to stabilize the spacecraft at a high roll rate (260°/s at canard deployment) but would be effective at a 20°/s roll rate, the limit of the Saturn EDS.

A-004, January 20, 1966. The final test of the escape system aboard a Little Joe II vehicle was also the first flight of a Block I CM; it was designed to test that the vehicle could successfully orientate itself correctly after experiencing a high rate of tumble during the powered phase of an abort and that the CM would retain structural integrity when loaded to the design limit under test conditions. The propulsion on this the fifth and final Little Joe II was supplied by four Algol and five Recruit motors. The RCS was deleted from the Attitude Control System, though a ground command pitch-up maneuver was required to initiate the tumbling of the vehicle. This time a modified Block I CSM (002) was used along with a modified Block I Launch Escape System to more closely resemble the characteristics of an operational flight vehicle. To ensure tumbling took place the centre of gravity in the CM and the thrust vector were changed. After several postponements for technical problems and weather constraints the flight was launched and the pitch-up command successfully signaled when the telemetry indicated the desired altitude had been reached. The abort was initiated 2.9 seconds later, and immediately the vehicle stated tumbling, with pitch and yaw rates peaking at 160°/s with roll rates recorded at −70°/s. All other systems and operations performed nominally at a peak altitude of 23,829 m above mean sea level and traveling 34,631 m from the launch pad in 410 seconds, with the vehicle tumbling four times before stabilization.

From these test launches it was determined that, had the tests been manned, the crewmembers would have landed safely; therefore, the Little Joe series qualified the Apollo LES (and Earth landing system) for manned flight. They tested the performance of an Apollo spacecraft under LES conditions that were not expected for manned spaceflight operations.

Saturn 1 unmanned test program

A series of ten launches of the Saturn 1 launch vehicle demonstrated the compatibility of the Apollo system using boilerplate Apollo spacecraft and production-type launch escape systems on the Saturn 1 launcher. Of the ten Saturn 1 launchers two included the LES in the configuration.

Apollo AS-101, May 28, 1964. This was the first flight to demonstrate the compatibility of a boilerplate spacecraft in a flight environment that was similar to that expected for the manned Apollo Saturn V missions. The LES was also tested in a nominal firing of the tower jettisoning motor much as on a nominal mission. For this flight the launch escape and pitch control motors were inert. The spacecraft used was BP-13 on the sixth Saturn 1 mission (SA-6). All objectives were successfully met for the LES phase of the mission. Planned for ejection at 158.5 s into the flight, the operation occurred at a GET of 161.2 seconds.

Apollo AS 102, September 18, 1964. This time the back-up mode of tower ejection was successfully demonstrated using the launch escape and pitch control motors to jettison the tower instead of the primary system. BP-15 was used as the CSM on launch vehicle SA-7. Separation was scheduled for a GET of 159.2 s, but occurred at 160.2 s with all systems operating as designed.

Unmanned Apollo missions. There were four unmanned launches of the Saturn 1B to qualify the vehicle and systems, two of which (AS-201/CSM 109 launched on February 26, 1966, and AS-202/CSM 011 launched August 25, 1966) featured launch escape systems that were ejected in the normal tower jettison mode. The first flight of a Saturn V (Apollo 4/CSM 017) occurred on November 9, 1967 and its LES was also ejected normally by the tower jettison mode. Apollo 6/CSM 020 launched on April 4, 1968 also demonstrated a nominal tower jettison mode. All was now clear for placing men onboard the Apollo and ensuring both their launch safety via the LES and modes of abort on the long haul to orbit.

Manned operations

To NASA's credit, the five manned launches of the Saturn 1B (Apollo 7, Skylab 2, 3 and 4, and Apollo "18"/ASTP) and the ten manned launches of the Saturn V (Apollo 8 through 17), despite some failures *en route* to orbit, never reached the point of calling an abort situation although incidents on two manned launches challenged the technology, engineering and astronauts.

When the escape tower was ejected the first sight that the crew inside the CM were presented with was (apart from the night launch of Apollo 17) the sunlight streaming into the windows, the noise and flash and in some cases the amount of debris (probably parts of the explosive bolt system); however, there was no doubt about the action happening in front of them.

During the unmanned Apollo 6 launch in April 1968 the first stage suffered from the effects of longitudinal vibrations called "pogo" which had a crew been aboard could have made their trip quite uncomfortable. By pumping helium into the propellant lines the problem was solved; the situation was not bad enough for the launch escape system to activate. The decision to shut down the S-II early from Apollo 10 in

May 1969 to avoid the pogo effect seemed to solve the problem, until the vibrations on the April 1970 Apollo 13 mission increased to such a rate that the switches installed to detect improper thrust were inadvertently activated, resulting in an earlier than planned shutdown of the central engine. The single third-stage engine, the S-IVB, made up for the difference and Apollo 13 made it safely to orbit.

The closest the November 1969 mission came to a launch abort occurred during Apollo 12 when the stack was struck by lightning 36 seconds after lift-off into an overcast cloud layer. The lightning traveled down the Saturn and the trail of ionized gas to the launch tower. The crew were soon reporting loss of power and guidance in the CM. The Saturn was operating fine but the spacecraft was flying on back-up battery power until the crew were able to restore the systems once in orbit. Controllers considered ordering an abort using the launch escape system, sending the crew to the Atlantic and then detonating the Saturn by range safety. With the spacecraft on back-up battery power, ground controller John Aaron who was monitoring the electrical system remembered a similar event occurring during a training simulation over a year before. He recalled a switch position that resolved the problem. This was relayed to the crew and fortunately Al Bean was able to follow instructions to configure it. Safely in orbit the crew were given the go-ahead for departure to the Moon. It had been a close call and one that Pete Conrad later recalled almost allowed him to "fly" the Saturn manually to orbit, a facility only just installed from Apollo 11. This would have suited the former Navy test pilot just fine.

Summary

In 11 manned launches of Apollo the launch escape system was not called upon once, nevertheless the extensive program of ground and atmospheric testing had placed significant confidence in the system. Events in the flight of Apollo 6, 12 and 13 had also underlined the consistency of redundancy systems that enabled the vehicles to overcome and compensate from in-flight mishap without resorting to the last-ditch abort command, thus saving the mission and ensuring the safety of the crew. This system, though unused on Apollo, is now being used as a baseline design for the Constellation Program and the Orion manned spacecraft from 2014.

SOYUZ LAUNCH ESCAPE TOWER

By far the longest serving operational launch escape system is the escape tower used on the Russian Soyuz spacecraft. After unmanned testing the first manned Soyuz flew in April 1967 and the emergency escape tower has featured on every manned variant since that time, an operational period of over 40 years with no clear signs to the end of the venerable program.

Soyuz launch.

Development

It had been determined that the escape rocket would only be required in the lower dense layers of the atmosphere, where loss of launch vehicle control would lead to atmospheric disintegration of the vehicle and explosion. If a problem occurred in the higher reaches of the atmosphere, velocity could be such that a nominal entry and landing could be initiated (Soyuz 18A). Therefore, the escape tower ejection system was adapted for Soyuz as the primary method of launch abort (Shayler, 2000, pp. 151–167; Hall and Shayler, 2003, pp. 188–192; 303–307).

The Vostok seat ejection system and personal parachute descent profile had its limitations in safely propeling a cosmonaut clear of the launch vehicle in the event of a malfunction, due to time to respond, distance traveled before the explosion and altitude at which it could operate. In addition, with the introduction of a multi-crew spacecraft, Voskhod (a modified Vostok), it was impossible to include separate ejection seats for each crewmember, as was the case for the American two-man Gemini spacecraft. For the Apollo Program and a three-person crew, the Americans had devised an improved escape tower similar to that used in the one-man Mercury Program. The Soviets were also looking at an escape tower for their new spacecraft, later known as the Soyuz.

In 1961 studies of an adequate launch escape system for Soyuz were addressed by OKB-1's Department 11 (Zak, 2008). Part of the original specifications for the new spacecraft included the requirement to provide an adequate and reliable method of complete crew escape (up to three cosmonauts) all the way up to orbital insertion. As the most difficult escape situation would be an explosion of the rocket on the launch pad, a three-person ejection seat system was impractical, and no guarantee that it would clear the cosmonauts away in time or completely escape the associated debris of the ruptured launch vehicle, let alone the huge fireball of the burning propellant. It became clear that the most suitable design would have to be an escape tower system. This was a fortuitous decision, as in 1983 a Soyuz crew were saved by such a system during an actual launch pad explosion, which created a fire that burned on the pad for 20 hours after the event and rendered the launch pad out of action for some time while repairs were effected.

By 1962 it was evident that provision of a dedicated solid-propellant rocket to pull the crew compartment clear of the Soyuz from the launch vehicle would be enough to clear the determined range of an exploding R-7 launch vehicle, in time and to an altitude to enable nominal recovery by using the Earth landing parachute system. In cooperation with OKB-1 evaluation studies had been completed by the LII flight research institute under the direction of N. S. Stroev. Tests conducted at the LII confirmed that the concept was suitable; such work was led by G. I. Severin.

The most likely places for an exploding rocket were on the pad or in the lower reaches of the atmosphere where dense cloud layers could add to the loss of control, aerodynamic break-up and loss of the vehicle. In the upper layers of the atmosphere it was determined from the data available at the time that the chances of a rocket stage exploding were minimal and therefore the LES would not be required, and the Soyuz could be separated from the stage and complete a ballistic-type re-entry and land by parachute in the normal way. Again this was a valuable option in that such a situation occurred in 1975 when this contingency situation was resolved after ejection of the Launch Escape Tower.

The following year, 1963, saw the development of a structural design known as the "Jettisonable Emergency Escape Head Section" (Russian OGB SAS or *Sistema Avariynogo Spaseniya*). This work was coordinated by OKB-1, Department 3 and 11 supported by a team led by K. D. Bushyev and S. S. Kryukov from Department 15. In addition, the city of Kuibyshev (now Samara) branch of OKB-1 was responsible for the development and production of the R-7 launch vehicle. This was a sensible move as the launch escape system had to be incorporated into the R-7 design and flight profile. The major feature of the system, the solid rocket motor, was developed by a team headed by Chief Designer Ivan I. Kartukov, at Design Bureau No. 2, Plant 81, based in Moscow.

The effectiveness of the launch system was tied to a reliable diagnostic system which had to detect a variety of possible failures of the launch vehicle; again, experience with the operational flights of the R-7 (both manned and unmanned) provided sufficient evidence for where a vehicle might go wrong. At the same time engineers realized that amendments to the parachute system had to be incorporated into that intended for use on the Soyuz to enable it to perform successfully and safely after an aborted launch where high *g*-rates and accelerations were significantly different from nominal profiles.

By the end of 1963 the technical specifications for the escape system and amendments to the parachute landing system had been formulated. In case of launch failure on the pad the escape altitude had to be no less than 850 m with a range of no fewer than 110 m. The crew could not experience more than 10*g* during a pad abort and up to 21*g* for an altitude abort. The system also had to be able to cope with an abort up to $T + 400$ seconds into the flight. To achieve sufficient velocity the maximum thrust of the escape system was 76 tonnes and the total mass of the escape package was 7,635 kg. Separation of the system for a nominal mission was set at 157 seconds ground-elapsed time (GET) for the escape rocket and 161 seconds for the fairing, for the original variant.

In 1964 the design of the escape system was completed. That same year K. D Bushyev headed a meeting with some of the leading members of OKB-1 to determine the most likely failures during the launch of an R-7 which would require the use of the escape system. This team included B. Chertok, S. Kryukov, E. Shabarov, S. Okhapkin and V. Timchenko. They arrived at five possibilities for the use of the escape system from their investigations.

- gyroscopes and sensors aboard the R-7 detecting a loss of vehicle control by deviations to the flight trajectory
- premature Stage 1 booster separation (the four strap-ons)
- combustion chamber loss of pressure
- velocity decrease
- loss of thrust.

To identify these situations a range of sensors and instrumentation was fitted throughout the R-7. Onboard Soyuz the design incorporated a "special sensor of weightlessness" which if activated early enough (i.e., before the loss of thrust) would initiate the escape rocket.

The design was approaching the final stages, which Sergei Korolyov would personally approve, but in 1965 a new problem appeared. The design of the Soyuz featured a three-section spacecraft. The cylindrical Propulsion Module (Russian PAO or *Priborno Agregatny Otsek*, instrument aggregate compartment) contained instrumentation, correction engines and subsystems and supported the solar wing arrays. The headlight-shaped crew compartment known as the Descent Module (Russian SA or *Spuskaemyy Apparat*, descent apparatus) was in front of the PM; also in front of this was the "egg-shaped" Orbital Module (Russian BO or *Bytovoy Otek*, living compartment). The whole spacecraft was enclosed in a launch shroud for ascent through the atmosphere to protect it from aerodynamic forces. This design meant that in a launch abort the crew were inside a module that included the re-entry heat shield and Earth landing system (all within a shroud) and between two other modules. The escape rocket on top of the shroud was capable of ejecting the crew, but they had to get out of the shroud and separate from the rest of the spacecraft before effecting the Earth landing system. This was in marked contrast to Apollo where the crew rode in the Command Module atop the Saturn directly beneath and attached to the Launch Escape System. Apollo's Service Module was behind the CM and the Lunar Module was stored in an adapter section on top of the third stage of the Saturn V, under the Service Module.

To solve the Soyuz configuration problem at the time of an abort the Service Module would be electrically and mechanically severed from the Descent Module and left behind with the rest of the R-7 as the escape system shot away. Then at the appropriate safe distance the Descent Module would separate from the fairing and slip out of the open base of the shroud to follow its parachute descent profile. It was while studying this action that a new problem revealed itself. Completely escaping collision with the SM as the shroud surrounding it was pulled away by the escape tower was seen to be impossible. After evaluating several options it was decided to reduce the payload fairing so that it split in two upon the abort command. The top part would be separated with the escape system; the lower section would remain with the SM on top of the failing R-7. To improve the aerodynamic stability of the escape tower/shroud/spacecraft configuration during the abort profile, four stabilizers folded against the shroud during nominal ascent could be deployed to aid in the abort recovery process using the reserve parachute due to the lack of height at descent.

It was also decided to allow launch control to have the ability of initiating the abort command should the need arise. Based at Site 23 at Baikonur Cosmodrome the Kvant ground station could transmit radio signals to command the emergency escape system to activate the system on the pad.

Testing the system

Two test-firings of the escape system were used to human-rate the system: the first in 1966, the second in 1967. These were the equivalent of the US beach aborts simulating off-the-pad aborts. Post-flight analysis revealed that on one of these tests, sections of the shroud covering were ripped away by excessive acoustic loads, exposing the structural bearing of the shroud. To prevent this happening again, thermal protection was added to future fairings.

The first launch of a Soyuz (unmanned and under the codename Cosmos 133) occurred on November 28, 1966. It was safely placed in orbit, though problems soon developed in orbit and a planned second spacecraft designed to dock with it was cancelled. By December 14 the problems with the first Soyuz were resolved and the second was planned for a solo test flight rather than an ambitious automated docking profile. However, things did not work out exactly as planned as a result of an unplanned demonstration of the escape system.

Shortly after ignition the R-7 engines were shut down. Once it was determined the pad was safe from fires the launch crew returned to make the vehicle safe. Suddenly, with pad workers all around the rocket, the escape system suddenly activated with a flash and a whoosh; as a result the third stage ignited and workers scrambled for their lives before the stage exploded. Miraculously, only one life was taken in what could have been much more tragic.

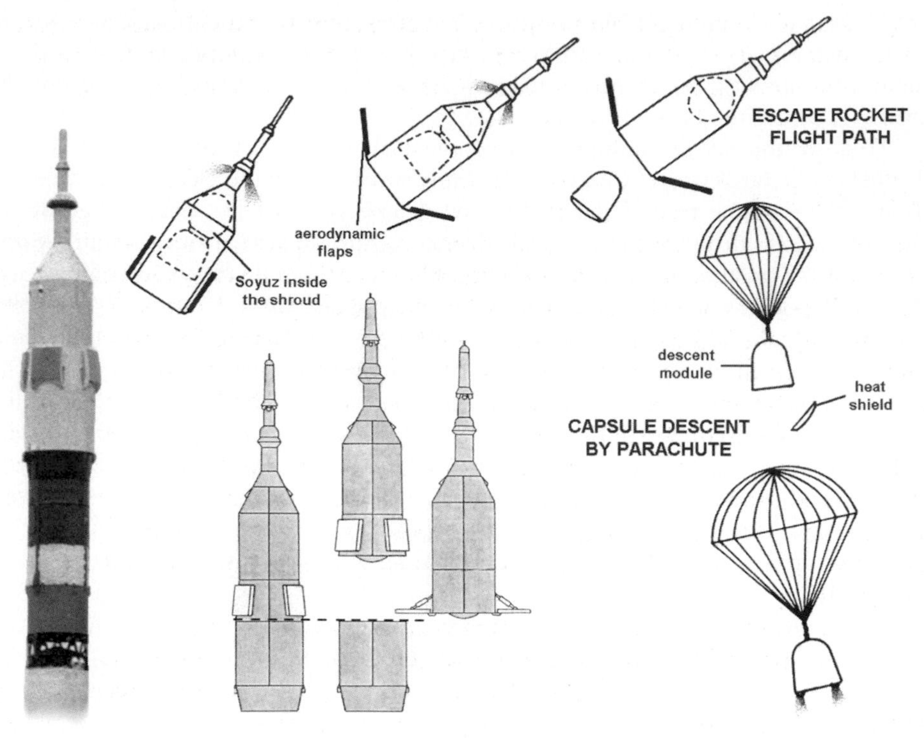

Soyuz abort sequence.

The subsequent investigation focused not only on why the R-7 failed to launch but also why the escape system had activated when none of the parameters requiring activation had been infringed, and also why the system resulted in the rocket catching fire culminating in the explosion that claimed one life. Just three days before this launch attempt (December 11) the escape system had undergone a further test at Vladimirovka, near Kapustin Yar, when a similar event had occurred. This time, though, the Propulsion Module had not be fueled and there was no explosion. It was found at the Baikonur incident that when the Descent and Propulsion Modules had been severed the Thermal Control System lines were also cut and its iso-octane coolant had leaked out and ignited; from the exhaust flames of the escape system this spread to the PM and then the upper stage of the R-7. The Director of Cosmonaut Training Nikolai P. Kamanin was dismayed to learn of the event at Vladimirovka only after the explosion at Baikonur that had nearly killed him, near a building some 700 m away from the pad on which the vehicle exploded.

As for the R-7, it was determined that an ignitor on one of the strap-on boosters was at fault. This problem took little effort to rectify. Why the escape rocket ignited when clearly it should not have done, Kamanin initially theorized that the gantry surrounding the R-7 had somehow shifted its position when replaced after the abort, but after more analysis Boris Chertok, Deputy to the OKB-1 Chief Designer

Valeri Mishin, thought the gyroscopes in the core stage of the R-7 had accidentally activated the system after power-down. In any event adjustments had to be made before entrusting the system to a flight crew, though it did work as designed (Hall and Shayler, 2003, pp. 125–127).

System upgrades

The Soyuz Launch Escape System has over the years undergone upgrades to match that of the R-7 vehicle and the Soyuz itself. Work continued on improving the escape and landing system, with development of an updated version starting as early as 1968. This was initially led by a team headed by V. A. Timchenko, Department 241 at OKB-1, then in 1974 a special laboratory, No. 179, took over this development which led eventually to inclusion in the Soyuz T spacecraft from 1979.

The new solid rocket motor was developed by the Iskra plant located in the city of Perm. It was capable of much higher reliability and could operate at greater altitudes and produce greater separation distances with the ability to use the main parachute for recovery instead of the reserve used previously. The change in wind characteristics to select the most favorable escape trajectory was also a feature of the new design.

A second set of solid rocket motors was placed on the tower system above the original set to provide additional thrust so that separation following a pad abort incident could be done much more quickly; this second set was used during the September 1983 pad abort event. At higher altitudes this second set would add valuable stability to the vehicle.

In addition to upgraded flight control systems, two pairs of solid motors were added on the fairing. This allowed for an escape option after separation of the escape motor but prior to the jettisoning of the payload shroud in the upper reaches of the atmosphere. Previously there had been no such provision for this region during a powered ascent. To compensate for the added mass of the escape system, the escape rocket could now be ejected, if not required, 123 seconds into the flight instead of the previous 160 seconds.

This new system was tested during ground simulations. A special test simulated a launch vehicle failure after separation of the escape tower, known as a Phase 1A escape. Here a test vehicle with two engines replaced the R-7 and fired a test spacecraft to about 2.5 km where the new solid rocket motors on the fairing were evaluated.

A further modification by the Iskra plant occurred for the Soyuz TM variant introduced in 1986. This was basically another weight-saving measure. To reduce the mass of the escape rocket the engines were redesigned with a single one-chamber design replacing the earlier two-core engines (a central and an add-on engine). In the new design, though, both changes would burn along the same profile they would use with one nozzle. In addition to the weight of the new design the overall diameter was reduced which in turn added to improved aerodynamic characteristics. All of this meant that the system, if not used, could be jettisoned even earlier at about 114 to 115 seconds. The total 60 kg mass saved in this redesign also allowed a flight trajectory in which both the jettisoned escape rocket and the four first-stage boosters of the R-7 could come down in the same drop zone.

Description and operation

The emergency SAS includes the re-entry capsule holding the cosmonaut crew, the orbital module, payload shroud and the solid-propellant escape rocket. Prior to the launch the center of gravity would be calibrated for that particular vehicle. Should the rocket deviate from its planned trajectory and prescribed altitude the escape module gyroscopes and dedicated thrusters could be modified to compensate. The system was activated 15 minutes prior to the planned lift-off and was capable of being used up to 157 seconds into the ascent (on the original Soyuz variant 1967–1981). It was then separated, if not needed, followed four seconds later by the payload shroud. To ensure a smooth separation from the Soyuz, three "floating" struts on the inside of the fairing followed the structure of the spacecraft within it.

In an emergency situation then, these struts would be fixed to the lower structural ring of the DM, whereupon all loads would be transferred from the payload fairing, which separated at that point during the escape profile. The system was fully automated, and when a launch failure was detected a red light would illuminate in the crew compartment and the Launch Control Station.

When activated the escape motor burned for between 2 and 6 seconds pulling away the upper section of the shroud and the combined habitation module and decent module. As the vehicle ascended it accelerated to 50 m–150 m per second and during a pad abort it would loft the combination to about 1 km–1.5 km, thus enabling a nominal parachute landing. Once a safe distance was reached from the failed R-7 a separation motor fired and the DM separated from the OM and dropped out of the open end of the shroud, an action often likened to a projectile leaving the barrel of a gun or cannon. On a nominal mission the escape motor is used to separate

Monitoring the September 1983 Soyuz pad abort.

the tower from the ascending rocket at such a velocity, angle and distance not to incur re-contact as the R-7 sped the spacecraft towards orbit.

A combination of three propulsion systems was incorporated in the design. The central rocket engine was the primary method of DM escape from the failed R-7 and was used to attain the necessary altitude for parachute recovery. The altitude control rocket thrusters were used to follow the preset spacecraft escape trajectory from the pad during the first few seconds of the ascent. To ensure a successful evasive separation trajectory was achieved after normal jettisoning, separation rocket thrusts were provided. These were also used to separate the aerodynamic shroud from the crew module during an abort.

In addition there were thrusters located on the shroud as stated above for additional thrust on the pad or at altitude following separation of the escape tower or the shroud. The escape subsystem of automated equipment worked in conjunction with the spacecraft and the R-7 to issue commands to the escape system, during either pad or altitude aborts.

The operational phase was subdivided into six portions.

1. From the moment of arming (15 min prior to lift-off) to the moment of lift-off—in this phase the emergency signal could only be initiated by the Launch Director via the radio system from the Launch Control Room. Upon initiation of the "emergency signal" the command to sever connections with the propulsion module and ignition of the main central rocket engine were issued. Just 1.4 seconds later a signal was sent to the attitude control rocket thrusters to fire in accordance with data about the current wind direction and speed. This was followed 2.6 seconds later by the shroud thrusters. Once the escape trajectory had peaked the automatic equipment issued the command to jettison the cosmonaut visual system periscope and separate the descent module and orbital module, with thrusters on the shroud firing to boost them and the habitation module away from the descent module preventing possible collision. The descent module then followed a parachute recovery in a reduced time frame.
2. From lift-off to the 20-second point in the flight—this was the low-altitude abort mode in which the propulsion system of the R-7 was not terminated, thus enabling the thrust to propel the vehicle as far from the launch pad area as possible prior to ejection of the escape tower and recovery of the crew in a reduced time period.
3. From a GET of 20 seconds to escape tower propulsion system ejection—during this time a launch vehicle propulsion system emergency situation could be commanded. The sequence followed that of the first portion with the exception that only the first chamber of the central rocket engine was ignited, as the altitude was sufficient for nominal recovery.
4. From the programmed ejection of the tower to the jettison of the shroud—here the shroud thrusters were used to assist the escape of the crew. Upon receipt of the abort command, the spacecraft would be separated at the DM/Propulsion Module interface and two shroud thrusters ignited. Just 0.32 seconds later the second ring of shroud thrusters were fired to separate the crew module from the

failed booster. The spacecraft would then be jettisoned and the recovery of spacecraft by parachute initiated.

5. From the programmed jettisoning of the shroud to the preliminary separation command from the upper stage of the R-7—now out of the shroud and almost in orbital insertion position the spacecraft had no active aids to move away from a failed launcher so onboard Soyuz separation aids had to be employed. Upon issuing an abort signal the automatic equipment onboard the vehicle issued a command to perform an emergency cut-off of the launch vehicle propulsion system. At the same time separation signals for the spacecraft modules were issued and the crew module performed a ballistic re-entry and nominal parachute landing.
6. From the preliminary separation command to the shut-off command of the third stage of the R-7—this scenario foresaw the Soyuz separate from the launch vehicle but make it to an off-nominal orbit. In this mode the life support requirements for the crew (up to 30 minutes) were ascertained and a crew rescue would follow. On the abort command the automated system transferred command for spacecraft separation to the vehicle's third stage. Almost immediately the profile for descent was computerized and a nominal capsule separation and parachute landing would follow.

Operational experiences

In the four decades of manned Soyuz flight operations there have only been two instances where the emergency systems had to be employed on an ascent to orbit. In April 1975 the intended second resident crew to Salyut 4 experienced a high-g ballistic recovery due to the failure of their launch vehicle. Then in September 1983 the crew of what should have become Soyuz T-10 and the fourth resident crew of Salyut 7 were the unplanned participants in the world's first manned pad abort using a tower-based launch escape system (Shayler, 2000, pp. 151–167; Hall and Shayler, 2003, pp. 188–192, 303–307).

In the April 5, 1975 incident following a perfect lift-off and separation of the four strap-on boosters, the core stage shut down and two sets of pyrotechnic charges should have separated the latticework holding it to the upper stage. However, some of the charges fired prematurely partially separating the stage but not completely. Now the flight was in deep trouble with the vehicle being diverted 10° off-course and unable to do much about it from inside the Soyuz. The abort system on the vehicle detected the deviation and initiated the Soyuz separation maneuver and separation of the modules for a parachute recovery. The crew had reached an altitude of 192 km and landed 320 km inside the Soviet border near China, and on the side of a mountain; they had experienced as much as 20.6g during the ordeal. The mission had been planned for 60 days but lasted less than 22 minutes. This type of abort fell under the Portion 5 scenario listed above.

The September 27, 1983 pad abort started at $T - 80$ seconds prior to lift-off when a propellant valve failed to close and a fire at the base of the booster broke out. As flames licked up the side of the vehicle, this damaged the primary abort system and ten seconds elapsed before the ground teams realized there was a serious problem at the pad. As the Soyuz toppled, the abort system finally activated via a back-up system and shot the two cosmonauts away from the pad as the R-7 exploded. The DM hit the ground 3.2 km away after a short and rough ride and parachute landing after a 5 min 13 s ride of what should have been a 90-day residence on the Salyut 7 station. Both cosmonauts were administered a stiff drink of vodka by the recovery team but did not need to attend hospital; the pad burned for over 20 hours. This type of abort fell under the Portion 1 category of the above list.

Summary

In almost 100 manned launches the escape system has been used just once during a pad abort, and contingency procedures that are also part of the launch escape scenarios have been called on once. For four decades the Soyuz Launch Escape System has given Soyuz crews a method of escape from an errant launch vehicle based on a design began almost half a century ago. The added benefit of putting one's trust in the system is that it has actually been used to save cosmonauts lives in a real emergency situation, and that the procedures incorporated into the Soyuz design had also safely returned a second crew from a high-altitude and high-*g* mission.

OTHER SOVIET ESCAPE TOWERS

In addition to the Soyuz Escape Tower the Soviets also employed "escape tower" technology in its manned lunar program and the military space station program Almaz.

TKS ferry for Almaz

The military-based manned space station called *Almaz* (Diamond) had its origins in 1964 and was developed by the OKB-52 design bureau headed by Vladimir Chelomey. Designs envisaged one configuration where the crew would be launched on the same vehicle as the orbital station. Riding in a crew capsule into orbit on top of a Proton launch vehicle the crew could access the space station via a hatch in the heat shield similar to that planned for the US Manned Orbiting Laboratory program using a modified Gemini spacecraft. This design featured a powerful de-orbit engine on top of the crew module and on top of that was a long cylindrical escape tower featuring two sets of solid-propellant rocket engines which would be used to recover the three-man crew in the event of a launch abort. This design continued to evolve into the Transport Supply Ship, the TKS. This was approved in 1970 to deliver crews and cargo to the Almaz space station. The upper part of the tower-shaped structure on the apex of the vehicle comprised the launch escape rockets for escape and the de-orbit retro-rockets for descent.

In order to human-rate the system the decision was made to test two unmanned VA spacecraft on each test flight to provide maximum information on the system and hardware more quickly. Between 1976 and 1979 four pairs of the return apparatus (VA) were tested in part to evaluate the launch escape system and the de-orbit and landing systems. These were Kosmos 881/882 launched December 15, 1976, Kosmos 997/998 launched March 30, 1978, and Kosmos 1100/1101 launched May 22, 1979. The fourth pair (launched August 5, 1977) were lost due to a failure in the Proton launch vehicle. In addition launches from Site 51 at Baikonur were used to evaluate the escape system's ballistic trajectory between 1974 and 1977; one of these tests apparently ended in failure.

Though no manned crew were launched on these spacecraft, its design became the base for add-on modules flown to Salyut and Mir and subsequently evolved into the core module for ISS (Zarya).

Cosmonauts for the Moon

The Soviet lunar program was divided into separate but integrated steps proposed by Sergei Korolyev in September 1963, all identified by the L for *Lunik* (Russian for Moon) prefix.

- L-1 was a circumlunar mission based on the Soyuz complex (later termed Zond)
- L-2 automated lunar rovers to explore the lunar surface
- L-3 manned landing missions
- L-4 lunar orbital research and mapping
- L-5 manned lunar roving vehicles.

Launch vehicles for the manned phases of the program were planned as Proton for the L-1, and N-1 for the manned lunar landing, originally envisaged as a launch vehicle for Mars missions. Despite challenges from Chelomey's proposed UR-700 and Mikhail Yangel's R-56, it was the two Korolyev launch vehicles that were chosen to be the powerhouse for the Soviet manned lunar program (Harvey, 2007; Siddiqi, 2000).

Launch escape

The Proton/Zond combination featured an escape tower system similar to that used on Soyuz. An improved launch escape tower was employed on the N-1 vehicle.

As the system was being used in the manned Soyuz program the ground test and flight test program also qualified the system for use on the lunar program. Though no cosmonauts ever flew a Proton or N-1 launched mission under the lunar program the ability to use the launch escape system in the event of a launch mishap to theoretically save the crew was demonstrated several times in the unnamed role.

The launches under this program have been previously described in other titles in this Springer/Praxis series of books (Hall and Shayler, 2003, pp. 23–33; Harvey, 2007). Zond 1P was used in ground tests and Zond 2P was a boilerplate flown on Cosmos 146 as an Earth orbital test mission in March 1967 with no recovery planned.

In April of that year 3P was launched on Cosmos 154 in a partially successful second Earth orbital test. From then on the launch escape systems were fired on the following missions.

L-1 Zond circumlunar program

- September 28, 1967—what was planned as a circumlunar trajectory ended almost as soon as it left the launch pad. Just 60 seconds after lift-off the Proton launch vehicle began to deviate from its planned trajectory; its malfunction initiated the use of the escape tower which pulled the L-1 capsule (4L) away from the Proton. Though the mission failed it was clear that the emergency escape system operated as designed.
- November 22, 1967—this time the first-stage operation performed as designed, but just four seconds into the firing of the vehicle's second stage the Proton again veered off its planned trajectory activating the launch escape system to recover the 5L capsule. The next launch in the series was successful and designated Zond 4 (6L); this was a deep-space test of the spacecraft and recovery system.
- April 23, 1968—capsule 7L was launched on time but at 3 minutes 14 seconds into the mission the escape system suddenly activated and carried the crew compartment for recovery 520 km downrange from the launch pad. This time it was not the Proton that was at fault but an erroneous signal from the spacecraft that activated the escape system. The intended next launch was planned for a July 19 launch. However, five days prior to that (during processing) an unfueled Blok D oxidizer tank exploded due to an electrical fault over-pressuring the tank, killing one launch technician and injuring a second. The vehicle toppled over and leaned against the launch tower but the ejection system was not activated. Though it was reported that the spacecraft, 8L, could have been launched it was never flown under the L1 program.
- January 20, 1969—the next launches (Zond 5/9L and Zond 6/12L) occurred as planned, but 8 min 21 s into the flight of spacecraft 13L a deviation from the flight plan was enough to initiate an abort scenario. It was determined the No. 4 second-stage engine had shut down 25 seconds early and this was followed three minutes later by the third-stage failure preventing compensation for the lost second-stage thrust.

The remaining two Zond launches (Zond 7/11L and Zond 8/12L) occurred without launch incident and did not require the escape system.

L-3/N1 launch aborts

- February 21, 1969—the first N1 launch vehicle was planned on this date to launch a Zond spacecraft to the Moon. However, just 70 seconds into the mission with two of the 30 first-stage engines already shut down, all 28 remaining engines were shut down causing the vehicle to deviate from its flight plan and once again the launch escape system to activate. Apparent incorrect ground testing of the engine management system had resulted in an erroneous signal on engine No. 12

and its direct opposite No. 24. Then excessive vibrations ruptured the fuel line of engine No. 12 causing a fire that shut off the system and all engines.

- July 3, 1969—the second launch of the N1 fared worse; despite all 30 engines igniting as planned, it rose only 200 m off the pad before foreign objects were sucked into the No. 8 engine initiating an explosion that destroyed several other engines depriving the vehicle of the requisite lift-off thrust and seriously damaging the pad as well. Weighing 250 tonnes the whole vehicle exploded throwing chunks of hot molten metal towards Earth with debris falling 10 km from the pad due to the velocity and force of the explosion. A disastrous incident, thankfully free of fatalities, was only partially balanced by the success once again of the launch escape system which at $T + 14.5$ s into the mission pushed the Zond Descent Module clear of an impending fireball and landing the module safely 2 km away; had a crew been onboard they would probably have survived the traumatic event. It was another bitter blow to the Soviet manned lunar program at the time of the historic success of the Apollo 11 mission two weeks later.
- November 23, 1972—the third N1 was launched on June 27, 1971 with a dummy spacecraft and launch escape tower, but again the vehicle failed in flight at 51 seconds into the mission, as the escape tower was a mock-up and could not be fired. The 1972 launch attempt involved an upgraded N1, and this time it managed to fly higher and faster than any other vehicle of the type, but at 90 seconds into the flight it too ended in a ball of fame. Apparently, engine No. 4 had caught fire followed by an explosion in the tail of the vehicle, resulting in a huge explosion just seconds before the ignition of the second stage. The escape system fired; once again, had a crew been aboard they would probably have survived.

Though another launch was planned, improvements to the vehicle and changes to the Soviet manned lunar program in light of the American successes were not enough to prevent suspension of the program in 1974 and its complete cancellation in 1976. The focus of Soviet manned space ambitions were now centered on the creation of permanent manned space stations. Soviet cosmonauts may not have reached the Moon but their safety had been clearly demonstrated by the escape system which had ejected the crew module no fewer than seven times over a period of five years; 16 launch escape systems were launched under the L-1 and L-3 programs.

SHENZHOU LAUNCH ESCAPE SYSTEM

The launch vehicle used for Shenzhou manned flights is the Long March 2F (*Shenjian*, which translates as Magic Arrow) upgraded from the Long March 2E version. To make the system acceptable for manned flight a series of 55 engineering changes were required to human-rate the system. The primary difference is the addition of the manned Shenzhou system and launch escape tower on top of the rocket. This system had been described as one of the most direct applications by the Chinese of Russian Soyuz technology.

The launch escape system can pull the spacecraft crew compartment clear of a pending explosion of the launch vehicle either from the pad up to 15 minutes prior to launch or in the first 160 seconds of powered flight when the system is separated from the launch vehicle. Duplicating the Soyuz escape system the launch escape tower pulls the Shenzhou clear of the upper stages inside an aerodynamic shroud. After only a few seconds (i.e., once the tractor rocket fuel is exhausted) separation commands drop the crew compartment out of the base of the aerodynamic shroud for parachute descent and landing.

For Shenzhou there are two emergency profiles using the escape system: low altitude up to 39 km and high altitude above that distance. A fault-monitoring system is also incorporated into the systems aboard the Changzheng 2F launch vehicle. To aid in aerodynamic stability, just like Soyuz four flaps are located on the upper shroud.

The complete escape system measures 15.1 m long with the tower structure itself measuring 8.35 m; it is 3.8 m in diameter and has a mass of 11.26 tons. It can be operated by the crew inside the spacecraft, by mission control or by the automated guidance system which can detach, a departure from normal flight sequencing or trajectory.

The propulsion system features six solid-fuel motors There are four control motors, a low-altitude separation motor (with eight nozzles) and a low-altitude escape motor (with four nozzles). The upper part of the shroud features six motors which comprise four high-altitude escape motors and two separation motors.

The fault-monitoring system can sense a problem with the launch vehicle and can then activate the escape system. Low-altitude escape requires separation of the upper and lower parts of the shroud, together with the top two parts of the spacecraft. The escape system motors ignite to pull the combination clear of the exploding rocket.

For a high-altitude abort the escape tower is not used. In this scenario escape motors on the side of the shroud are used to propel the vehicle to a safe distance from the launch vehicle.

For aborts after 201 seconds into the mission at altitudes above 110 km the Shenzhou itself would be separated from the upper stage of the launch vehicle and then use its own propulsion system to place it in a low Earth orbit; re-entry can then be performed after the first, second or non-standard fourteenth orbit (Chen Lan, 2004).

The system was first flown on Shenzhou 3 to test the effectiveness of flying it on a manned vehicle. Shortly after the launch of the third Shenzhou spacecraft the escape tower was reported to be separated, apparently qualifying the systems for further testing on Shenzhou 4, prior to the first manned launch on Shenzhou 5. The system was not installed on the first or second Shenzhou testing in order to prevent erroneous signals activating the system during a nominal ascent. It was revealed that the "designer" (chief engineer?) for Shenzhou 7 (September 2008) rescue system was Zang Shuting.

ORION LAUNCH ESCAPE TOWER

The Constellation Program is the new initiative within NASA to create a new generation of manned spacecraft to replace the Shuttle fleet after its retirement in 2010. The main manned spacecraft called Orion will be used as the primary U.S. mode of transportation to Earth orbit (and the ISS), to the Moon, and support activities to develop a manned lunar base and, hopefully, manned flights to Mars. In light of the tragedies within the Shuttle program it was an early decision to develop an effective launch escape system for the whole crew module, one that returns to the capsule-type design of the Apollo era rather than the spaceplane design of the Shuttle. The system chosen for the new spacecraft employs a launch escape tower. The tower carries a set of three rockets. The primary rocket will be used to pull the Orion crew module clear in the event of pad or launch abort and clear of the Ares launch vehicle. The smaller engines will be used to jettison the tower, if not needed, and to aid in attitude control.

Plans for testing the system included developing and constructing a new test launch pad at the US Army White Sands Missile Range in New Mexico (which

Launch of Ares.

Impression of an Ares abort test.

commenced in November 2007) to support a series of launch abort tests similar to those conducted during Mercury and Apollo.

The initial test for Pad Abort 1 is scheduled for September 2008. Of the ten test flights qualifying new systems and hardware under the Constellation/Orion/Ares program at least five are based at White Sands. There will be a second pad abort and three in-flight altitude trials between 2009 and 2011. These are designed to measure the effectiveness of the new system at subsonic and supersonic speeds and during tumbling notions. In 2012 a high-altitude abort test is planned from Kennedy Space Center to test the system at the upper limits of its design envelope. These will help qualify the Orion for manned test flights from 2013, then operational flights initially to ISS from 2014, but eventually to the Moon leading to a manned landing around 2020. The technology and flight operations will be used to support expanded activities beyond the Moon at a later date.

The development of the Orion Launch Escape System is very much centered on the experiences and results from the Mercury and Apollo Launch Escape Systems, bringing the cycle of the manned spacecraft launch escape tower full circle. The program is constantly under development but in a recent American Institute of Aeronautics and Astronautics (AIAA) paper (Williams-Hayes, 2008) some details of the program were expanded upon.

The Crew Exploration Vehicle (CEV) is the main (parent) module of the Constellation system and is called Orion. The Orion spacecraft comprises a Crew Module (CM) that can accommodate up to six crewmembers, a Service Module (SM) and a Launch Abort System (LAS). Resembling the Apollo Command and Service Modules it is significantly more advanced, is larger and has a greater mass. Due to its design, adequate escape options for the whole crew had to be incorporated into its systems for the final moments on the pad and in powered flight up to Earth orbit. After a program of consideration and study a new launch escape tower system was chosen as the most effective method to address this requirement. As a result a program of test flights qualifying the abort system for manned flight was also developed.

The Flight Test Office

To investigate and develop a system that would, in the event of a catastrophic failure on the launch pad or during the early stages of a powered ascent to Earth orbit, safely pull the crew module to a point where a parachute descent could be made to rescue the crew, the NASA Constellation Program Office established the Flight Test Office (FTO). The FTO's primary objective is to conduct a program of unmanned flight tests of the Launch Abort System chosen for Orion in a series of pad, mid-altitude and high-altitude aborts prior to committing to manned launches.

Assembling experts from NASA field centers and contractors, the FTO team consists of members from four NASA field centers

— Dryden Flight Research Center, Edwards, California
— Johnson Space Center, Houston, Texas
— Langley Research Center, Langley, Virginia
— Glenn Research Center, Cleveland, Ohio

and three aerospace companies

— Lockheed Martin Space Systems Corporation, Denver, Colorado
— Orbital Sciences, Dulles, Virginia
— Orbital Sciences, Chandler, Arizona.

NASA retains overall responsibilities as the test organization, while the Orbital Sciences facility at Virginia is responsible for the LAS and Orbital Sciences in Arizona for developing the Abort Test Booster 9 (probably similar to the Little Joe II system used during Apollo) to be used for the flight tests. As prime contractor the Lockheed Martin team in Denver is responsible for the design, construction and development of many key systems, especially the avionics, operation flight instrumentation and the flight software.

The test program

There are currently six planned unmanned flight tests of the abort system. At least five will be conducted out of the White Sands Test Facility in New Mexico, the sixth may be from the Kennedy Space Center using the Ares launch vehicle. There will be two pad abort (PA) tests which do not require an external booster yet should demonstrate the system capability to eject the crew module from the launch pad. The four ascent aborts (AAs) will use a government-supplied test booster. Termed the Abort

Test Booster (ATB) it is being developed by Orbital Sciences in Arizona. This will provide data about the abort system under different test conditions. The final AA may be launched from KSC.

In chronological sequence the test program envisages

- PA-1 An abort from the launch pad using a NASA-provided boilerplate CM and a flight test–specific LAS. This test has more of an engineering development nature and involves the collection of flight and performance data rather than certification towards an operational system.
- AA-1 A maximum dynamic pressure abort profile will be flown again using a NASA-provided boilerplate CM and flight test–specific LAS.
- PA-2 A launch pad abort but with a more flight-like abort trajectory. A Lockheed Martin (LM) operational CM will be used in this test along with an operational LAS.
- AA-2 A transonic abort profile. From this test all future CMs will be LM-produced operational CMs and each LAS will be of the operational design.
- AA-3 Off-nominal maximum dynamic pressure abort.
- AA-4 High-altitude abort. Here the government-supplied ATB could be replaced by an Ares launch vehicle and launched from KSC. This is still to be determined at the time of writing.

The vehicles used in the flight tests are described as Flight Test Articles (FTAs) and will be configured specifically for the tests that are intended to be performed and the data intended to be collected. Of course, there are limitations in testing a system that

Ares abort profile.

is intended to be on an operational vehicle that cannot be fully determined until the test results are known. Therefore, the FTAs are based more on operational vehicle design studies, but will be crucial in the gathering of flight data that can be used in the final design confirmation of the operational Orion spacecraft and its Launch Escape System. Therefore, all the systems and configuration of the operational Orion cannot be incorporated in the FTAs, and the time constraint between each abort test precludes major changes to the vehicles on subsequent tests unless major anomalies occur.

Test vehicle configurations

Each flight test vehicle is composed of a Crew Module, a Launch Abort System, a Service Module subsection and a separation ring. For pad aborts a test booster is not required, but for ascent aborts the use of a launch vehicle is required.

Initially, two boilerplate CMs will be fabricated by NASA's Langley Research Center. These will be used for the first pad abort profile and the initial ascent abort test. These two vehicles have the exterior mold line of the final vehicle but do not represent the final design. For all other tests the CMs will be supplied by the primary contractor Lockheed Martin.

The operational vehicle is still under development, and therefore the final mass quantities for the vehicle and those for the test program are still being defined. As the date approaches for the first PA test the data have been collected, though there are constraints and exceptions for PA-1 that will change as the series and program develops.

For the mass of PA-1 the mass property chosen was a "reasonable reflection of the mass properties of future operational vehicles." For PA-1 it was a balance between the known weight limit of the first-generation recovery parachutes and the center-of-gravity (cg) value of the later operational CM. This resulted in an under-drogue parachute mass of 7,787 kg for PA-1. Added to the fact that the LAS on this first test is heavier than the operational design due to manufacturing lead time issues, the cg of the first CM is further forward than it will be on the operational vehicle allowing for a more stable vehicle during the test. In future tests both the operational constraints at the time, combined with developments in parachute design and dynamic pressure limits, will affect the overall mass properties of PA-2 as well as for the four AA flight tests.

Computer simulations of the abort profiles of the Crew Exploration Vehicle/Crew Launch Vehicle (CEV/CLV) combination have been developed by NASA JSC. This 6-degree-of-freedom nonlinear simulation program is called ANTARES and has been modified to support nominal and dispersed trajectories flown under the abort test program, providing mathematical models of what could be expected during the planned program, and thus creating an overview of altitude and distance trajectories. In simulations the data was sampled resulting in a nominal flight for PA-1 which pitched the vehicle over to a 15° angle of attack, achieved a 3,050 m altitude and a downrange distance of 2,100 m, and reached Mach 0.7 with a dynamic pressure of 600 psf. These computerized simulation altitude and downrange predictions for PA-1 should be met during the actual flight test. Other simulations have reproduced the preliminary plans for the PA-2 test, simulating a transonic abort profile. Under the ANTARES program the command for an abort is given at Mach 0.99 and a dynamic pressure of 610 psf imparting maximum drag on the simulated vehicle. After reaching a peak altitude of 11,094 m a downrange distance of 4,267 m was simulated.

Launch Abort System

Development of the Launch Abort System for Orion is being headed by Orbital Sciences of Dulles in Virginia. Designed to be manufactured from lighter materials, the units supplied for the first two flight tests turned out to be heavier due to the added lead time to manufacture with lighter materials and added time constraints in the delivery of the test vehicles.

The most recent design of the LAS features three motors. There is an abort motor with four canted nozzles (25° from the main housing case) with the fuel being extinguished in just 4 seconds resulting in an acceleration of 15*g* for the crew from either the pad or a failed launch vehicle. The attitude control motor (ACM) is fired at the same time as the abort motor to stabilize the flight trajectory. Eight nozzles are equally spaced 45° apart around the housing towards the top of the casing. These supply motion control in both the pitch and yaw axes burning for about 20 seconds with the resulting thrust directed through the eight nozzles to achieve controlling movements. At abort motor burn-out the vehicle is stabilized by two canards on the tower that will deploy which, combined with the ACM, are used to re-orientate the crew module and the LAS for entry and landing. The heat shield at the base of the CM now faces the direction of flight and the air stream. Now that the vehicle is orientated for landing, a jettison motor is ignited to provide the impulse to separate the LAS from the CM. This final motor includes four nozzles canted at 35° from the main casing. In just 1.5 seconds a total of 1,860 kg/0.3 m thrust is generated separating the tower and Boost Protective Cover from the descending CM allowing for nominal parachute recovery. The Boost Protective Cover (BPC) is a fiberglass structure covering the CM providing protection during the early stages of ascent or abort. A similar device was incorporated in the Apollo Launch Escape System.

Pad Abort 1

Currently, the first test flight in the series is Pad Abort Test 1 which is planned to be imitated from a site near to Launch Complex 32 at White Sands. Trajectory is to be in a due north direction relative to the pad. For this first crucial test a number of primary objectives have been set both for the test and several subsystems. The primary objectives for the PA-1 are "to demonstrate a ground initiated abort as well as to demonstrate the capability of the LAS to propel the crew module to a safe distance from a launch vehicle."

Escape rocket test stand for Ares.

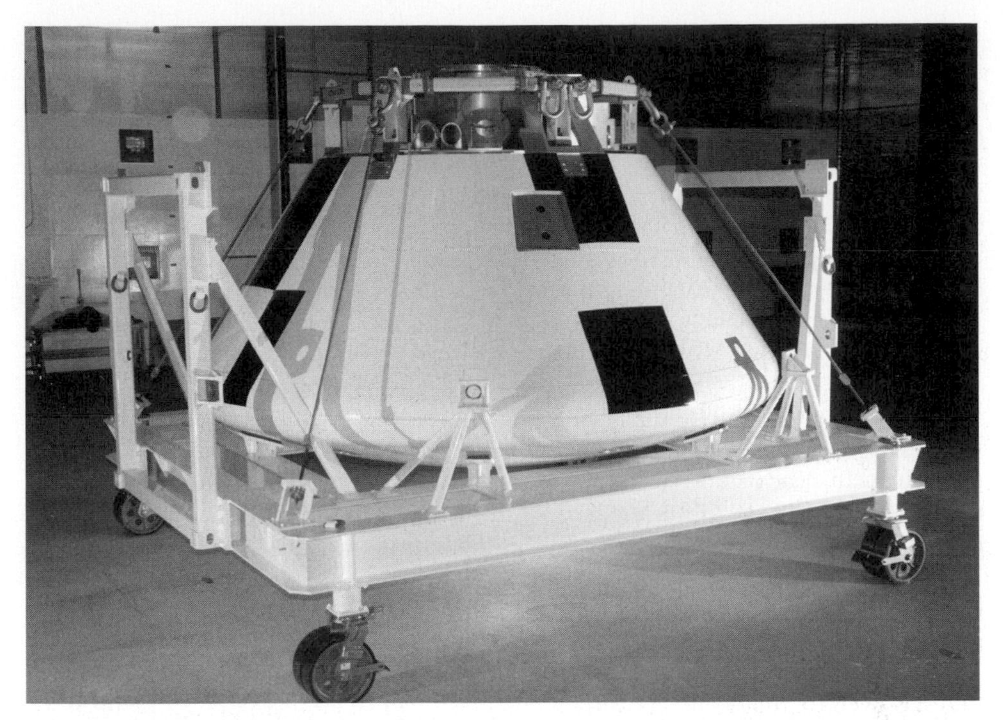

Ares abort test capsule.

Subsystem evaluation and objectives include a demonstration of the stability and control of the LAV during ascent and acquisition of structural load data on the interface between the LAS and the Crew Module. The abort and attitude control motor performance of the LAS will be measured along with demonstration of the ability to jettison from the CM. In testing the separation events, evaluation of the abort sequences will be demonstrated in addition to the separation mechanism of the LAS Crew Module, and jettisoning of the forward bay cover. Additional data on the ground impact locations for the hardware will also be of help in planning future operations and procedures.

Additional objectives include development of the parachute recovery system and landing system events. Environmental issues to be studied include the external acoustics before separation of the Boost Protective Cover and the quality of transmissions of telemetry from the CM through the BPC. Another area being evaluated is the ground support network during the test. The functional performance of the command control monitoring system will be featured as well as the range of transportation devices, special tools and test equipment used in the test.

Test stand prepared

In June 2008 NASA announced that a vertical test stand used to support full-scale test-firings of the Orion Launch Abort System was being prepared for use later in the summer. The support structure has to be large enough to restrain the 5.2 m high and 0.9 m diameter abort motor that will be used to clear Orion crews from an Ares I launch vehicle in the event of a failure up to 91,440 m.

A full-scale inert motor was installed without the oxidizer in its propellant upside down on the test stand with its four nozzles pointing skywards. The test stand was located at the ATK Facility in Promontory, Utah where a series of bench tests and check-outs were completed over the summer leading to the first firings. The ignition assembly was also scheduled for tests. The ignition motor was the assembly inside the abort motor which provides an ignition source for the propellant. Upon ignition the propellant burns extremely quickly and is consumed in about five seconds resulting in 2,200,000 N of thrust. Most of the high-impulse propellant is burned in just the first three seconds, the so-called "critical time frame", to ensure the crew is pulled from danger as soon as possible. Using a reverse flow design, hot gas is forced through the manifold's four nozzles, creating the required quick pulling force to drive the crew compartment clear of danger. In this design the hot gases with four plumes each three times the length of the motor are located at the tip of the structure in order to allow clearance of the crew compartment during operation.

AA-1 and beyond

The first pad abort test is currently scheduled for the fourth quarter of 2008. Once that has been completed the remaining flight tests can be more accurately planned. In the latest *Multi-Program Integrated Milestones* (NASA, 2008) the plan is as follows:

AA-1 third quarter 2009
PA-2 second quarter 2010
AA-2 third quarter 2010
AA-3 first quarter 2011
AA-4 ? to be determined.

It remains to be seen how accurately these and others developments unfold as the Shuttle Program is wound down and Constellation efforts hopefully are increased to meet the Vision of Space Exploration milestone of returning humans (Americans) to the Moon by 2019/2020, the 50th anniversary of the first Apollo lunar landings. Though this time scale is likely to change, the testing of the abort system leading to the first manned flights of the Ares/Orion system will bring that goal much closer to reality.

SUMMARY

One or another emergency escape tower for manned spacecraft has been in operational service constantly since the late 1950s. Initiated on Project Mercury (up to 1963) they were briefly considered for Gemini, re-entered service on Apollo until 1975 and reinstated for Constellation in the forthcoming Orion Program. In the Soviet Union escape towers were introduced for the Soyuz Program, and have supported its manned operations (with variants) since 1967 (some 40 years later). The Chinese too have decided to employ escape towers in their manned spacecraft (Shenzhou) which have supported the manned flights in that program since 2003. Though employing different designs to match different requirements and capabilities their objective remain the same: rapid and failsafe separation of the Crew Module from an errant launch vehicle from the pad to a safe altitude for parachute recovery.

In almost 50 years of operations they have only been used, as designed and intended, once in 1983. A remarkable record of unwanted employment but a necessary facility should the need be required for their use, they must work for the first and only time in order to save lives. Other systems have been evaluated but it will be some time before a more effective method of rescuing multiple crewmembers from an impending exploding launch vehicle can be developed that is as cost-effective, simple in design and dramatically effective in operation. They are also probably the ultimate white-knuckle ride one could experience—though, given the choice, a ride you would not wish to experience unless you had to—as the crew of Soyuz T-10A can testify.

REFERENCES

John Catchpole (2001). *Project Mercury*. Springer/Praxis, Chichester, U.K.

Chen Lan (2004). *Inside Shenzhou Spacecraft*. Available at Go Taikonauts! website *http://www.geocities.com/capecanaveral/launchpad/1921/story-8.htm?200830* (last accessed 30 June 2008).

Rex Hall and David Shayler (2003). *Soyuz: A Universal Spacecraft*. Springer/Praxis, Chichester, U.K.

Brian Harvey (2007). *Soviet and Russian Lunar Exploration*. Springer/Praxis, Chichester, U.K.

NAA (1968). *Apollo Spacecraft News: Launch Escape*. North American Aviation, CA.

NASA (1963a). *Mercury Project Summary*, NASA-SP-45, October. NASA, Washington, D.C.

NASA (1963b). *Project Mercury: A Chronology*, NASA-SP-4001. NASA, Washington, D.C.

NASA (1966a). *This New Ocean*, NASA-SP-4201. NASA, Washington, D.C.

NASA (1966b). *Manual Control of High Altitude Apollo Launch Abort*, NASA-TN D-3433. NASA, Washington, D.C.

NASA (1975). *Apollo Program Summary Report: Section 10 Launch Site Facilities, Equipment and Prelaunch Operations—10.1 White Sands Missile Range*, NASA-JSC-09423. NASA, Washington, D.C.

NASA (2008) *Multi-Program Integrated Milestones*, Document #MPIM, rev-FY2008-Q3, revised 29 April 2008. Available at NASA Human Spaceflight website *http://www.nasa.gov/mission_pages/shuttle/nes/index.html* (last accessed 10 June 2008).

David Shayler (2000). *Disasters and Accidents in Manned Spaceflight*. Springer/Praxis, Chichester, U.K.

Asif Siddiqi (2000). *Challenge to Apollo*, NASA-SP-2000-4408. NASA, Washington, D.C.

Peggy S. Williams-Hayes, Aerospace Engineer, Control and Dynamics Branch, NASA DFRC (2008). *Crew Exploration Vehicle Launch Abort System Flight Test Overview*. NASA Dryden Flight Research Center, Edwards, CA.

Anatoly Zak (2008). *Emergency Escape System of the Soyuz Spacecraft*. Available at Russian Space website *http://www.russianspaceweb.com/soyuz_sas.html* (last accessed 30 June 2008).

6

Launch escape, 2: Ejection seats

A second mode of escape during ascent has been used on two capsule programs—the Soviet Vostok and American Gemini spacecraft. In addition, the development of ejection seats for rocket research aircraft and spaceplanes has been utilized for the Space Shuttle designs. In the American Shuttle Program this only supported the Approach and Landing Test and Orbital Flight Test program, was made inert for the first operational shuttle fight (STS-5) and removed altogether after the flight. They have never been reinstalled on any of the Orbiters. There were plans for the Soviet Buran shuttle and European Hermes spaceplane to feature ejection seats but these never progressed to operational use. In addition, a unique series of launch abort modes have been developed to support Shuttle flight operations where ejection seats and escape systems are not viable.

EJECTION SEAT HISTORY

As military aircraft were evolved in the 1930s to fly higher, faster and farther so the need to provide a rapid escape option for the pilot emerged beyond opening the canopy and bailout (or falling) to open a parachute. A more rapid exit had to be devised and the solution was found to be an ejectable seat and jettisonable canopy.

In 1939 Germany, Karl Arnold, Oscar Nissen, Rheinhold Preuschen and Otto Schwarz of the Junkers aircraft works received the first patent for an aircraft ejection seat. An additional patent was awarded to Erich Dietz for adding a power cartridge to the seat. The same year the prototype Heinkel 176 received the first working ejection system, followed by installation of ejection seats on the Heinkel 280 and 219 on the production lines. The dubious honor of performing the first live emergency ejection on January 13, 1943 fell to Herr Schenke while flying a Heinkel 280.

Work in developing the ejection seat system in modern jet aircraft continued throughout the Second World War and in the post-war years. Some operational ejections were recorded during the conflict, the Luftwaffe having recorded about 60 emergency ejections, though how many pilots survived is not so clear. In January 1945 Bernard Lynch made the first live static ejection test. Development expanded around the world after the end of the war. In 1955 the first live runway-level ejections were made by RAF pilots. In 1961 the first live static rocket-assisted ejection was made by Doddy Hay followed the following year by the first in-flight rocket-assisted ejection seat test by Peter Howard. The Martin Baker Company in England has been a pioneer in seat development since the mid-1940s and is now the world's longest established manufacturer of ejection seats and associated support equipment including survival equipment, location devices and recovery techniques. Thousands of lives have been saved using ejection seat technology, and as exemplified in the space program utilizing that experience and skill in associated and follow-on programs adds to the reliability and confidence in such systems to ensure lives are saved. Around the world leading aircraft companies and their contractors have developed reliable systems and hardware for rocket planes and spacecraft. Though the selection of contracts in the U.S. and Soviet Union was vastly different, their aeronautical design experience was fundamental in adapting ejection seat technology for use in manned space programs.

Despite their widespread use in military aircraft and incorporation into the rocket research planes program they have not been widely used for manned space launch systems and included only in six Vostok, ten Gemini and four Shuttle flights.

ROCKET PLANES

The use of ejection seats in the pioneering military aircraft programs of the 1940s and 1950s was duplicated in the important research work being conducted on pioneering research aircraft both by the military forces and NACA, the forerunner of NASA. Here development included breaking the sound barrier, increasing the speed of aircraft to hypersonic flight, developing new technology in power plants, aircraft design, sweep wings and lifting bodies that were the forerunners of today's military leading-edge technology of vertical and short take-off and landing, and stealth technology as well as providing a valuable database for spaceplane design that led over many years to the NASA Space Shuttle and a crew escape proposal for ISS. Similar research in the Soviet Union led to the Buran shuttle. The high-performance cutting-edge technology in test-piloting work is linked with manned space exploration as are related developments in missile technology, unmanned spacecraft, remote sensing, miniaturization, micro-surgery, communications and navigation. The development of new technologies in the military have forged forward pioneering techniques, some of which have been mirrored in the space program, and others in everyday life. The cross-over technology of rocket research is one program that demonstrates this (Miller, 2001).

X-1, breaking the barrier with a parachute (1946–1958)

There had been studies to include an ejection seat on the X-1, the aircraft that broke the sound barrier on October 14, 1947 with pilot Chuck Yeager at the controls. It was decided that the extra weight would be a heavy penalty and the seat would have been of little help at such high speeds, an area being investigated and little understood. Therefore, in an emergency a pilot would have to disengage the control column by removing a securing pin so that it could be hinged out of the way while he tried to remove the right-hand door panel, and then simply fall out using a conventional backpack parachute for landing. Ejection seats were finally installed in second-generation X1s (X-1A, X-1B, and X-1E) which further explored the regions around and beyond Mach 1. It was installed only after the series had completed part of its flight research program. The ejection seat installed on X-1E was a surplus seat from the second X-4 aircraft. The series investigated dynamic stability, air load investigations, heat transfer research and refinements in airframe construction materials and engine technology reaching speeds up to Mach 2.5. Ground tests on the seats included using dummies to provide evidence that no part of the pilot's anatomy would come into contact with fixtures and fittings, including the canopy, which would be previously ejected, hinged open, a comforting factor for any pilot wishing to fly the aircraft and know that his ejection seat will indeed normally save his life—not take it.

Skystreak and Skyrocket (1947–1956)

Along with the X-1 these aircraft investigated the transonic research area up to and beyond Mach 1. They had a *zero–zero* escape capability which meant the aircraft could be sitting at the end of the runway standing still (zero speed, zero altitude), push a button and eject. There was also the option to jettison the nose section, then by activating a second handle the backrest fell away allowing the pilot to exit the cockpit and descend by parachute (Hunley, 1999, p. 39).

X-2, a different approach (1952–1956)

The X-2 program was designed to push the limits of speed and altitude far beyond that possible with the X-1 family; in addition, the study of sweep wing technology was also a primary objective as was provision of flying laboratory information on aerodynamic heating. Paramount in designing the X-2 was pilot safety and attempting to improve the chances of survival for the pilot at high speed and high altitude. It would also assist in investigating the idea of an alternative type of crew escape that might have applications in future military aircraft. As a result the designers at Bell evolved a jettisonable crew compartment and nose section. In this design four explosive actuated gas pistons were located 90° from each other around the perimeter of the rear of the bulkhead which included the pilot seat. Upon ejection gas-powered pistons literally pushed the nose of the aircraft and pressurized cockpit section away from the rest of the vehicle. Following separation a 2.4 m diameter ribbon parachute would be deployed automatically from the center of the aft bulkhead and used to stabilize the

Bell X-1.

section at about 150 mph as it approached lower altitudes to allow the pilot to perform personal egress. The audio warning device indicated when the correct altitude of 4,600 m was reached. The original altitude of 6,000 m was lowered when studies of the pilot's oxygen requirements revealed the tower altitude was more suitable.

The pilot would then have to manually open the hatch canopy and egress the crew seat using the seat pack parachute that he would also have to operate himself. In case the pilot was semi-conscious or injured an audible barometric warning device was sounded indicating low-altitude ejection limits. Of course, this looked good on paper but when tested in August of 1948 questionable results were achieved.

A 411 kg dummy X-2 nose fitted with data collection equipment was attached to the top of a captured Second World War V2 and launched from White Sands Missile Range under the code name *Blossom III*. When the 4,600 m altitude was reached some 101 seconds after launch and at a velocity of 61 m/s, the nosecone was released and the nose section descended as planned. Stresses recorded on the experimental hardware included 15.84g at separation. However, the altitude was far higher than envisaged for the actual X-2 flight profile (whose ceiling was 38,100 m), the recovery parachute behaved poorly during deployment and other data were inconclusive. No further tests of this system were carried out. The system was used once on the final X-2 flight by pilot Mel Apt on September 27, 1957. During the flight Apt lost control of the vehicle at the end of powered flight bringing about an increase of roll rate, the resulting g-forces and changing angle of attack threw Apt about the cabin as recorded by onboard film recovered later. Apt regained some control to eject the capsule at 12,200 m but for some reason Apt was not able to release his seat lap belt and was unable, or ran out of time, to jump clear of the rapidly descending cockpit. He died when the cockpit impacted the ground, his body found partially ejected from the cockpit with the cockpit intact but severely crushed. Impact velocity was determined to be about 190 km/h (±8 km/h) with the deceleration forces on the pilot seat at $90g$ ($\pm 40g$).

X-3, stiletto (1952–1956)

This "exotic" looking X-plane was designed to investigate high-speed flight at Mach 2 for up to 30 minutes and was the first X-plane to take off and land under its own power—not air-launched. The ejection seat provided was fin-stabilized and unconventially left the aircraft pointing downwards—not upwards. This was also the method used to get into the cockpit and out of in the normal flight profile. Sliding on rails the seat section was lowered to the ground, then with the pilot strapped in the unit was mechanically hoisted back up to the fuselage and locked in place. Tests of the system featured a range of scale models and test configurations, designed to investigate the radically new concept of pilot ejection at high speed. This was soon outdated by rapidly developing technology, but was pioneering for its day.

X-4, tailless research (1948–1953)

This turbojet-powered research aircraft was designed to investigate semi-tailless or tailless configurations at transonic speed to about 0.85 Mach. The X-4 was also the first X-Plane to be designed and fitted with an integral ejector seat.

X-5, swept-wing and variable swept-wing research (1951–1955)

This aircraft ejection seat was provided with a cordite cartridge and was guided by a rail system until it cleared the cockpit area.

X-13, vertical take-off and landing research (1955–1959)

This X-Plane was a pioneer in US VTOL technology and utilized a specially manufactured transportation, launch and retrieval trailer to allow the aircraft to be hoisted up on to its tail after conventional take-off and landing were evaluated. Ironically, with the aircraft on its tail it resembled the position an astronaut would find himself in on the launch pad; if seated on an ejection seat an astronaut would experience ejection sidewards and upwards instead of directly upwards as a nominal aircraft ejection seat would perform. To provide safety the ejection seat could swivel (with 45° of movement in the pitch axis only). Having a zero–zero capability it was fitted initially with only one actuator but later with actuators on both sides, manufactured by Stanley Aviation of Colorado Springs, Colorado.

X-14, VTOL technology development (1957–1981)

This was a long research program in which an aircraft was upgraded twice, but due to weight restrictions was not fitted with an ejector seat. Tests therefore were restricted to 3.5 m–4.5 m or lower or at 760 m where it was considered safe for single-engine failure recovery or pilot egress and parachute descent.

X-15, pushing the envelope at the fringe of space (1959–1968)

This was the most famous X-Plane next to the venerable X-1 and proposed X-20 Dyna Soar. Three X-15s performed 199 free flights for almost a decade pushing the limits to Mach 6 and 76,000 m or above, to the very fringe of space. With such speeds and altitudes an improved ejection seat was required. Early studies featured a capsule or ejection seat and significant input into this was completed by North American Aviation's Scott Crossfield. Eventually, an ejection seat was chosen which could be operated at 36,600 m and up to Mach 4. To stabilize the seat a pair of telescoping booms with attached wings were fitted. In the seat plan were two oxygen cylinders to feed the pilot's full-pressure suit in normal flight and in the event of ejection (NASA, 2008).

Operation featured a rapid sequence of events. The seat was rocket-propelled and designed to perform a safe ejection at 167 km/h; wind blast could kill a pilot not adequately protected so the seat was designed with protection as well as safety in mind. The seat footrest included ankle restraints which were activated by the pilot by pushing his feet backwards which activated the shackles, and deflectors in front of his boot toecaps. As the ejection handles were raised so thigh restraints were activated and the elbow restraints rotated inwards towards the body. At the same time the emergency oxygen supply safety pin was pulled activating the system.

Bell X-2.

Automatic canopy ejection was achieved as the ejection handle was pulled to 15° of full travel. At canopy jettison the seat ejection motors were fired and ejected along guide rails. During this period the timer was activated to determine seat separation and parachute deployment, and the rocket motor ignited to increase separation distance over the aircraft vertical tail in almost any circumstance within the operating performance envelope. The telescopic booms and side wings were also deployed as the seat left the aircraft to provide stability. When the 4,570 m altitude was reached the onboard sensors would initiate pilot/seat separation by jettisoning the headrest, releasing the seat belt and quick disconnect of all personal lead connections (on the left of the seat bucket). Descent by personnel recovery parachute would follow. Manual separation was available should the automatic system fail (Godwin, 2000).

As activating the ejection handle could inadvertently arm the system and lock it in the firing position—though inhibited to fire only when the canopy was attached—it was the responsibility of ground crews to check through the windows that the handle was in the safe position prior to each research flight. An external handle on the right side of the cockpit could be used to sever the catapult initiator connection to prevent this igniting in an unplanned firing. Another safety feature was providing the pressure suit with a face heat battery used to prevent ice build-up on the faceplate and allowing the pilot to estimate seat separation when he could not see his correct altitude reading. In the program several in-flight critical situations occurred but did not require the use of the ejection seat system. The one fatality on November 15, 1967 occurred at altitude, with the pilot Michael Adams, possibly disorientated or suffering from vertigo from excessive roll movements, when the vehicle began to spin and aerodynamically broke up, killing the pilot who was unable to eject from the vehicle. Structural loads had increased to $\pm 15g$, far beyond the design limits of $+7.33g$ and $-3g$.

The decision about the best method of escape from the X-15 was an early consideration. A fully ejectable cockpit was more complicated, and heavier, while ejection seat and personal parachute recovery would be lighter, quicker and use existing technology. Early recommendations were for the ejector seat in 1955, but more work would be required to support this proposal. As meeting after meeting discussed the pros and cons for a seat, so the testing continued. By November 1956 wind tunnel testing at the Massachusetts Institute of Technology was completed with encouraging results. With the loss of Mel Apt in the X-2 which featured an ejectable capsule, support for the ejection seat option increased. Between 1958 and 1959 a series of sled tests of the proposed X-15 ejection system were conducted at Edwards Air Force Base in California and despite some early setbacks work progressed satisfactorily. Parachute drop tests were not considered necessary for the development program (Houston, 1959, pp. 139–157).

X-20, Dyna Soar (never-reached-flight stages)

The objective of this program was to demonstrate a manned, maneuverable vehicle for experiments involving hyperspace and orbital flight-testing equipment to determine future military uses of piloted spacecraft. Before cancellation in 1963 this

Valuable information on high-g load was gathered by manned sled tests (top). Ejector seat tests for lifting bodies (bottom).

program was a development from several earlier USAF proposals for manned spacecraft which could be launched vertically into space, operated in orbit as a spacecraft with a completely controlled re-entry and landed on conventional (military) runways. This was of course the profile the NASA Space Shuttle would follow 20 years later. Studies included pilot ejection systems that ranged from no escape provisions at all to fully encapsulated crew compartments. It was decided that the single crewmember would probably have a conventional rocket-powered ejection seat optimized for subsonic use only. Access on the launch pad would be via an overhead hatch but this was quite restricting with just a pressure suit on, let alone used in an emergency egress situation. Clearly, the ejection capsule design was a more viable option had the program progressed to flight status. Dyna Soar was canceled in December 1963.

Lifting body technology, flying bathtubs (1963–1975)

A series of lifting body vehicles operated between 1963 and 1965 developing the technology and experience in low-speed operations and control of blunt-body vehicles. The M-2, HL-10 and X-24 were used to investigate these regimes. At first the initial lifting body, the M2-FI, had no provision for crew ejection in an emergency, but later a lightweight Weber rocket-propelled zero–zero seat was installed. The M2-F2 featured a modified zero–zero ejection seat from an F-106 *Delta Dart* aircraft. For the X-24 series (its A and B variants), a rocket-propelled zero–zero ejection seat was provided for safe pilot ejection in an emergency situation during nominal flight envelope operations.

From the sound barrier to the Space Shuttle and beyond

For almost 30 years aeronautics research programs at Edwards Air Force Base in California have supported the development of high-speed, high-altitude and new technologies in support of not only military aircraft technology but also the theories of flying winged vehicles into space, aircraft safety and commercial improvements, all of which have benefited not only military but also civilian aerospace technology and design. This research supported the development of the NASA Space Shuttle (see p. 207). With the Shuttle due to retire in 2010, NASA plans a return to the capsule-like configuration of Apollo. However, there are plans for private access to space under the Virgin Galactic program. Evolved from the pioneering flights of SpaceShipOne and White Knight, which in 2004 took the X-Prize, crew escape from SpaceShipOne was not via an ejection seat. These were judged much too heavy and expensive. Instead, the crew would have bailed out once the front hatch was cranked open and fell away. Fortunately, they were not called upon to be tried out in flight. Clearly, satisfactory crew escape will have to be developed to support private space ventures in SpaceShipTwo.

VOSTOK, SINGLE-SEAT EJECTION

History records that the first human to enter space was Soviet cosmonaut Yuri Gagarin on April 2, 1961 aboard the spacecraft *Vostok* (East). He completed one orbit of the Earth in 108 minutes and survived the mission to prove that humans could indeed be launched into space, perform useful observations while there and endure the rigor of re-entry and landing. The method of ensuring he indeed survived the launch and the landing was solved by the Soviets using the same system for launch abort and landing escape: an ejection seat.

Developing the system

The theory of human spaceflight was in part realized by Russian academic Konstantin Tsiokolvsky and brought to realization by pioneers such as Sergei Korolyov, one of the leading spacecraft and rocket designers of the Soviet Union. The development of the initial Soviet manned space program called Vostok has already been described in a companion volume in this series and complementary volumes (Hall and Shayler, 2001; Siddiqi, 2000; Abramov and Skoog, 2003), and are suggested for more in-depth research.

During 1955 no fewer than five different versions of "spacecraft" for vertical (suborbital) flight were being considered by the OKB-1 design bureau headed by Sergei Korolyov, continuing the work pioneered by Mikhail Tikhonravov on a two-man cabin for suborbital flight in the late 1940s. Studies for manned suborbital flights in the smaller R-5 rocket were abandoned in favor of placing a human crew on the much larger R-7 ICBM vehicle modified for carrying a manned spacecraft. Designs of various spacecraft included winged spacecraft and a vehicle with a rounded nose and spherical base (capsule) were rejected in favor of a spherical design (termed OD-2) which reduced the stresses on the vehicle during entry and provided a much simpler design hurdle to overcome. Recovery of the capsule was not prioritized as it was envisaged the lone occupant would be ejected from the capsule after re-entry and make a separate parachute-supported landing.

The manned spacecraft termed Object 3K was one of four variants designed to explore manned and automated scientific research and development as well as automated military reconnaissance. The Descent Module (termed SA) would feature a lone occupant seated in an ejection seat apparatus that could be used for launch abort situations as well as end-of-mission recovery. This was decided upon after studies revealed providing a suitable soft-landing system to support a crewmember inside the descent module for land recovery (the preferred option in the Soviet Union) was too complex and time-consuming for this first spacecraft. That option was incorporated into the more advanced spacecraft under planning, later termed Soyuz. The size of the descent module was determined by the launch capability of the R-7 launch vehicle and its dimensions. Therefore, the mass of the Vostok descent module was about 2,400 kg and some 2.3 m in diameter. The ejection seat with a human payload represented about 7% of the total mass of the whole Vostok spacecraft.

There were studies supporting the provision of an escape tower, similar to the U.S. Mercury system, for Vostok, but these were abandoned in favor of an ejection seat system. Ivan I. Kartukov, Chief Designer at OKB-81 in Moscow had been pressurized by Korolyov to develop an escape tower for the Object K manned spacecraft. Kartukov was the leading designer of solid-propellant rocket accelerators used in Soviet missiles of the day. The design submitted was found to be too heavy for the launch mass of the spacecraft on the R-7. Kartukov was afraid that lightening the system could place added stress on it and therefore increase the chances of failure and add to the risk of the crew unfortunately having to use it. As a result, despite a plea from Korolyov to continue with weight reduction, the escape tower for Vostok was rejected in favor of an ejection seat system that was effective up to the first 40 seconds of spaceflight. Should a problem occur then ground command could initiate blowing the outer spacecraft hatch and initiate the ejection of the cosmonaut on his seat for parachute descent. OKB-81 continued in participating on the design of the escape system by supporting the development of the complex outer-hatch system which had to be separated allowing the ejection seat to pass without injuring the occupant (Siddiqi, 2000, p. 196).

Operation

As there was no launch escape tower on Vostok the cosmonaut rode in an ejection seat all the way to orbit, which could have been used, should a problem occur, within the first 40 seconds of powered flight resulting hopefully in a nominal parachute descent, similar to that completed at the end of every one of the six manned Vostok missions.

With the cosmonaut strapped into the seat, final checks on the suit and system were completed before the hatch was sealed. Inside the Vostok the cosmonaut would lie in a semi-reclined position 65° to the horizontal. This design ensured both the correct attitude during a possible ejection to shoot the occupant up and away from a potentially exploding launch vehicle, and to minimize the acceleration forces encountered during the ride into orbit. In the event of deviation from a planned flight trajectory the sequence would begin by explosive separation of the outer hatch of the descent module. Two solid fuel–powered rockets under the seat would then be initiated for ejection, hand-pulled by the cosmonaut. The seat and cosmonaut would then ride on rails through the opening in the side of the aerodynamic shroud which covered the spacecraft up to about 180 seconds into a nominal flight. After a sufficient height had been achieved the cosmonaut would separate from the seat and perform a landing using his personal parachute.

During tests of the system the angle of ejection of the seat from the spacecraft was adjusted from sideways to slightly forwards in order to clear the slipstream of the ascending rocket as well as increasing the safe separation distance. A special "gun" was installed on the seat whose "cartridge" was a tether connected to the apex of the seat's recovery parachute to increase inflation of the recovery parachute for stabilization and deceleration at low altitude.

Integral to the escape system was the Vostok pressure garment termed SK-1 for male cosmonauts and SK-2 for the female cosmonaut (Valentina Tereshkova) who rode the Vostok. It provided protection for the cosmonaut in ejections up to 8 km and an oxygen supply during parachute descent from up to 10 km.

A variety of tests both ground-based and airborne were completed prior to the first manned flight. These featured unmanned tests of the ejection system using tower-supported test rails, from the backs of high-flying aircraft. Mannequins, animal test subjects as well as human testers were used in this evaluation program to determine the ejection profile of the seat and the recovery system. Considerable experience had been gained from ejecting smaller unpressurized or pressurized containers from high-altitude sounding rockets or suborbital research flights in the 1950s, and this experience was beneficial to the development of the Vostok ejection seat system at comparable speeds and altitudes.

The Vostok ejection seat and pressure garments were developed by Plant 918; the chief designer was Semyon Aleksaeyev. Located in Tomilino, it was established in 1952 to develop high-altitude, full-pressure garments for military aviation crews. It was a logical step to become involved with pressure garments for spaceflight and crew rescue. This company became more commonly known as Zvezda.

No escape from Voskhod

The Vostok launch escape system was not called upon during any of the six launches involving a human crew, though each mission utilized the system under normal flight conditions during the landing phase. Vostok was adapted to become the multi-crewed Voskhod spacecraft that supported the first space crew (Voskhod 1 in October 1964) and the first spacewalk (Voskhod 2 in March 1965) before being grounded in favor of developing the Soyuz system. With two or three cosmonauts aboard Voskhod there was no way ejection seats could be provided for each crewmember; in addition, it was impractical (1) to add a launch escape tower to the new variant so late in the design phase and (2) due to the shortage of time Voskhod was needed to be launched before the Americans commenced their more advanced Gemini spacecraft which, like Vostok, employed ejection seats for its two astronauts. Voskhod was the first and to date only manned spacecraft where there was no provision for crew escape or recovery from launch accidents or launch vehicle failures. Though other missions were planned, the two crews who flew the Voskhod into orbit were fortunate that the launch vehicles worked as designed and placed them in a safe Earth orbit.

GEMINI, TWO-MAN EJECTION

As discussed earlier the original design for the two-man Gemini spacecraft, called Mercury Mark II, featured an escape tower but was rejected in favor of dual ejection seats to be used in the event of a launch abort. In Mercury it was the sequencing of the escape system added to its mass, which on a successful launch was not required and was essentially irrelevant to the functioning of the manned spacecraft itself, that helped focus attention on an escape seat system for Gemini. In addition, the inclusion of an ejector seat also fitted well with the new vogue of modularizing systems.

Human-rating the Titan

The problem was that the fuels used on Atlas (liquid oxygen and a mix of hydrocarbons termed RP-1) to launch Mercury were a highly explosive mixture in the event of a break-up of the launch vehicle. The automated systems and escape tower on Mercury addressed the need for rapid separation from that potential fireball. No

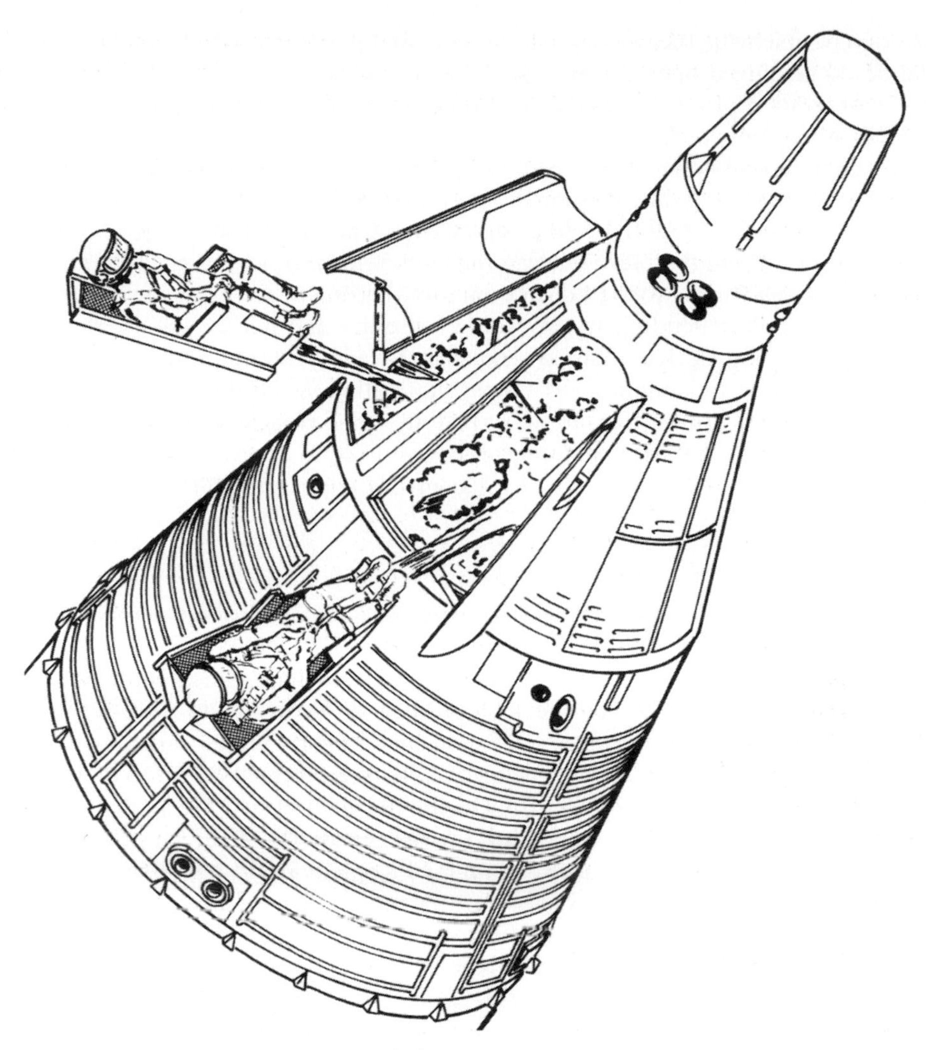

Gemini ejection seat escape.

ejection seat could supply enough power to eject an astronaut away in time. A new booster called Titan II was being prepared for service in the USAF, and it had potential for use in the space program. It was more powerful and could easily lift the new Gemini vehicle. The Titan's fuel was hypergolic, therefore less violent and did not require an ignition system—it burned upon contact. This therefore required fewer internal relays and could accompany redundant systems to improve the safety of the vehicle, and its human-rating capability. There were still disagreements as to how much force an explosion on a Titan would entail compared with an Atlas, but the idea appealed to the developers and managers undertaking the new Mercury Mark II program which eventually became Gemini. It also provided compatibility with the idea of a flexible wing recovery system, instead of a parachute system, and of hatches that could be hinged open for in-flight EVA activities.

In order to improve the reliability of the launch vehicle the decision to abort was handed to the crew, making the system much more simple; reliability was increased by including a redundant guidance dual-control system in the launcher. The most serious potential for a failed Titan II was a first-stage "hard-over" where a failed guidance and control system would push the engine nozzle over to one side deflecting the line of thrust and therefore sending the launcher out of its designed flight trajectory—essentially veering to the side or even cartwheeling to aerodynamic destruction. An automated abort-sending system would sense this and initiate the abort in time, but would astronauts? Therefore, a second first-stage guidance and control system was added offering redundancy and almost eliminating the need for automated abort systems. The Titan II test program was certainly by no means trouble-free but these problems were resolved; in the flight program it proved to be a reliable and successful launch vehicle for manned spacecraft (Shayler, 2001, pp. 168–174).

Gemini ejection system

The objectives of the Gemini escape system envisaged escape from pad aborts at zero velocity and 30 m "zero" altitude to heights in excess of 12,190 m at a velocity of over 305 m/s. This range, from almost a standing start to high Mach number, posed challenges to the designers of the system such as evolving an effective testing program to human-rate the system to fly on operational Gemini missions. Gemini ejection seat system guidelines were defined at 30 m zero velocity (on top of Titan on the pad) to 4,510 m flying at Mach 0.75. During the main recovery system, the escape seat system is used as a back-up feature for the remainder of the launch phase (to 12,190 m) as well as during the re-entry and landing phase of the mission (see Chapter 8). The final selection of the ejection seat system for Gemini was made after consideration of launch mass limits, the performance of Titan II, and the need to provide both crewmembers with adequate and reliable escape options in both primary and back-up modes during ascent and re-entry.

System description

The launch escape seat for Gemini offered a dual role of primary restraint during the nominal dynamic phases of ascent from the pad through to orbital insertion, orbital flight and nominal entry and landing, as well as the primary method of crew escape during ascent and landing. Each seat combination comprised a number of components arranged in subsystems many of which were designed to operate in series. The major subsystems of the ejector seat for Gemini were

Hatch actuator system—the system of opening the crew compartment hatch. This would have been initiated by either crewmember pulling a handle located in between their legs, thus activating a series of mildly detonating fuses that were attached to actuators which drive rods upwards to operate the hatches and lock them open. This

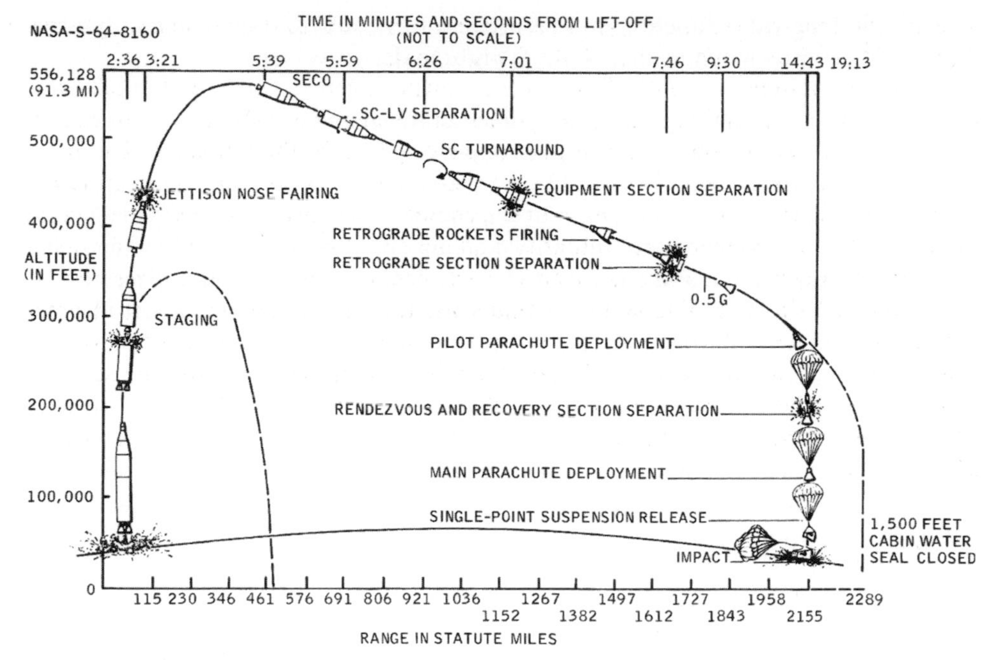

Gemini abort profiles.

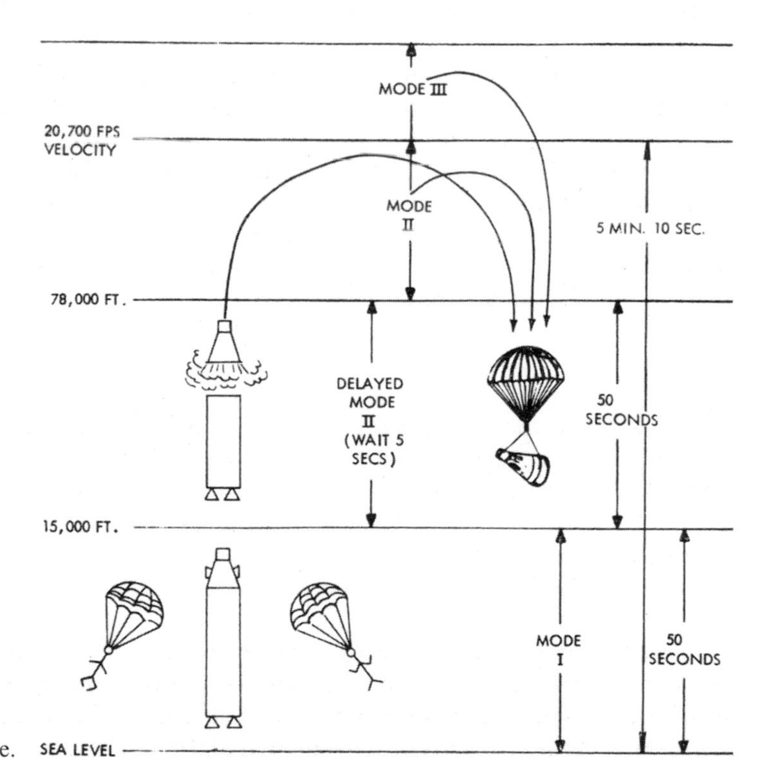

Gemini abort detail. Mode I: eject after shutdown. Mode II: salvo retros after shutdown. Mode III: shut down, separate, turn around, retro-fire.

system was designed to function up to altitudes of 21,340 m and sustain aerodynamic forces on the open hatch at that altitude and the design velocity.

Seat assembly. Resembling a military aircraft–type design the seat was fabricated mostly from titanium and aluminum, which provided an excellent weight-to-strength ratio. The back of the seat area was fitted around the rocket catapult structures with the center of gravity of the astronaut at a minimum distance from the thrust vector of the rocket motor. Structural integrity was also provided by the side panels that support the arm section, seat pan and seat back. It was attached to the spacecraft by a torque box rail assembly sliding through six titanium blocks on the back of the seat supporting a range of g-loads during operation.

To provide the propulsion required to eject the seat and astronaut occupant a rocket catapult (ROCAT) was designed. The limits of this design was defined by a launch vehicle failure on the pad—the most critical escape trajectory. Using half-scale tests on launch vehicle explosions and other data, a calculation was made of the maximum fireball radius expected, and from this and studies of heat flux the criteria for ROCAT were determined.

To support the astronaut in the seat an adequate restraint system had to be provided. This was to be used in dynamic flight and zero-g restraint as well as in the ejection mode. With an astronaut properly restrained, a nominal center of gravity could be determined. However, outside forces (such as an explosion or involuntary head movement) during ejection could affect this and had to be compensated for in the thrust vector of the ROCAT. Restraints provided included a contoured fiberglass cushion, foot stirrups, elbow guards, leg straps, lap belts and shoulder straps. A V-block headrest also helped to restrict side movements of the astronaut's head.

To locate a variety of support equipment a backboard was included in the seat design. Here the parachute, the ballute, survival kit and associated pyrotechnics were located. The shoulder strap inertia system was also located in the backboard. The straps were at the tip of the structure which moved along inertial reels to the lower back. By use of a control this could be operated in manual lock or automatic modes. Manual lock meant that though an occupant could be reeled in they could not be extended. In automatic mode the straps were automatically locked when extended at a rate of between $2g$ and $3g$. The reel straps could extend to 48 cm.

Also included in the seat structure was an egress kit, located in the seat pan, which provided the astronaut with oxygen and pressurization during an abort ejection at altitude. Pressurized at 124 bar with gaseous oxygen, it was reduced to a flow of 3 bar by means of a regulator at 0.016 ± 0.001 kg/min and a check valve in the suit. Upon leaving the suit it passed through a shut-off and relief valve dumping the oxygen overboard maintaining suit pressure at about 0.25 bar at ejection above 10,670 m. It was activated by a lanyard pulled by seat motion at ejection.

Parachute system. Between separation of the seat and astronaut at 2,290 m altitude a balloon-shaped "ballute" was used to stabilize the descending crewmember. Fabricated from coated nylon fabric, four reed-type vents inflated it by ram air. With a minimal diameter of 1.2 m, it eliminated tumbling motions induced at seat separation and inhibited any future tumbling possibilities. It also provided the feet-down attitude required for nominal parachute deployment. Decreasing the rate of descent by 9 m/s, it also limited a flat spin rate of no more than 45 rpm. The personnel parachute of Gemini was a standard 8.5 m diameter flat circular military-operated C-9 canopy. A pilot chute was included to help in main chute deployment, and a harness of Dacron webbing formed in a double "figure 8" was fitted to the astronaut. These were individual fittings and could be adjusted for comfort and fit by the chest strap. It also featured quick disconnects for ease and speed of unharnessing the unit.

Survival equipment. This was stored in two containers, one on the backboard assembly on the left front of the assembly, the other on the rear.

Sequence of events

Upon the decision to abort, and with a force of at least 12 kg one of the astronauts would pull the ejection control handle between their legs only 13 mm to activate the firing mechanism that released the dual firing pins and detonated the booster charges and hatch actuators igniting the hot gas that fired the propellant cartridges to push gas into the piston forcing open both hatches and locking them in the open position. The whole sequence from pulling through the ejection handle to opening the lock hatch took 0.3 seconds.

The seat ejection cycle began with the gas pressure of 35 bar being received through a ballistic hose from the hatch actuator. This activated through relays and charges the seat-firing mechanism propelling the rocket motor, and thus the seat, through the length of the catapult housing and out the open hatch. Dual igniters on the rocket assembly provided redundancy. For 108 seconds after rocket burn-out the astronaut was still strapped to the seat. This allowed dynamic pressure to decay in the case of high-*g* ejections. A cartridge then fired a rod-shaped movement-initiated mechanical linkage and released the astronaut from the seat.

Stabilization was then performed by the ballute which was deployed by an aneroid control device. Above 2,290 m there was a 5-second release delay allowing dynamic pressures to decay, therefore reducing additional parachute opening loads. Below 2,290 m the release firing pin was not ignited and the cutter separated the ballute allowing main canopy deployment. At 1,740 m and after a 2.3-second delay a main charge fired generating gas pressure that propelled a 10-ounce "slug" away from the crewman attaching a wire to the top of the drogue chute and pulling one-third of the main chute out of the pack. When inflated the drogue pulled out the rest of the main chute. High-pressure gas from the drogue gun was also used to initiate the equipment jettison sequence, after a 5-second delay. Three charges separated the restrain strap cutter, lap belt disconnect and jetlox release that separated the oxygen hoses and wire cable from the egress kit. The backboard and egress kit then fell away, with only the survival kit lanyard remaining connected. The kit and lanyard were now below the descending astronauts as the rest of the equipment freefell to Earth. The time from the ejection handle being pulled to the astronaut being free from his seat

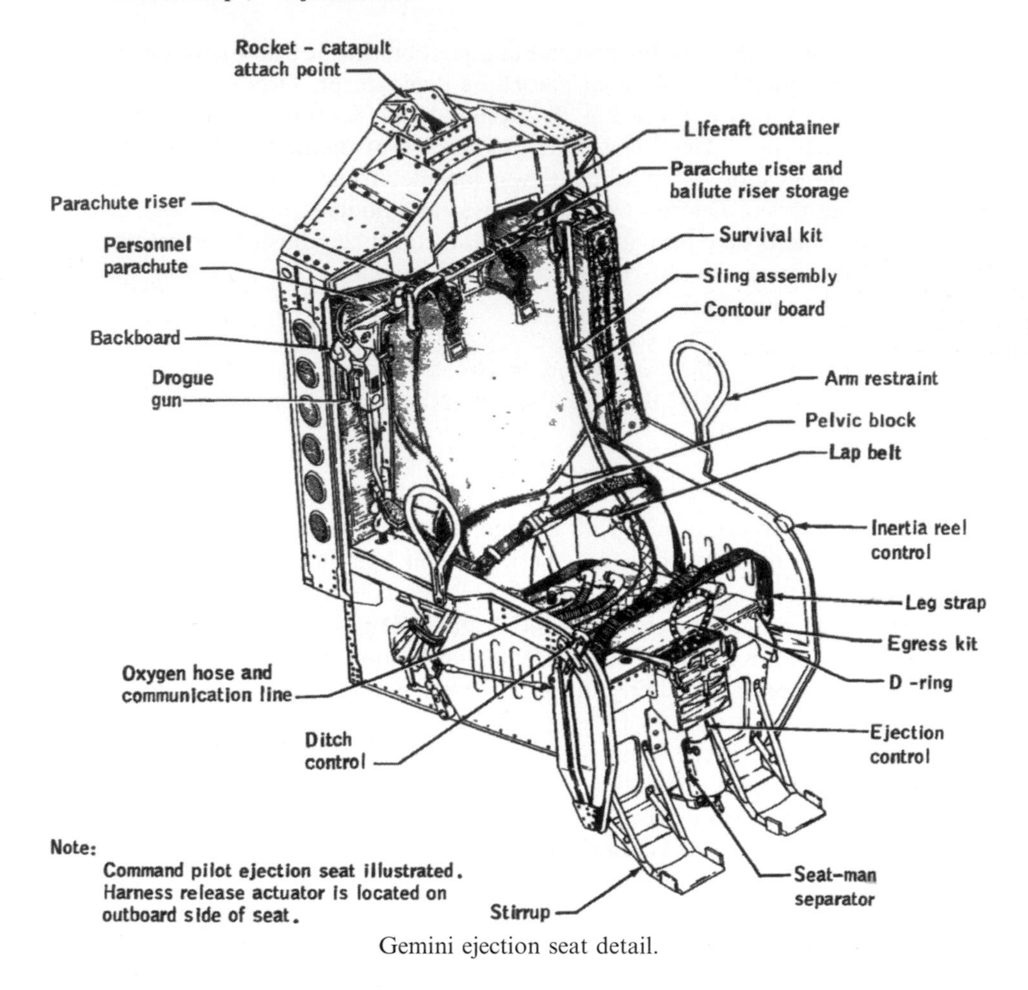

Gemini ejection seat detail.

was just 1.50 seconds for ejection at 1,740 m and less than another 7 seconds elapsed from the astronaut being free to equipment jettison. This changed to 5 seconds above 1,740 m, but depended on altitude.

Test program

A range of tests and evaluation of the Gemini Escape System were performed to human-rate the system. These included wind tunnel testing, static tests, air drops, body slump tests, simulated pad ejection tests, track vehicle tests, and a range of qualification tests that encompassed sled tests, simulated pad ejections, recovery and survival system aircraft drop tests, ground qualification tests and high-altitude ejection tests (Ray and Burns, 1967). This program is summarized below.

Wind tunnel testing. These featured tests with Mach numbers ranging between 0.5 and 5.8 on the seat–man configuration. This objective was to gather aerodynamic

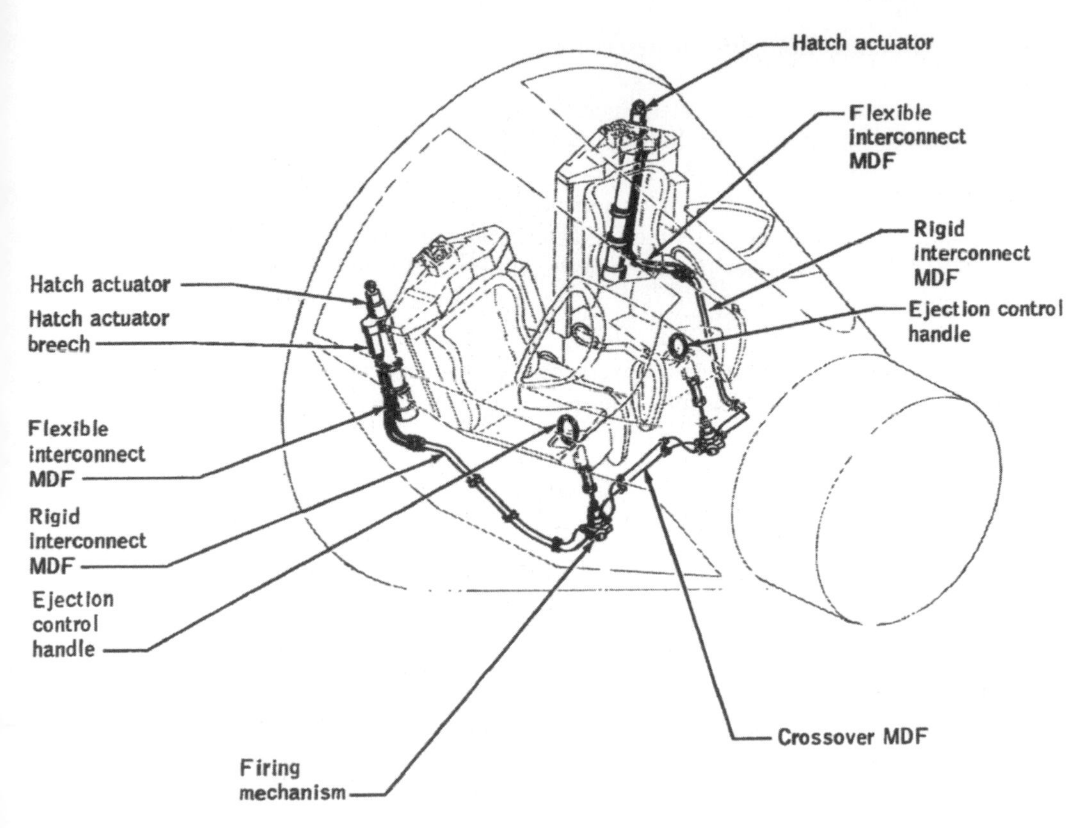

Gemini capsule details.

characteristics of the combination at varying altitudes that could occur following ejection. From this, data was used to determine trajectory calculations and flight characteristics, adding the scant information available on a similar configuration at such extreme altitudes and Mach numbers. This provided a more defined database to work from. Testing was completed in three series using models. A 20% model was used in the free-stream tests while the 10% model was used in the rocket exhaust effects and proximity test.

Series I used a 20% model to determine various pitch angles, while in Series II various yaw angles were evaluated. Both series gained additional data on sideslip at 90°. In the second series a 10% model was utilized to determine the effects of the rocket plume and the close proximity of the spacecraft with its hatches open. The types of tests conducted in the wind tunnel included separation of the seat from the rail, the hatch seat clearance point and the midpoint between both positions. Series III tests were carried out to determine what modifications needed to be made after the initial tests had been completed. Here the changes incorporated into the design included shortening the vertical height of the seat and raising the foot position for the astronaut, deducing the best armrest angle and lowering the armrests, and adding a larger headrest. The tests carried out on Series II ranged from Mach 0.5 to 3.5.

Ballute wind tunnel tests determined the opening characteristics and drag effects of the design. Ballute–man stability parameters were also investigated. Low-speed tests used a one-fourth scale model and a one-sixth scale model; the initial supersonic test utilized a one-fourth scale model; a second supersonic test used both a one-fourth scale model and a one-sixth scale model. Two supersonic tests were also completed on a full-scale model. A series of low-speed tests were also performed expanding the range of data.

Static tests. There was an extensive program of bench test and individual component testing prior to a range of static tests designed to verify one or more events in the sequence of ejection as the design evolved.

Separation tests were conducted to ensure that any modifications made to the system would not hinder clean separation of the astronaut from the seat, and to demonstrate this an anthropomorphic dummy was used during the development and testing period. The test consisted of placing the dummy on the seat and then activating the separation system. This phase also confirmed that the sequences of separation actually occurred as designed.

A series of rocket catapult (ROCAT) tests were conducted to explore the capabilities and range of the system beyond that envisaged for Gemini. Therefore, a group of static rocket tests were completed to investigate the limits of the operating envelope providing data for possible use in future programs, thus saving time and expense in additional testing should the Gemini ejector seat system be adapted for

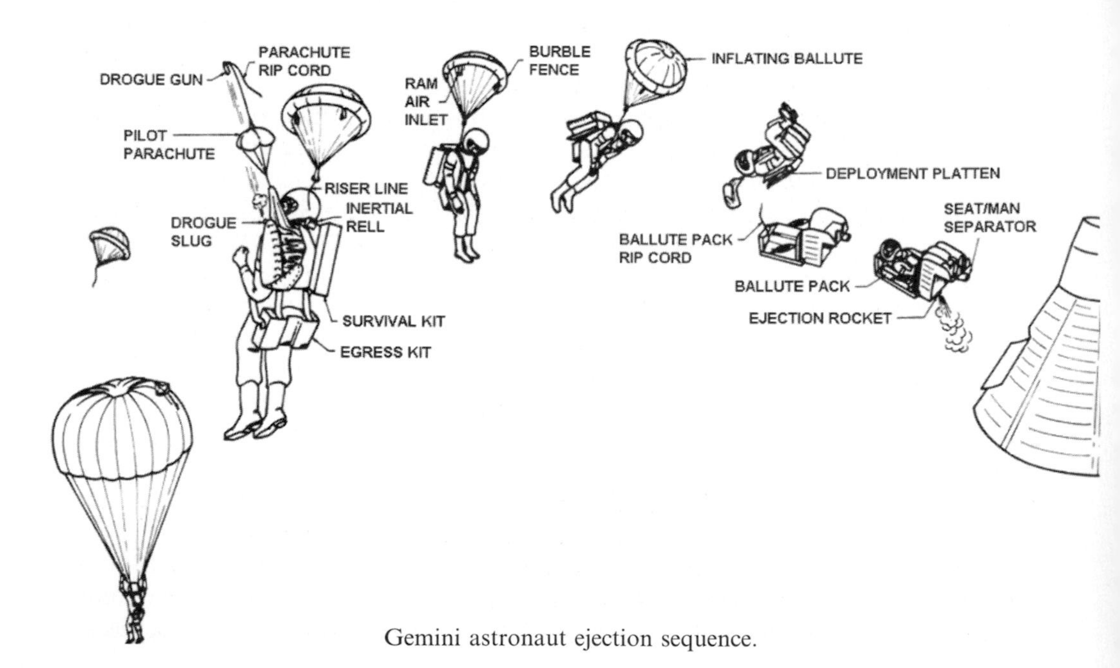

Gemini astronaut ejection sequence.

these new programmes. In the early tests structural failure in the thrust pad area was experienced on the bottom of the seat and as a result design modification was required. To test the changes a dummy was mounted on a seat on a concrete pad between track rails. The track catapult fired and propelled the sled down the track by reactive force. The rocket motor was then fired to determine the force vector of the new design. Other static firings were completed in altitude chambers to determine performance at various altitudes.

In addition, three flight tests were completed during 1963. The first on April 19 was carried out to obtain the optimum trajectory as well as evaluating further design modifications. The unit landed 217 m downrange and 15 m right of the centerline. The second and third tests on September 17 and 18 were to gather details on the trajectory of the seat with predicted maximum eccentricities. Therefore, the second test landed 149 m downrange and 5 m left of the centerline path, while the third test landed 168 m downrange and 70 m left of centerline.

Harness tests. Initially a torso dummy was used to evaluate both the structural load and the comfort of the personnel recovery parachute harness. At first, static pull and dynamic drop tests were completed adding forces as the tests continued, moving on to a tower later in the program. Early results found that the Navy canopy release was insufficient to prevent premature release, so it was changed to an Air Force unit which provided more resilient acceleration and loads, and so this was the version integrated into the Gemini Program. There were nine dynamic drop tests in which a 136 kg dummy was dropped from a tower. The fall distance ranged from 3.5 m to 4.8 m with peak forces from between 0 kg and 567 kg.

Comfort tests were also made with human jump subjects replacing the dummy but using the same tower and expected nominal opening loads. This verified the harness under opening loads and revealed there should be no unexpected problems if the system was used on an operational Gemini mission. In this phase, seven tests were completed ranging from a fall distance of just 0.60 m and a peak force of 394 kg up to (on Jump 7) a fall distance of 2.1 m and peak forces of 789 kg.

Air drops. These were employed throughout the development program to not only verify the functioning of the system, procedures and equipment but also the various modifications incorporated into the design as the development continued. The make-up of these tests varied depending on the test required of the item of equipment to be evaluated. There were drips used with weights only, though most used the anthropomorphic dummy. To test the ballute, humans were used to evaluate the system.

During June and July 1962 the first parachute drops were completed and tested the parachute with weights attached; two systems were tested in parallel at three velocities, 108 km/h, 162 km/h and 216 km/h. System A featured a standard C-9 canopy, a 106 cm drogue, spring pilot parachute pyrotechnics, automatic opener and a rectangular daisy chain pack with B5-type pilot parachute flaps. The B system was similar except the daisy chain was locked by a steel pin and attached to a drogue gun. The 106 cm pilot parachute featured vanes instead of the spring design. The results revealed that System A varied with altitude but System B was constant; this, combined with an early tower test of System A, led to the decision to proceed with System B. There were a total of 19 drops completed, which ranged from 305 m to 396 m, and featured 13 tests of System A and 6 tests of System B.

A second series of 14 air drops was completed between March and May 1963. These again demonstrated the configuration using a dummy. There were five drops from a helicopter or a two-engine plane. In the helicopter drops the dummy was suspended from a cable 61 m below to avoid downwash. From the two-engine aircraft a lanyard reeled out the test unit and also activated the drogue gun to start the test. The other 9 of the 14 tests in this phase verified the various air loads on the unit. There were three helicopter drops and two twin-engine aircraft drops completed in March 1963, each of which were from the 305 m altitude. The nine aircraft drops from between 914 m and 3,764 m were completed during April and May using B-66, H-21 and C-130 aircraft.

High-altitude parachute tests.

Rocket slide ejection test rig.

There were eight tests to investigate the inclusion of deflation pockets on the personnel recovery parachute. These automatically trap water and deflate the chute on impacting the ocean. During a pad abort there was little time to determine whether these pockets affected the deployment time. Using dummies suspended from a helicopter with parachutes with or without pockets, results indicated an increase in opening with the pockets, and so pockets were removed and the extra time of 0.75 s meant full inflation could occur outside the parameters allowed for a pad abort ejection.

In August 1962 a dummy was dropped on a complete seat system from an aircraft flying at 305 m at 162 km/h. This evaluated the seat–man separation and parachute system prior to simulated off-the-pad ejection (SOPE) test No. 6. All system functions were as designed.

Investigation of the use of a larger ballute to stabilise an astronaut (a follow-up of the early wind tunnel tests) used a 91 cm ballute and 45 cm single riser, which provided conflicting results on the ability to stabilize the astronaut. In the new test a 91 cm ballute was used with a human test subject. Special equipment was used along with a 10.5 m HALO parachute. The jumpers were deliberately confused and disorientated. Several tests were completed on the 91 cm ballute and 45 cm riser, as well as on a 91 cm ballute and a 1.5 m bridle. Ballutes were increased to 107 cm then 122 cm. A series of 24 jumps was made, six of which featured dummies, the rest human test subjects. Lower rotation was experienced but it was decided to proceed with a 1.5 m dual bridle in order to reduce changes-of-stowage problems and entanglement.

Body slump tests. A program of 40 tests (27 human and 13 dummies) was completed to determine body slump tests. The purpose of using a dummy was to compare the effects with the human body. Using photographic evidence, records of acceleration forces on specific points on the body were recorded. Using a Gemini seat mounted on a tower 20° from vertical and tilted back 34°, a braking system stopped the seat as it moved up the tower structure. Live slump tests were completed between May and August 1963.

Simulated pad ejection tests. A program of 12 development ejection tests to simulate a pad abort situation using a complete seat system and a 45 m tower was completed. There were eight single ejections and four dual ones simulating a two-man Gemini crew escaping from a Titan on the pad. Results were obtained both from telemetry and photographic evidence. These tests took place between July 2, 1962 and July 16, 1963. The first six were single ejection; the first dual ejection occurred on Test 7 on September 26, 1962. There were varicd results, and changes had to bc incorporated accordingly. On the first test the dummy did not separate because of failures of the separator system, on the second test the parachute wrapped around the dummy. On Test 3 the dummy did not separate from the seat and impacted downrange with the seat. Test 4 saw the parachute fail to deploy. Test 5 again experienced a failed parachute deployment and required addition of a second drogue gun for redundancy. Test 6 was the first normal operation of the series. The first dual-seat ejection on Test 7 also recorded successful results. On Test 8, though, only one parachute deployed. Test 9 was originally planned as a dual-seat test, but was changed to two single tests, both were successful. Tests 10 and 11 were dual-seat ejection tests and demonstrated changes made to the system as a result of the earlier test. Both of these were successful.

Track vehicle tests. There were four tests (six runs) on a rocket-propelled track. This track had been used during the Mercury Program but was modified for Gemini where a boilerplate spacecraft was installed. Hatches were fixed open, and seats were initiated by electric power supplied trackside. Propulsion for the sled was solid-propellant rocket motors with the number varied depending on the test. Braking was by water brake troughs after ejection. The dummy test subject was instrumented to record acceleration levels, rates and events and, coupled with telemetry from the sled and high-speed documentary photography, provided a clear record of the sequence of events.

During the first test on November 9, 1962 twin seats were installed. The test was designed to study velocity drag capabilities and structural integrity—no ejection was planned. A failure of the pusher motor third stage resulted in the motor penetrating the boilerplate spacecraft, severely damaging the spacecraft though it still remained on the track and was subsequently repaired. Data were obtained and a new pusher vehicle designed and manufactured. A maximum dynamic pressure test was planned for Test 2 on June 20, 1963. Though ejection velocity was slower than targeted, it was successful and both dummies were recovered safely.

Test 3, designed to test ejection under 15° yawed conditions, was aborted when it was detected that test velocity was out of tolerance. Rescheduled as Test 3A on August 9, 1963 all objectives were successfully met. Test 4 was designed to simulate an ejection during descent after re-entry and was very successful. Test 5 was a rerun of Test 2 due to the lower velocity encountered on the earlier run. One dummy was recovered by parachute, the other had a low trajectory and impacted the ground prior to deployment of the drogue. This test was not rescheduled due to time constraints in the qualification program.

Qualification tests. Sled tests were completed between June 4 and December 11 1964 and comprised four tests. They qualified the system during high dynamic pressure and during simulated re-entry conditions. Tests 6 and 7 were highly successful. Test 8 resulted in the left (command pilot) dummy ejecting normally, but the right (pilot) dummy did not separate from the seat due to a faulty left armrest/side panel separating from the seat resulting in the malfunctioning of the seat–man separator. After correction Test 9 was a success.

SOPE Tests 12, 13 and 14 were completed between January 16 and March 6, 1965, to qualify the off-the-pad abort region. Test 12 resulted in a successful command pilot left-side ejection and recovery, but not the right-side pilot dummy. Failure caused by premature ignition of the ROCAT resulted in a redesign allowing Tests 13 and 14 to be completed successfully.

A Recovery and Survival System air drop test was completed between January 11 and March 13, 1965 to qualify the system and sequence, as well as its performance and integrity under the dynamic conditions that followed separation of the seat and man; 20 dummy drops and 18 human subject test drops were also conducted. The dummy tests were conducted between 1,740 m and 13,650 m, the human tests were completed between 1,740 m and 9,450 m. Despite some problems and failures, all tests were completed successfully, meeting all the objectives in this program.

Ground qualification tests to certify the system under the most severe conditions and loadings expected were carried out, as were tests on the ballute loads on the attachments and backboard assembly, personnel recovery parachute system and ejection during re-entry from abort. All objectives were met.

An F-106B aircraft was used for high-altitude ejection tests. Three tests were utilized: a static ground ejection on October 15, 1964 proved compatible with the F-106B of a Gemini ejection seat, a subsonic ejection at Mach 0.65 at 4,785 m on January 12, 1965 and a supersonic ejection the same day at Mach 1.72 at 12,192 m.

Gemini flight operation

The last test under the escape system qualification program occurred on March 13, 1965, just ten days prior to the launch of the first manned Gemini mission, Gemini 3 on March 23. Between that date and November 11, 1966, ten manned Gemini missions were launched successfully without the need to utilize the Gemini escape system. The closest a mission came to an abort ejection was during the Gemini 6 launch attempt of December 12, 1965, when (as subsequently discovered) a small electrical plug in the tail of Titan II dropped out prematurely and a small plastic dust cover had obstructed the oxidizer flow inlet line of a gas generator. Resulting in an engine shutdown after 1.2 seconds prior to lift-off, the commander/pilot assessed the situation and determined it was safe to remain in the vehicle. His decision, based on data received and his "awareness" of the situation, allowed the vehicle to be remanifested for a launch to meet the Gemini 7 spacecraft three days later (Shayler, 2001).

As can be seen by this extensive program of development and testing for the escape system, it was a long drawn-out process over three years, all for a system that was not used operationally. This clearly demonstrates the efforts and background work required to design, develop and qualify a rescue system for a spacecraft, which hopefully will never be required, but if it is needed is tested to a level that test data demonstrate an above-average chance of survival within the design envelope of rescue systems—given enough time for it to operate, of course.

GEMINI AND MANNED ORBITING LABORATORY

While the Gemini spacecraft was being developed, the USAF had given the go-ahead in 1963 for a Manned Orbiting Laboratory launched on a Titan IIIM vehicle and crewed by two military pilots flying a modified Gemini spacecraft. The USAF had a long interest in utilizing Gemini for its own program, and the resulting MOL program was the apex of that desire. For crew safety, thoughts were turned to an escape tower initially but the decision was made to go with the twin ejector seat combination used on the NASA system as an easier and cheaper option that was already qualified for NASA use, thus saving time and launch mass—the reasons NASA opted for seat ejection instead of a tower in the first place.

A study of the safety of a crew launched on a Titan IIIM established the baseline of the MOL Crew Escape System during ascent to orbit (MOL and Gemini B Office, various dates).

During ascent the fatality rate was not to exceed 3 crews in 1,000 flights. With failures during ascent less than 32,000 ppm (parts per mission) the Titan IIIM failures less than 30,000 ppm. There should be no violations in abort ceiling levels, and the capacity to provide thrust to terminate the launch at any point should be available.

Titan IIM Malfunction Detection System

The Titan IIIM MDS featured an automatic thrust termination of the solid rocket motors which came into play when angular rates threatened an uncontrolled vehicle, immediate loss of a solid rocket motor pressure or manifold pressure, or inadvertent SRM separation. There was also a redundancy guidance and control system with automatic switch-over for rapid malfunctions.

Information displayed on the Gemini control panel would initiate manual abort scenarios, as with the NASA Gemini. The Gemini B escape system included six retrograde rockets' salvo or ripple fire, manual separation of the spacecraft and escape operation, manual seat ejection, manual termination of thrust for Stages 0, 1 and 2, and crew monitoring of critical booster performance and manual switch-over of slower malfunctions.

Controllers at the USAF Mission Control would monitor the launch from Launch Complex 6 at Vandenberg and monitor the trajectory for slow-drift malfunctions. Wherever possible, they would avoid launch abort constraints, perform guidance systems analysis, and predict where impact points would be as the ascent continued and monitor basic subsystem performances.

Gemini B escape modes

There were four defined escape modes for Gemini B/MOL launches. Mode A was on the pad and up to 30 seconds into the flight, after which water impact was imminent, thrust on the launch vehicle was terminated, salvo-fire of six retrograde rockets and seat ejection from the spacecraft. Mode B covered a GET of 30 seconds to 7,315 m/s velocity. With the thrust terminated a "ride-out" was initiated with 2, 4 or 6 bursts of retro salvo-fire, jettisoning of the adapter section followed by re-entry and deployment of the recovery system. Mode C covered 7,315 m/s with a 488 s flight time (±20 s) during which the spacecraft would separate and, following staging, a specific burst of retro-fire would be required to achieve the desired safe touchdown, followed by retrograde, re-entry and recovery. A Mode D abort was an abort to orbit for at least two orbits whereupon the spacecraft would be separated from the Titan IIIM by manual command from the crew, the spacecraft onboard systems would provide additional acceleration to achieve orbital velocity again manually operated by the crew and where possible a degraded mission (alternative/contingency) could be performed, followed by normal entry and landing.

The survivability of a crew during a Titan IIIM launch deviation was also evaluated. The launch vehicle failure mode and its effects were assembled from known data, and from these failures the causes of aborts were identified. By simulating these failures, associated warning times could be established. This resulted in the probability that an abort might be required at any time during ascent. By assessing the failure modes of the Titan IIIM, the capability of the Gemini B escape system to cope with them could be determined. From this—assuming an abort was required—the data could indicate whether an escape was possible. Then, by determining the equipment reliability of the Gemini B escape system (by using such data as presented above under qualification of the Gemini escape system for NASA) and from NASA flight performance, the possibility that the escape system would function as designed—assuming an abort was required—and whether an escape was possible could be determined.

Some evaluations had the purpose of improving blast protection by undertaking an appropriate test and failure analysis program, especially around some of the cryogenics on the MOL in the event of a Titan IIIM explosion, which would also destroy the laboratory module as the Gemini crew escaped.

Escape tower on Gemini B?

The MOL safety studies also re-evaluated the use of an escape tower over the current ejection seat system. A Mercury-type design was suggested with a 386 cm tower structure supporting an escape rocket with a thrust of 29,484 kg and burntime of 1,375 s with a 40,540 kg/s total impulse. An aerodynamic spike was also incorporated in the design attached to the re-entry module by a segmented clamp ring.

The study found that the escape tower utilized during Stage 0 only improved pad abort mode and ride-out mode escapes, but required additional blast protection for Stages I and II where retro-rockets would be used for escape. There was, however, a high risk in development both in terms of scheduling (in an already much delayed program) and cost (already escalating). An extensive test program was also required, something that NASA had not done on its Gemini series. Therefore, a significant impact would be felt in the Gemini B, Titan IIIM and MOL programs. The total weight penalty of equivalent payload loss and blast protection added to 118 kg and the estimated cost added $52 million to the budget. Importantly, crew fatality in such a system had been estimated as 1.6 crew/1,000 missions.

Alternative choices

From these studies it was determined that blast protection was crucial as it could reduce potential fatality from 10.7 to less than 3.0 per 1,000 missions. An escape tower *as well as* blast protection should be eliminated due to the high risk of development penalties. Of the remaining alternatives, the baseline ejection system would be improved with the addition of a thrust vector control, a blast shield and a "pop-gun" action in which adapter compartment pressures rise during retro-rocket action which increases velocity during abort by venting following capsule separation.

The conclusion reached was that blast protection alone was unacceptable as it afforded no pad escape provisions. A full blast shield was too costly and heavy. TVC with blast protection was a good compromise and provided the best survivability at pad abort. Adequate Mode B survivability and both overall added cost and mass penalties were not excessive. The pop-gun option suggested the laboratory section would always fail but the crew would survive; however, it would be better to improve technologies to ensure the laboratory did not fail all the time. The pop-gun effect is good for the high max Q region but contributed little to pad escape. Therefore, the TVC system was recommended as a baseline for Gemini and implementation of blast protection on laboratory cryogenic tanking and improve burst strength limits on the pressurized shell structure and forward dome of the laboratory.

In the event Gemini B/MOL was canceled by the USAF in June 1969, with only a single unmanned launch taking place. By then Apollo was flying and so was the Soviet Soyuz, both employing escape rocket technology rather than ejection seats. However, on the design boards was a new generation of manned spacecraft that featured multi-crews, heavier payload capability and a reusable technology to "shuttle" cargo and crews to and from space. In the question of crew safety and rescue systems, escape towers, multiple ejector seats, crew-recoverable modules and provisions in the launch profile were all under study. The advent of the Space Shuttle era added new challenges to crew rescue both on the pad, in orbit and on landing.

SHUTTLING INTO SPACE

The development of the NASA Space Shuttle system evolved from studies in the 1960s into a reusable space delivery system. Work on developing such a system had occupied space planners and designers for years, indeed the idea of fully reusable space "planes" can be traced back as far as the early part of the 20th century. The complicated and long gestation of the NASA Shuttle is best recorded in Dennis Jenkins' book on the history of the national STS program (Jenkins, 2001) and the reader is encouraged to review this volume to appreciate the changes that have been incorporated into the Shuttle design and equally the safety systems and rescue procedure.

NASA SHUTTLE LAUNCH PROFILE

Plans were made to launch the Shuttle from both the NASA Kennedy Space Center in Florida and, for DoD missions in polar orbit, the USAF Vandenberg Air Force Base in California. The idea of launching a Shuttle from California was abandoned in 1986 in the wake of the Challenger accident, and all launches have taken place from LC 39 (Pads A and B) at KSC, which was modified from Apollo Saturn to be compatible with that of the Shuttle, a similar transition is currently underway to convert launch facilities at the Cape from the Shuttle to Constellation.

In launch configuration the Orbiter and two Solid Rocket Boosters are attached to an External Tank in the vertical (Shuttle nose up) position on the pad. Facilities for pad rescue have been described in Chapter 4. At launch three main engines ignite, fed by liquid hydrogen and liquid oxygen from the ET. When verification of the correct thrust levels is confirmed, the twin SRBs ignite and the vehicle must leave the pad. Should the engines fail prior to SRB ignition a Redundant Set Launch Sequencer (RTLS) abort takes over shutting down the SSME and aborting the launch (see pp. 100–102). From lift-off to entry into orbit the vehicle is under powered flight and aborts during this phase fall under different regimes depending on the configuration of the vehicle, its altitude and distance from the launch pad and speed.

On a nominal ascent, as the Shuttle clears the tower it is rolled to heads-down attitude to reduce atmospheric pressure on the vehicle. Maximum dynamic pressure is reached at a Ground Elapsed Time of 60 seconds; 1 minute later (and two minutes into the flight) the twin SRBs have used their propellant and are separated for parachute descent and ocean recovery. The three main engines continue the powered ascent fueled from the ET. About 8 minutes from launch the ET fuel is exhausted and the SSMEs shut down. The tank is then discarded and re-enters the atmosphere and is destroyed. The Shuttle fires its twin Orbital Maneuvering Engines (if required) once or twice depending on the mission profile to complete orbital insertion and circularize its orbital path.

Orbital operations now take over for up to 14–18 days, and this is followed by a retrograde burn to low Earth orbit and re-entry for land landing on a runway. The orbital safety element and entry and landing provisions are described in the final two chapters. Here we focus on confidence operations and rescue during launch.

Early developments

The X-Plane and lifting body research programs contributed to the development of high-speed, high-altitude flight and runway landings of a reusable vehicle that also provided the groundwork for the development of what became the Space Shuttle. Reusable space vehicles, both manned and unmanned, had been studied for decades before the Shuttle evolved into the program at NASA that followed Apollo. With regard to crew escape and rescue some designs featured ejectable crew compartments, or even whole nose sections, that would be jettisoned from the main vehicle to make its own landing back on Earth. Some designs also featured deployable fly-back vehicles with air-breathing engines which could be used in the event of a failed main rocket engine. These were incorporated both in the designs for a manned booster vehicle and the "Shuttle Orbiter" design. Studies under Phase A looked at a range of advanced conceptions from the late 1960s through to the final decision to award a contract to North American Rockwell on July 26, 1972. In the previous four years there had been a wealth of studies, designs and proposals mainly focusing on the overall design and flight profile of the vehicle. As stated earlier, the complicated genesis of the NASA Space Shuttle has been extensively detailed in Jenkins (2001); here we summarize these developments.

Under the Phase B Project Definition program, commenced in 1970, the aerospace companies bidding to receive the formal contract had to look at the safety of the configuration, its design and integrated systems of recovery, especially in crew escape during launch. With a proposed multi-crewed vehicle the escape tower option used on Mercury and Apollo was rejected in favor of studies focusing on three areas of escape. These were an encapsulated ejection seat, which was similar to the type used on the North American experimental bomber the XB-70; a completely separate crew compartment, which was the concept already used in the General Dynamics F-111 and the early Rockwell B-1A aircraft; or ejection seats and pressurized suits, which was the option for the Lockheed A-12/SR-71 triple-sonic aircraft. The only design suitable to rescue more than two crewmembers was that of a separable crew compartment which was found to raise not only development costs and mass, but would also require far more studies and effective advanced warning.

In April 1971 a revision was added to the definition document which established the requirement for "rapid emergency egress" during the planned development flights. These development flights would grow to include an atmospheric flight

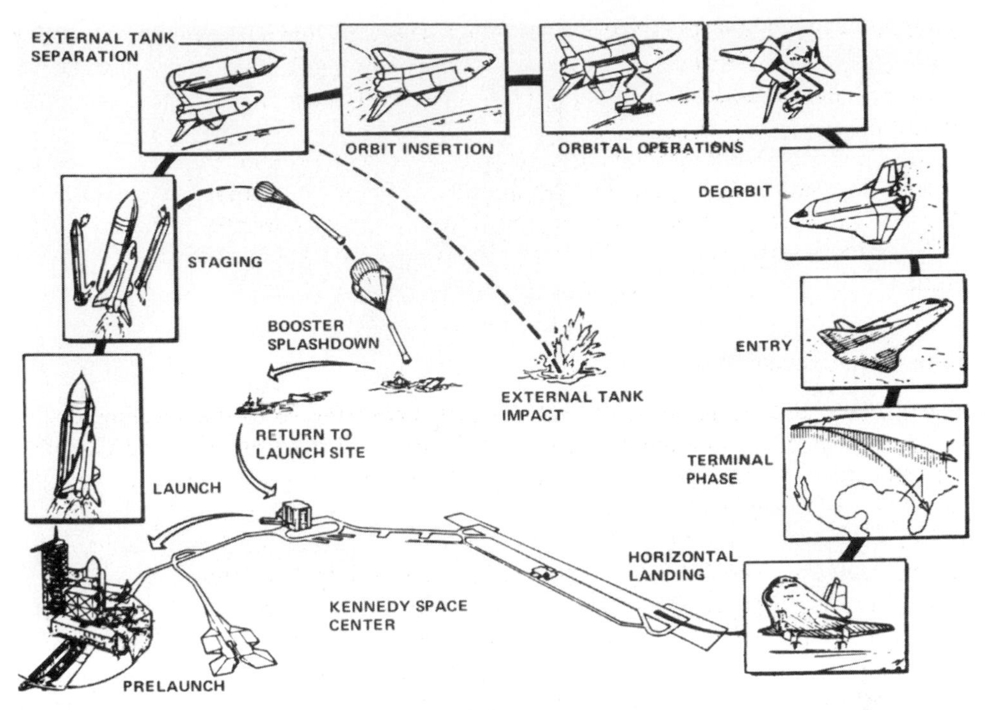

Shuttle mission sequence.

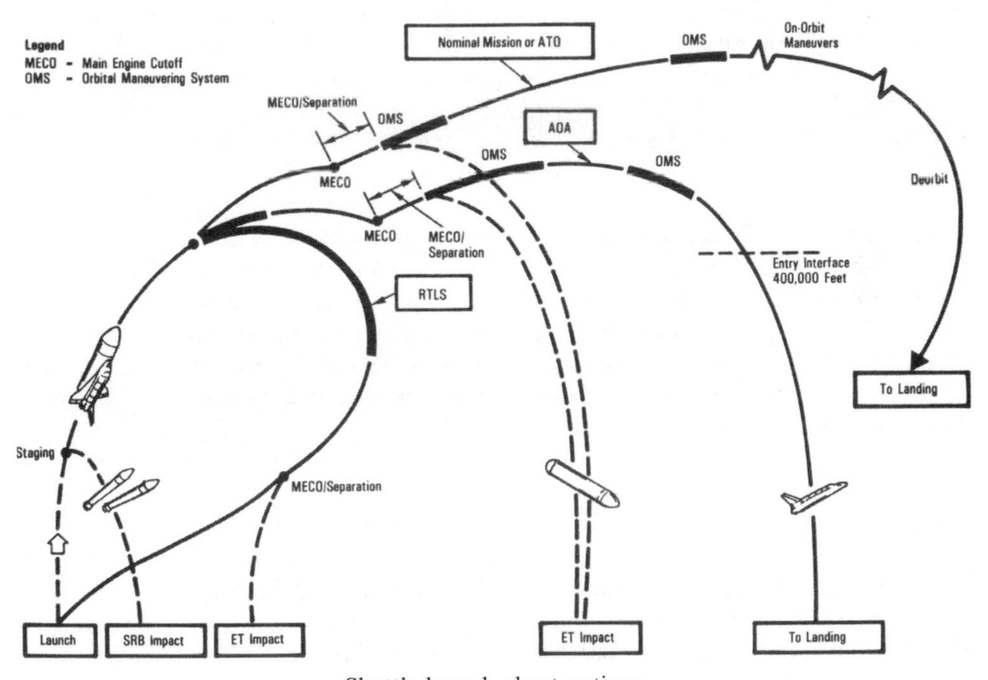

Shuttle launch abort options.

Escape suits for Orbital Flight Tests.

test—initially powered by the onboard air-breathing jets, but later off the back of the converted Boeing 747 aircraft—and early orbital flight tests (up to six). It was eventually decided to use modified SR-71 ejection seats for the OV-101 Enterprise airborne flight tests and OV-102 Columbia orbital flight tests. The philosophy was to provide the planned two test flight crewmembers with the added protection, albeit limited, of escape from potential disaster. It was decided that upon completion of the test flights most—if not all—unknowns would be solved leading to classing the Orbiter similar to that of an "operational airliner" which does not have ejection seats, escape modules or parachutes for everyone onboard.

Ejection seats installed

The Approach and Landing Test program, completed by the Orbiter Enterprise during 1977 demonstrated the Shuttle Orbiter capability to perform an unpowered approach and landing on a runway, the final phase of a Shuttle mission. Both pilots were provided with modified A-12/SR71 Lockheed zero–zero ejection seats. These were also installed in Columbia for the first four Shuttle flights under the Orbital Flight Test Program (STS-1 through STS-4), flown between April 1981 and July 1982. For these four missions the ejection seat could tilt forward enabling the pilot to position himself closer to the instrument panel while on the back position in the seat on the launch pad. Following the last flight they were disabled for STS-5 and removed completely for Columbia's next launch on STS-9 in 1983.

Ejection on the Shuttle was capable of initiation by either the commander or pilot if the vehicle became uncontrollable, there was a fire onboard, or the landing was to be on water or unprepared surfaces away from the planned runway landing. In total it required about 15 seconds for the situation to be recognized, the crew to initiate the ejection sequence and to clear the vehicle. Operationally, ejection seats were available for use during launch during the first stage or during a gliding flight below 30,480 m. However, subsequent studies revealed that had a crew ejected during ascent they would have been exposed to the plumes from the SRBs and SSME, though landing escape options below 30,480 m remained a possibility. From STS-5 through STS-51L there was no provision for launch escape for the crews other than the defined launch abort options.

The options detailed below were the most viable options to recover the vehicle and crew during the mission. Aborts on the pad would be prior to SRB ejection and handled by pad emergency egress procedures (see Chapter 4). For the early part of the ascent it was theoretically possible to return the vehicle back to the launch site and land on the landing facility's runway, depending on the velocity, altitude and timing of the abort situation, this is known as a "return to launch site" abort. As the Shuttle flew over the Atlantic (or Pacific from Vandenberg) and should a problem occur beyond the range of return to the launch site and before it could limp to orbit, then a landing on a ballistic trajectory may be possible, either in Europe or Africa from KSC, or the Pacific Islands from California. This became known as a "transoceanic abort landing" capability. If a problem developed at a time when the vehicle could make it to orbit, but not remain there for long and land after one revolution of Earth, then the "abort once around" capability was devised. Finally, a situation may arise which sees the Shuttle making it to a safe but lower temporary orbit from where it could be raised in altitude, perform an amended mission or indeed come home early.

This would be an abort-to-orbit situation allowing more time for the controllers and crew to assess the situation.

It was the loss of Challenger in 1986 that focused minds on the difficult issue of crew escape during launch. A wealth of proposals for ejection seats, escape compartments or pads, and slide poles were evaluated. The escape pole concept was finally chosen as the most suitable system to incorporate in the vehicle without a major redesign and prohibitive development costs and to avoid a complete grounding of the fleet. This escape system is described in Chapter 8; calls for ejection seats and escape poles were again revisited after the loss of Columbia in 2003. Reinstalling two ejection seats for the commander and pilot along with a rescue pod with separate escape engines and recovery systems for the rest of the crew were proposed in the aftermath of the Columbia tragedy. The rescue pod would be located in the payload bay and would require the remaining crewmembers (up to five) to move from the flight-deck and mid-deck through the airlock hatches into the pod, seal it and separate it, possibly without opening the payload bay doors, allowing time to activate the rescue system before the vehicle disintegrated or was too close to the ground to effect the recovery system. To re-engineer the front section of the Shuttle to land separately was simply not practical, and it was deemed easier to build a whole new Shuttle to incorporate such a design from the start. In the end it was decided to finally retire the Shuttle after completing the International Space Station, and develop a new vehicle to return to the Moon, and support U.S. ISS operations, with an integrated rescue capability resembling the launch escape tower from the Apollo Program. For the remaining Shuttle flights, the abort options, escape pole, added photo-documentation, in-orbit inspections, provision at ISS of a safe haven and opportunity to launch a second rescue Shuttle remains the only option. It was clear from the STS-51L accident that in reality no system of crew rescue or escape was available for the rapid events that unfolded on January 28, 1986. From STS-5 through STS-51L the crew operated in what was termed a "shirt-sleeve environment" with no protective suits worn, only flight overalls. A clamshell-type helmet was worn, however, to provide a supplementary oxygen supply in the event of cabin depressurization during ascent or re-entry (but below 15,240 m).

A portable oxygen system was available inside the Orbiter called the Personal Egress Air Packs (PEAPs) which, along with the early Launch and Entry Helmet, could also be used by an EVA crewmember in flight to pre-breathe prior to EVA denitrogenization of the circulatory system. The PEAPs were located on each crewmember's seat and supplied with oxygen from the atmospheric revitalization system. Oxygen was supplied through flexible quick-disconnect hoses, and could supplement each crewmember with up to 6 minutes walkaround (independent) capability when seated away from the main oxygen supply. There were ten oxygen system locations installed on the Orbiter: four on the aft center console on the flight-deck, four on the mid-deck ceiling and two in the airlock (*Space Shuttle ...*, 1984, pp. 302–303). When the crew compartment of Challenger was finally recovered from the Atlantic Ocean it was found that some of the PEAPs had been activated on the flight-deck of Challenger following the explosion of the vehicle 73 seconds after launch.

Launch protection 1985, clamshell helmet and flight suits (top). Shuttle Payload Specialist receives ejection seat instruction for the T-38n aircraft (bottom).

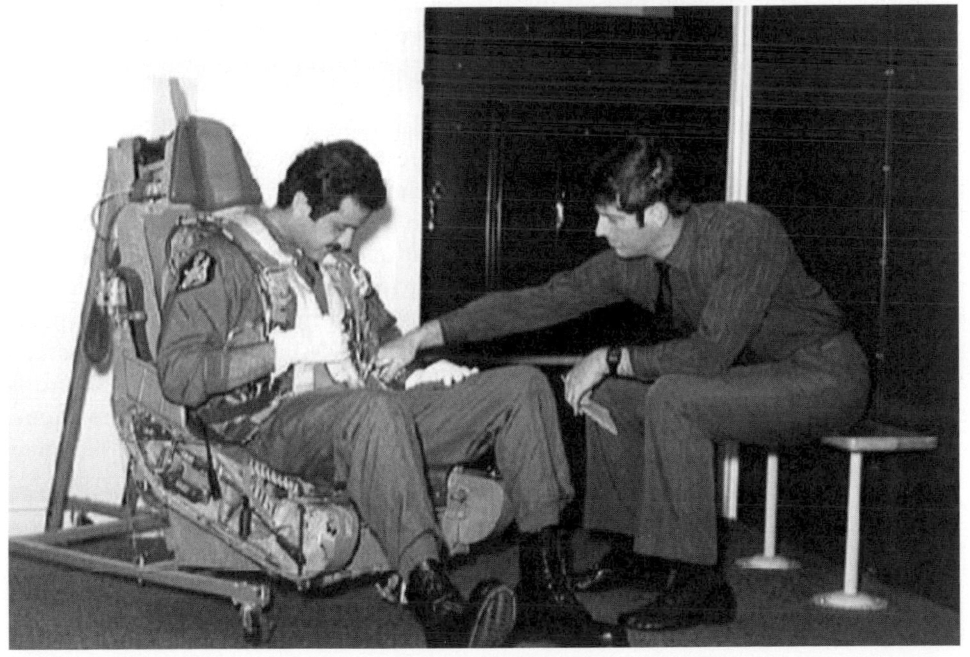

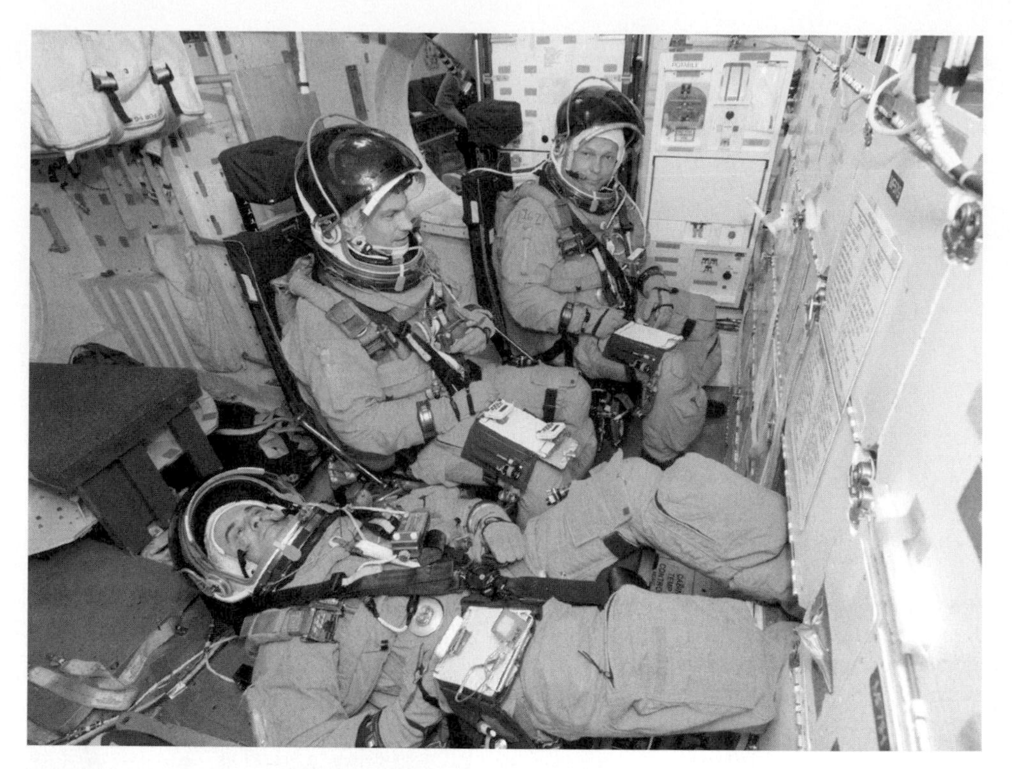

Post-insertion/de-orbit training session in one of the Shuttle full-scale trainers at JSC. (Foreground) ESA astronaut Leopold Eyharts, (center) NASA astronaut Stanley Love and (rear) ESA astronaut Hans Schlegal. All three are attired in (2007) training versions of Shuttle launch-and-entry suits.

For the first four orbital missions the two-man crew wore the David Clark Company Model S1030A Ejection Escape Suit (EES), which provided crew protection. It was a full-pressure garment for ejection up to 24,384 m and at velocities of Mach 2.7. The modification for the NASA astronauts involved integration of an anti-*g* suit. The original USAF S1030 Pilot Protective Assembly (PPA) was worn operationally by SR-71 pilots. The contract for these suits was awarded in 1978 and included parachute and flotation support equipment. Early photos of the first six Shuttle crews reveal them wearing the ejection suit for publicity photos although only four crews worked with them operationally, the original six OFTs being cut to four missions. Interestingly, Mission Specialists Allen, Lenoir, Musgrave and Peterson (Thomas and McMann, 2006) were photographed modeling the suits, though no provision for them wearing such ejection suits was available on STS-5 (Columbia, OV-102) or STS-6 (Challenger, OV-099).

It is worth recording that during the ALT program the crew onboard the Shuttle Carrier Aircraft (a modified Boeing 747) also had a method of escape. This featured a 4.87 m long escape slide that connected the flight-deck with the forward cargo

Shuttle launch escape suit (1988) (top). On-orbit escape suits are stowed until entry day (bottom).

compartment where an egress port was located. In an emergency situation and a need to rapidly escape the aircraft a crewmember would activate a handle either side of the autopilot control pedestal. This in turn would explosively fracture 30 fuselage windows allowing rapid aircraft depressurization followed by explosive separation of the emergency egress hatch which would fall clear of the aircraft. In order to allow the crew to clear the engines, wings and landing gear, a spoiler was automatically extended. The four-man crew (two pilots and two flight engineers) would then slide down the chute, jump out the open hatch and descend by personnel recovery parachute. Each crewmember would take about 11 seconds to complete this, meaning that all four crew would be out, hopefully, in less than a minute from initiating the escape option.

To evaluate the Shuttle ejection seats for operational use, a complete upper forward fuselage was fabricated and used at Holloman AFB, New Mexico on October 26, 1976 for rocket sled testing. The test program commenced on November 18, 1976 and was followed by a series of unmanned ejections at speeds varying from 0 mph up to 450 mph. This series ran from January 11 to May 5, 1977 and qualified the system for use on Enterprise during ALT. A second program was required to qualify the system for the OFTs and this was completed in April 1980.

NASA SHUTTLE ABORT MODES

If a failure of a component threatens the integrity of the vehicle or safety of the crew a number of abort options are available to the Shuttle crew during ascent.

The nature of the design of the Orbiter vehicle, with two decks containing four upper seats and up to four lower-deck seating arrangements, prevented the use of an ejector seat capability for each crewmember. In addition, the option of an escape tower or a separating crew module was rejected early in the design studies. Failures in the main propulsion system, the structural integrity of the vehicle or atmospheric conditions need to be taken into account when initiating an abort situation.

Following lift-off, abort scenarios fall within three categories.

Performance—loss of one or more Space Shuttle Main Engines. Dependent on when the engine (or engines) fails during the ascent was the crucial point for what happens next. If the failure occurs early enough and the vehicle is still carrying a large amount of fuel it could make it to orbit if just one engine reduces its performance or is lost completely. This can result in no loss of speed, as telemetry received at Mission Control via the Abort Region Determinator (ARD) indicates whether an under-speed situation is predicted, which would lead to main engine cut-off (MECO). From these data a decision can be made immediately to continue or abort and at the same time which type of abort would be most suitable.

Systems—the failure of one or more major subsystem generates this type of abort. The controllers at Mission Control relay details of the problem and options to the

Shuttle pad abort (STS-41D) 1984.

flight crew from the data received. Again, depending on that data and predictions the appropriate actions can be instigated, should such a situation be required.

Range safety—despite planning and training for as many options and scenarios as possible during ascent, the remote possibility of a Shuttle becoming uncontrollable and heading towards a populated area has to be considered; moreover, how to deal with this potentially catastrophic event has to be planned for. This then is the area of aborting a flight that falls under the range safety–type of abort. The range safety of a Shuttle ascent is divided into the first stage where the US Air Force has the responsibility of detonating explosive charges inside each of the SRBs. This responsibility shifts to Mission Control during the second stage of the ascent based on the latest data from range safety with information relayed to the crew to act upon. Initiating a range safety abort is based on imaginary divisions around highly populated areas. In planning a Shuttle launch, range safety rules are devised by placing impact limit lines around populated areas. Inside these are drawn destruct lines. Under these rules potentially lethal items of hardware from a failed launch cannot land beyond the impact limit line. Due to the velocity of the vehicle or associated debris, termination of the mission has to be much earlier to ensure that the debris will not impact a populated area. Therefore, should hardware enter the destruct line, it is eligible for destruction.

The range safety system in the ET was deleted from STS-80 since performance history indicated that it was sufficiently downrange so as not to present a hazardous situation to landmasses. SRB range safety systems have been retained. Initiation of the destruct signal is by an encoded command to arm the system via a number of tracking stations, then a second signal to fire the charges. The arm signal is also illuminated on the flight-deck control panel for commander and pilot information. Though, it is a mission rule that no command to destruct the vehicle can be made until after the flight crew has tried to separate the Orbiter from the remainder of the stack to perform a contingency-type abort.

In some circumstances, during first-stage abort any deviations of trajectory could cause range safety to initiate destruction, but the Shuttle could still be under the control of the crew and therefore could be returned to a nominal trajectory or to complete a safe abort. In these circumstances the Flight Director and Flight Dynamics Officer at Mission Control use voice communications with the Flight Control Officer at Eastern Test Range (previously known as the Range Safety Officer). Any violation detected by the FCO is relayed to Houston and the FD and FDO. It is then up to the FD to establish if the vehicle is either controllable or uncontrollable. If it is determined by the FD that the vehicle is controllable then no action is taken by the FCO to terminate the flight just because the trajectory alone is deviating from the flight plan. The limiting factor in all of this is of course the time available to evaluate, decide and take action, and to recover the vehicle as safely as possible.

There are two types of Shuttle aborts: "intact", which is designed for the safe return of the Orbiter to a planned landing site, and "contingency", where the survival of the flight crew is assured following a more serious failure when an interactive abort is not possible, this would normally require ditching of the Orbiter and probable loss of the vehicle but rescue of the crew.

The four types of intact aborts are return to launch site (RTLS), transoceanic abort landing (TAL), abort to orbit (ATO), and abort once around (AOA).

Defining an abort

During ascent there remains a definite order of preference for choosing the appropriate abort mode.

- An RTLS abort involves the crew flying the Orbiter downrange to use up propellant and then turn around under power to return direct to the landing site (as close to the launch site as possible), preferably the Shuttle Landing Facility near the VAB and LC 39.
- A TAL abort is essentially a ballistic trajectory allowing emergency landing on one of several landing sites on the West European and African mainland. Here an OMS engine burn is not required.
- An ATO abort allows the Shuttle to enter an orbit that is temporary and lower

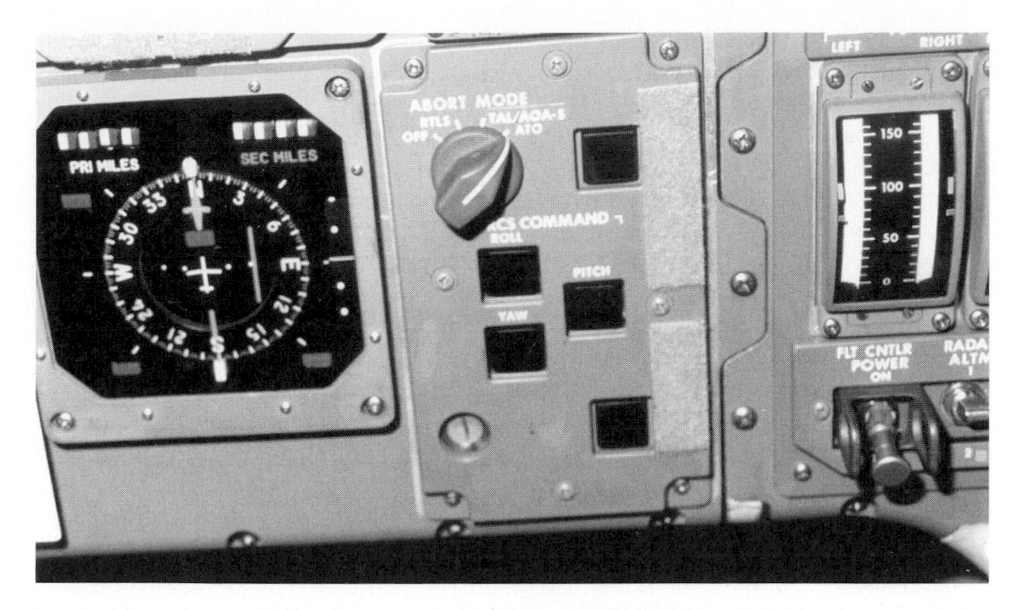

Shuttle abort controls (set to ATO) STS-51F 1985.

than that intended due to reduced performance from the propulsion system. This allows time to plan contingency mission operations or plan a more suitable de-orbit maneuver and early landing. This was the option take for the STS-51F (Mission 19) launch abort scenario in July 1985.

- An AOA abort allows the Orbiter to complete one orbit of Earth and a nominal entry and landing within 90 minutes of launch. This normally involves two OMS burns, with the second being the de-orbit maneuver. Entry and landing—normally at U.S. West Coast sites (Edwards AFB, White Sands New Mexico)—would be the preferred option in this scenario if a landing at Florida is not desirable or possible.

Depending on the type of failure and when it occurs determines which type of abort scenario can be used. If, for example, performance loss is the only factor in calling an abort, then the preferred sequence would be ATO, AOS, TAL, RTLS in order to preserve mission safety and still retain the maximum opportunity for completing a near-normal mission profile. Should an abort require the mission to end quickly in order to protect the lives of the crew or structural integrity of the vehicle, then the TAL or RTLS options may by more suitable over ATO or AOA. A mission rule is that contingency aborts are never selected if an intact abort is possible and available.

The primary instigator of aborts, based on received telemetry and visual evidence, is Mission Control in Houston who take over the control of each mission after the vehicle clears the tower. The team of flight controllers on the ground, backed up by support rooms, have direct knowledge of every system on the Orbiter from downlink data. Their knowledge is far greater than that of the flight crew riding the vehicle. As the vehicle ascends so the astronaut ascent Capcom (Capsule Communicator) on duty at MCC-H calls up the next abort mode available or when it is not available—such as "press to ATO"—meaning press on to abort to orbit for the next phase of the ascent. In the early Orbital Flight Test program (STS-1 to STS-4), when ejector seats were carried for the two astronauts aboard Columbia, when the upper limit was reached for use of the seat the call became "negative seats", meaning the crew would have to refer to an intact abort if a problem developed.

Should communications with the ground cease or become unclear, then the flight crew have onboard other methods to help determine the current abort situation and their options. These include displayed information as well as a series of cue cards and malfunction data files and procedure books. During ascent, it is the responsibility of the MS2/flight engineer, sitting in between and behind the commander and the pilot, who monitors the options and sequence of events for each specific abort as required.

It was during the STS-51F launch abort in July 1985 that MS Story Musgrave was deep in thought about where they may be heading should a transatlantic abort be called that he was temporarily not really listening to other conversations around him. He was fully aware of his responsibility and his own desire to get the correct information promptly to the pilots when they called for it. When MS Karl Henize, seated next to him on Seat 4 in the flight-deck, casually asked where they were heading, hoping he would finally make it to space for the first time after waiting for 18 years, Musgrave confidently said "Spain ...". At which point both Commander Gordon Fullerton and Pilot Roy Bridges spun their heads around and glared at Musgrave, who waved them away realizing his error. They made it to space and recorded a highly successful mission after some adjustments and contingency actions by both ground controllers and the crew.

Deciding which abort mode to select is a factor in the nature of the abort call or the timing of it during powered ascent. Obviously, crew safety is paramount. However, the best options to achieve full or partial mission success for primary and/or secondary objectives are also considered.

Should a problem with one or more of the Shuttle Main Engines occur (as with STS-51F and STS-93) the combined team of the flight crew and Mission Control will determine the best option to follow, which may not be the fastest mode. However, should the failure be a systems failure which can threaten the safety of the vehicle, and thus the crew, then the fastest mode is selected to return the vehicle to the ground at the earliest opportunity.

The quickest Shuttle abort profiles are RTLS and TAL (lasting about 35 min), while the AOA option has a duration of at least 90 min requiring one complete orbit of the Earth. The ATO situation allows for extended time in orbit to assess the situation and find the best option available to provide the safest way to continue the mission, the scope of any "amended" mission and its duration. With three good main engines, decisions as to which option to follow are assessed at the time of the abort.

To select the abort mode in the flight software an abort mode switch is turned to the appropriate mode and the "Abort" button depressed to initiate the update. Despite improvements to the flight controls and displays on the Shuttle with its "glass cockpit' technology, the abort selection switch and initiation button remain intact, and it is the role of the commander to select and initiate an abort mode by selecting the mode and inputting it.

In order of occurrence during a nominal ascent the four intact abort areas are

Return to launch site (RTLS)

This mode is available for loss of thrust in one of the SSMEs from lift-off through 4 min 20 s into the mission where there is sufficient propellant aboard to affect a return to KSC of the Orbiter, payload and the crew. The RTLS profile consists of three separate stages: the powered phase, a separation phase and the glide phase. The powered phase consists of continued main engine thrusting, the separation phase includes detachment of the External Tank and the glide phase is the approach to the Shuttle Landing Facility at the Cape.

The exact time for the crew to select the RTLS option is highly dependent on the reason for selecting that type of trajectory in the first place. The earliest time to effect an RTLS decision is after separation of the SRBs at 2 min 20 s as a result of losing an engine at lift-off. Further considerations concern the fact that the SRBs have to burn out, as they cannot be turned off or separated prior to burn-out, and the Orbiter has to gather the momentum and distance to affect a separation and turnaround move to have enough lift to fly back under control to the Cape. The latest the RTLS mode can be selected is 3 min 34 s into the flight when a three-engine RTLS is chosen.

Whenever the decision is made, the vehicle continues downrange to use excess Main Propulsion System (MPS) propellant; the aim is to retain just enough onboard propellant to allow it to turn around, aim it back to KSC, cut off the SSMEs, separate from the ET and glide back to the SLF at the Cape. The trickiest maneuver is the pitch-around maneuver, dependent in part on the exact time into the ascent of the failure of one or more SSMEs. As the vehicle continues downrange the pitch-around maneuver is begun to configure the Orbiter and the ET to a heads-up attitude, pointing the nose of the Orbiter toward KSC, essentially doing a 180° turnaround. Although the vehicle is pointing towards the Cape it's actually continuing flying downrange with the remaining SSME providing a braking maneuver to null downrange velocity. In order to improve both the Orbiter's center of gravity and mass, for the glide and landing the twin OMS and RCS propellants are dumped by continuous burning, which adds to the deceleration.

When the propellant levels in the External Tank drop below 2%, then the desired point for SSME cut-off is reached. A powered pitch-down maneuver is initiated 20 seconds prior to SSME cut-off, which aligns the vehicle for a separation from the ET and points the tank at the correct attitude and pitch so that it is pointing away from the fiery descent through the atmosphere. Following the shut-off of the three main engines the now spent ET is released, the crew initiate a translation burn of the RCS, which ensures the Orbiter does not recontact the tank, and align the vehicle at the correct pitch attitude to commence the glide phase of the RTLS abort profile. From this point the glide to the landing area is similar to a nominal entry and landing approach.

Transoceanic abort landing (TAL)

This abort option covers that area of the ascent from the final point of an RTLS opportunity to when an abort-to-orbit situation can be called on only two engines. It is also available for major systems failures in the Orbiter such as a significant loss of pressure or failure of the cooling subsystem where it's too late to return to the launch site but too dangerous to proceed to orbit. It therefore becomes imperative to get the vehicle and crew back on the ground at the earliest opportunity, which in this case means on the other side of the Atlantic.

The vehicle follows a ballistic trajectory across the ocean to a predetermined runway capable of handling a Shuttle landing and rollout. These sites are available as close to the nominal ground track as possible to maximize the use of available propellant aboard the ET. Other criteria include the length of the runway (ideally 36,576 m), the local weather conditions and whether the site has the approval of the U.S. State Department (bureaucracy and politics are part of the planning). Landing would normally occur some 45 minutes after launch.

Initially, this was termed the trans-*Atlantic* landing but was amended early in the program when it became evident launches from Vandenberg in California could not abort across the Atlantic. Had any launches occurred from Vandenberg, then the TAL would have been in Hoa and Easter Islands in the South Pacific, where construction of support facilities were completed prior to the cancellation of Vandenberg Shuttle launches.

Selection of the TAL mode has to be prior to the shutdown of the main engine; if a TAL is initiated after SSME cut-off then the computers automatically select the abort-to-orbit mode. From initiating the abort, commands steer the vehicle towards the plane of the selected landing site, roll the combination to a head-up attitude (should it be required) and initialize a propellant dump from the OMS by burning the fuel through the twin OMS engines and Reaction Control System nozzles. This decreases the mass of the vehicle which in turn increases performance; it also redefines the correct center of gravity for correct control of the vehicle and reduces the landing mass of the orbiter. At this point the ET is separated and landing is completed normally. The vehicle would be returned to the United States, if feasible, on top of a Shuttle Carrier Aircraft (a Boeing 747).

As the chances of landing a Shuttle at these transatlantic sites are rare the equipment at these sites is in some cases minimal. The early TAL sites were Morón (southern Spain), Banjul (Gambia, West Africa) and Ben Guerir (Morocco, North West Africa). Dakar (Senegal, West Africa) was also an early site but is no longer cleared by the U.S. State Department. A second site at Zaragoza Air Base (eastern Spain) is also a TAL site option. Normally, rookie astronauts perform Shuttle mission support duties at these sites during Shuttle launch operations, returning to the U.S. once the vehicle is safely in orbit.

East Coast abort landing (ECAL). This a variation on the TAL option where the Orbiter is flying a higher inclination launch that follows the eastern seaboard of the United States. It was devised in the early 1990s after Shuttle operations out of

Vandenberg AFB in California were terminated. It is not considered a separate abort mode as it is available only for a few mission profiles and for a short time during the climb to orbit. This option replaced certain bailout options providing an opportunity to return the Orbiter towards land rather than abandon it in level flight to narrow the chances of impact in populated areas.

The East Coast abort landing sites are (northwards from the Cape) Myrtle Beach, South Carolina; Cherry Point Marine Corps Air Station, North Carolina; Oceana, Virginia; Dover Air Force Base, Delaware; Falmouth, Massachusetts; Portsmouth, New Hampshire; Halifax, Nova Scotia, Canada; Stephenville, Newfoundland, Canada; Goose Bay, Newfoundland, Canada; Gander, Newfoundland, Canada; and St. Johns, Newfoundland, Canada.

Abort once around

When the performance of the vehicle has been lost to such an extent where nominal orbit insertion is impossible or there is insufficient propellant in the OMS to conduct the thrusting maneuver to place the vehicle in a safe orbit and complete the de-orbit maneuvers, or when a major systems failure necessitates a land landing as quick as possible. Once the AOA mode is selected an initial thrusting maneuver is completed by the OMS which adjusts the orbit at SSME cut-off enabling a second OMS thrust sequence to be possible to de-orbit the Orbiter and perform a nominal entry and landing, with slightly less energy at one of the three primary AOA sites in the continental U.S.A., some 90 to 105 minutes after launch. These sites are White Sands, New Mexico; Edwards AFB, California; or Kennedy Space Center, Florida.

Abort to orbit

When full performance has been lost and it's impossible to achieve the planned orbit, then this mode is selected to place the vehicle in a safe orbit, using the additional thrust of the remaining engines and added boost from the OMS engines. By placing the Orbiter in a safe orbit an alternative mission can be accomplished, the vehicle can climb, depending on available propellants, towards a planned orbit or a return to Earth may be chosen after evaluation of the overall situation and the status and limits of onboard consumables, which would result in a full-term or minimum mission (54 hour) duration flight depending on the circumstances at the time. Re-entry is nominal. Only one mission has fallen under this category—STS-51F (Mission 19) in July 1985.

Contingency abort

These are aborts where the loss of one or more main engine or serious failures in other systems result in a decision to maintain the integrity of the Orbiter for an in-flight crew escape if a landing cannot be made at a suitable landing site. If only one engine is lost the vehicle and crew would probably be recovered normally; if more than one engine is lost then, depending on the timing, a safe landing may still be possible; but if all three engines fail then the Orbiter would have to be ditched and in-flight escape of the crew prior to ditching will be required (see Chapter 8).

SOVIET BURAN LAUNCH ESCAPE

Like the American Space Shuttle system the Soviet Buran shuttle had a long gestation in designs and plans for winged reusable spacecraft going back decades before it flew. Though only one mission was actually completed, which was unmanned, on November 15, 1988, the infrastructure and planning was for far more missions carrying crews to the new space stations, Sadly, the funding, support and infrastructure for the program diminished and the hopes and aspirations of those who worked on the program were never realized, the hardware associated with the program left in many cases to deteriorate and the files stored away—a story untold until recent years. An excellent history of the Buran system was authored by Bart Hendrickx and Bert Vis in this series in 2007 and is highly recommended for readers wanting to review the complete story of Buran and its launcher Energiya (Hendrickx and Vis, 2007). Like its U.S. counterpart there were plans and provisions for crew rescue and escape in the event of a launch abort situation and these are summarized below.

Origins of Buran rescue

Studies of reusable winged aircraft originated with Fredrikh Tsander in the 1920s and continued with studies of rocket-propelled aircraft and missiles during the next 30 years. Interestingly, one of these early designs in the 1950s, also called Buran, was a missile (M-40) designed to carry a 3.4-ton hydrogen bomb. It was designed by the aviation design bureau OKB-23 under the leadership of Vladimir Myasishchev. Improvements were planned to the design—at one point even adding a small crew cabin from which the lone pilot would eject before the missile impacted the ground. This was an early plan to investigate the extreme physical and psychological stress of rocket-powered flight. Buran was canceled in November 1957 prior to reaching the flight program.

Early studies for human flight beyond the atmosphere in the Soviet Union featured not only a variety of landing techniques, but also ejectable crew systems for separate parachute recovery, something that was adopted for the Vostok Program. This system of pilot ejection featured in several designs for "spaceplanes" in the 1960s, such as the VKA-23 (aerospace apparatus from design bureau OKB-23). In an emergency during launch the pilot could eject at altitudes up to 11 km. If the problem occurred higher than that, then the whole vehicle would have to be separated from the launch vehicle and—prior to landing automatically on skids—the pilot would eject and land separately, either in an emergency landing or a nominal one.

Spiral, designed by OKB-155 under Artyom Mikoyan (famous for the MiG fighter aircraft family), was an air-launched spaceplane primarily for military objectives, similar to that of the USAF's Dyna Soar X-20. In the event of a launch emergency the pilot could separate the pressurized capsule/crew compartment from the spaceplane and perform a landing by parachute. In this design the provision for an escape from orbit was also considered, and the small capsule featured its own heat shield and de-orbiting engine which could be used to abandon the rest of the vehicle in orbit should the need arise for a contingency or emergency landing.

The military capabilities of the U.S. Shuttle was the catalyst that drove the Soviets to pursue a design of their own, but the similarities between the Soviet Buran and the American Shuttle were remarkable; in addressing this Hendrickx and Vis (2007) observed:

> "unsure of what *exactly* the threat was, they had little choice but to stick closely to the American design, as the mission was to be similar, and to make sure they would be able to respond to whatever strategic missions the [U.S.] Shuttle would eventually perform. Buran was not built out of some fundamental need in the Soviet space program, but was an answer to *potential* military and other applications of the [U.S.] Space Shuttle. This would eventually become the root cause of its downfall in the early 1990s" (pp. 82–85).

The authors also made it clear that, although the Buran configuration closely resembled the American design and technology, the Soviet designers had to overcome many hurdles themselves in order to make Buran fly. To complete its only flight, unmanned, was a compliment to their skills and talent, as well as their dedication, in overcoming these hurdles. Although Buran looked like the Shuttle and was designed to complete similar missions, it was a different vehicle. Nevertheless, its flight profile and mission objectives were similar, and of course its provision for crew safety had to take into account the same distinctive flight profile and hardware limitations as did the American system.

Launch emergency situation

The significant difference in the use of liquid-fueled boosters rather than the Solid Rocket Boosters used on the American Shuttle afforded Soviet cosmonauts a better range of rescue options than their American counterparts had. They, of course, were able to fly the vehicle into orbit. As the 1986 Challenger accident tragically demonstrated, the loss of an SRB in the first two minutes of a Shuttle launch will almost certainly result in the loss of vehicle, and most probably the crew. Had one of the four strap-on liquid boosters failed on a manned Buran launch then it would not necessarily have led to a similar disastrous conclusion. Escape while the vehicle was on the pads has already been covered in Chapter 4, but there were other options available to the crew, as with the American Shuttle ride from pad to orbit.

Ejection option

For the planned manned test flights of the Buran, where two to four cosmonauts would crew the vehicle (similar to the U.S. orbital flight tests), ejection seats would be provided and the cosmonauts would wear a full-pressure rescue suit. Studies into ejectable cabins and crew compartments during the spaceplane studies pointed to the development of personnel rescue systems, at least for the early manned test flights to reduce the cost and complication of incorporating separate ejectable crew compartments into the vehicle, which also had a severe mass penalty.

In the four-person tests, the configuration of the mid-deck would have to be altered to enable two ejection seat hatches to be placed between the forward Reaction Control System and in front of the forward flight-deck window, necessitating relocation of equipment normally at the front of the mid-deck to the rear. Though four seats could be installed on the flight-deck, operationally on the test flight the two pilots would ride on the flight-deck, two other cosmonauts would ride to and from orbit in the mid-deck forward area, but all four would have the option of ejection during a launch (or landing) emergency situation. This was something not afforded mission specialists on the Shuttle from STS-5 onwards.

The seats used were modified K-36 seats constructed by the Zvezda design bureau which have become standard issue of the majority of Soviet and Russian high-performance military aircraft. Seats of this type have been produced and tested in their thousands and have saved literally hundreds of pilots in real emergency ejection situations. Their performance record therefore was excellent. The difference for Buran was that their need was mainly for launch and not so much during landing. Therefore, consideration had to be taken to provide enough velocity to allow the cosmonaut to clear not only launch pad facilities, but also a potentially exploding launch vehicle. The modified seat designed for Buran was designated the K-36RB (or K-36M11F35) and could propel a cosmonaut 300 m in altitude clearing the 145 m high rotating servicing structure, by means of a solid-propellant rocket attached to the seat's structure. A pad ejection would see the pilot landing 500 m from the pad about 10 seconds after initiating the system. Another benefit of the seat (being modular in design) was that additional items could be added as required, or removed if not needed for the vehicle it was designed to be used in. Installed beneath the seat was a stabilization system with a drag parachute that was used for ejections up to 1 km altitude. Twin booms with stabilizing parachutes would be deployed to stabilize the seat and pilot, then separated towards the apex of the seat's trajectory.

Operational limits on the seats were to be up to 30 km (Mach 3.0) and 35 km (Mach 3.5) which on a nominal launch would have seen a ground elapsed time of 100 seconds. The limit was determined by the operational limits of the pressure suits worn. The seats could also be used for landing in the region below Mach 3.0 and down to wheel-stop on the runway. These types of seats were also mounted in the atmospherics test vehicle (BTS-002) used in similar approach and landing tests that the U.S. Shuttle Enterprise conducted in 1977. The ejection systems could by initiated by the crew, by command from the ground or by onboard automatic systems.

Testing of the seats was accomplished by a variety of means such as ground testing using ground-based mock-ups of the Buran crew compartment and manikins dressed in pressure suits. These were static or sled-driven tests similar to those carried out on the U.S. Shuttle ejection seats. Airborne tests with manikins were completed in the aft cockpit of a converted two-seater MiG25RU training aircraft.

Higher altitude tests were completed in a novel way by placing an experimental variety (K-36M-ESO) as a piggyback payload on five unmanned Progress resupply missions to the Mir Space Station (Hall and Shayler, 2003). These were launched between September 1988 and May 1990 and featured the seat installed in an ejectable compartment placed, instead of the inert launch escape engine, on top of the payload shroud. On Progress launches a standard Soyuz launch shroud and escape tower is used but the escape engine lacks the powder charge, as an escape option is not required for an unmanned launch. The seats were ejected during the ascent at altitudes between 35 km and 40 km and speeds of between Mach 3.2 and Mach 4.1.

Such seats were not used in the Buran program due to its cancellation in the 1990s, but the idea and design were employed in studies of the proposed Soyuz replacement vehicle called Kliper, where Buran seats would eject the crew during test flights; though, for planned operational commercial flights with up to six crew-members there would be no option for ejection from Kliper, other modes would have to be developed. Equally, as the four-man test flights on Buran demonstrated, should a crew comprise more than four, ejection seats would have to be removed, which also would save launch mass, and escape would have to be afforded by flying one of the abort profiles or rescue from orbit (see Chapter 7).

Buran Strizh full-pressure rescue suit

The full-pressure suit to be used by the first Buran cosmonauts was called *Strizh* (Russian for "swift") and designed to provide maximum comfort and allow the wearer to still perform their duties while unpressurized and to support work in an emergency situation with the suit pressurized to 440 hPA. It could be easily donned and doffed unassisted in a weightless condition and provided good compatibility with the ejection seat, especially in its connections to the parachute harness, restraint harness and alignment with the headrest. Development work commenced in 1977 based on Soyuz Sokol suits, and in 1981 the garment received its formal name. There were to be few standard sizes but with the advantage of personal adjustment to fit a broader range of wearer, thus reducing down the number to be produced.

In its development, work focused on the restraint system attached to the garment and how this integrated into the seat structure to enable the wearer to still function while restrained and under pressure. Initially, six functional mock-ups were constructed and a further four more suits were completed in 1981. The development continued to search for the most effective method of getting into the suit (this was found to be similar to the Soyuz suit with a frontal opening). Tests continued in the laboratory to simulate parachute restraint effects, thermal tests for use in winter conditions, water emersion tests, and helmet ventilation and visor-misting tests. There was also an extensive program of wind tunnel testing of the suit both with and without the ejector seat. From 1982, the laboratory test suit program began in earnest. This included significant laboratory and in-house testing as well as standalone and integrated testing with the life support systems. There were the standard fit and function tests, mobility tests and a program of parachute descents some of which were at extremely low temperatures to demonstrate the conditions and protection needed at these high altitudes and severe temperatures. By 1989, laboratory testing had been completed and flight-testing in the Illyushin IL-96 flying laboratory commenced. Suit development was completed by 1991.

The requirement for these garments was protection of the wearer for ejection up to 30 km altitudes and speeds of Mach 3. As a result of high aerodynamic loads and the thermal conditions, special fabrics had to be incorporated into the construction and thermal protection panels added. In case of ejection the cosmonaut could have pressed the nape of his neck against the headrest of the seat during periods of extreme pressure during ejection.

During the unmanned Buran mission in November 1988 two suits were flown with manikins taking the place of the normal human wearer to further qualify the suit and portable life support system for operational use. In the back-up mode the suit could operate at 270 hPa and could be worn in a pressurized cabin with a ventilating airflow supporting two suits for up to 24 hours. In an emergency and closed-loop configuration, the operation could sustain 12 hours of constant use. Overall mass was 198 kg and in production there were 27 test and training models fabricated and four flight models (Abramov and Skoog, 2003, pp. 211–221 and 352).

Buran launch abort options

There were three launch abort profiles available for manned Buran launches (Hendrickx and Vis, 2007, pp. 152–153).

Emergency separation (Russian acronym EO). This was an option available if a problem occurred or all engines had to be shut down above the higher limit to use the ejection seats. In this case, in theory Buran would have been quickly separated from Energiya using four small solid-fuel motors. It was then down to the crew to stabilize the Orbiter as soon as they could in order to make an emergency landing on a suitable runway downrange. After first dumping excess propellant, the ejection option was still available to the crew should the problem in stabilizing the vehicle continue, or a suitable landing site not fall within range.

There were other potential problems with this type of abort. First, providing the suitable ground support and runway infrastructure would have proven very costly had the program continued. It was also determined that it was probable that the Buran would hit the core stage or not perform a clean separation, something that NASA had concluded in studies of a similar abort profile for the Shuttle. Aerodynamic pressure build-up and a late separation from the attachment point could theoretically pitch the vehicle up and result in aerodynamic break-up.

The NASA model suggested a "fast separation" option in which the Orbiter could separate from the External Tank just three seconds after an SRB failure. However, as the Challenger disaster demonstrated there were some failures using an SRB where even three seconds was not enough time to react to a pending disaster. It is possible that such an abort option was not a suitable scenario for Buran even with liquid boosters.

Return maneuver (MV). This was similar to NASA's return-to-launch-site abort for the Shuttle. It was used for a single-engine failure of one of the four strap-on boosters or the core stage. Should one of the strap-on engines fail, the opposite number would have been terminated to balance the trajectory, followed by dumping excess fuel to lighten the ascending mass and ensure characteristics at separation would be close to nominal ones and not add to the problems already being encountered. To prevent explosive occurrences at altitude, the kerosene aboard the boosters would not have been dumped.

The return profile was similar to that of the Shuttle during RTLS aborts in that Buran would have flown farther downrange to expend more propellant, then would perform a 180° pitch-around to point towards the launch site and the vehicle would be orientated to head-up attitude prior to separating from the core rocket stage. As a nominal approach and glide landing was flown excess propellant from the orbital maneuvering system would be used by additional firing for additional velocity or braking and then dumped, thus moving forward the center of gravity for a safer landing.

The limit of "negative return" preventing this maneuver occurred at 2 min 5/10 s into the flight if the failure occurred on a strap-on or 3 min 00/10 s for a failure in the core stage.

Single-orbit trajectory (OT). The was the Buran version of an abort-to-orbit profile. After 3 min 10 s into the ascent, if one of the core engines failed Buran could still limp into orbit. This depended like the Shuttle on when this occurred and how much propellant was left to burn through other engines. By using the orbital maneuvering engines (Russian DOM), or the remaining engines, or a combination of both, a low or temporary orbit could be reached to determine the best options to continue or bring the crew home. Should the option be for a short orbit, the DOM engines would be used to insert the vehicle into a single-orbit trajectory then punch the vehicle out of orbit for landing, in much the same way as the Shuttle's abort-once-around option.

HERMES, EUROPE'S SHUTTLE

In addition to the American Space Shuttle and the Soviet Buran vehicles there was one more "shuttle" (this time developed in Europe) that was proposed for regular manned access to near-Earth orbital space operation. This was the French-initiated Hermes spaceplane.

Hermes was the result of early design studies conducted at the French National Space Centre (CNES) in the mid-1970s. Despite a significant record of participation and cooperating in space exploration by countries across Europe there was no agency to handle the varied national and international programs, nor a firm desire for manned space activity. There had been interest in participating in the U.S. Shuttle program with a pressurized laboratory called Spacelab and discussions with the Soviets to fly European astronauts on Space Station missions, but independent launcher capability and manned spacecraft was not a primary goal in Europe for many years.

Origins of a spaceplane

A coordinated European Space Agency was formed in 1974, but there remained individual national programs that demonstrated the interest in manned space projects across Europe. The French national space agency CNES initiated studies of a manned transportation system in 1976, and termed the project Hermes. Industrial work continued in study contrasts with Aerospatiale and Dassutt Aviation. At the 1985 ESA ministerial conference in Rome, France announced its intention to pursue the Hermes spaceplane concept. Two years later in 1987 a ministerial meeting in the Hague reached the agreement to "Europeanize" Hermes as one of three steps to independent European manned spaceflight autonomy. Ariane 5 would be the high-performance launch vehicle that could support low Earth orbit–manned missions. Columbus was a versatile laboratory that could support unmanned polar orbit operations, a permanent laboratory attached to the then U.S.-led Space Station Freedom (later the International Space Station), and a free-flying human-tended variant, while Hermes would be the cargo and crew transport vehicle with docking and EVA capability. This was a small Spacelab just 18 m long, with an 11 m wingspan and a crew of two pilots and up to four "scientists" on 400 km 90-day missions docked to a space station or 30 days of independent flight. Irish author Brian Harvey has detailed the development and background of Hermes in his book on the development of a European space program in this series, which is suggested for further reading (Harvey, 2003, pp. 294–304); here we focus on the rescue capability of the proposed spaceplane.

Escape for euronauts

As with the Shuttle and Buran, a euronaut had to be provided with a system of rescue from the vehicle during ascent, or modes of recovery in the event of a major systems failure. Hermes would be launched by Ariane 5 out of Kourou in French Guiana (South America) on trajectories over the Atlantic Ocean.

In early studies the Hermes design focused on a range of assumptions that defined the explosion risk as being limited to the boost phase of the mission profile. If a launcher failed, then separation of the spaceplane was considered an option, and therefore its structural integrity had to be sufficient to withstand the aerodynamic and thermal loads placed on it in an emergency situation.

The "separation" of Hermes during the boosted phase was to be by four boosters (2.1 tonnes) that provided an acceleration of about $8g$ for 5 seconds. This, it was determined, would be sufficient to propel the Hermes spaceplane beyond the Ariane launch vehicle, but did not ensure survival as a result of a shock wave from a launch vehicle explosion. After solid rocket burnout the booster could have been used in stages to compensate for the mass penalty. For the rest of the boosted phase it was deemed safe enough to turn off the cryogenic engines then safely separate the Hermes

European Hermes. Courtesy ESA.

with or without its own propulsion system to effect a suitable landing, or in extreme cases ditching the vehicle by parachute into the ocean.

For orbital flight then, it was Hermes itself that was determined safe for the crew and for re-entry in the event of mishap. Installation of a lightweight personal escape system (ejector seat) was provided for low-level escape, and parachute recovery with life support capability and preferred ocean recovery or land landing. The setbacks in this system were doubts about Hermes' capability to separate safely in the first place, provision of enough volume to each crewmember for ejection seat capability and for leaving the vehicle safely and efficiently, escape from the shock wave and the debris being ejected from such an explosion, decompression and aerodynamic forces, excess temperatures and the rigors of parachute descent into wilderness, Arctic or water conditions.

By 1985 it was determined that the vehicle should have ejection seats for each crewmember and that landing could be made at the launch site or the French AFB at Istres, near Marseilles in southern France. With a maneuvering capability of 2,500 km during glide-back to Earth this afforded a potential for RTLS-type aborts to French Guiana, TAL to Istres or AOA back to French Guiana.

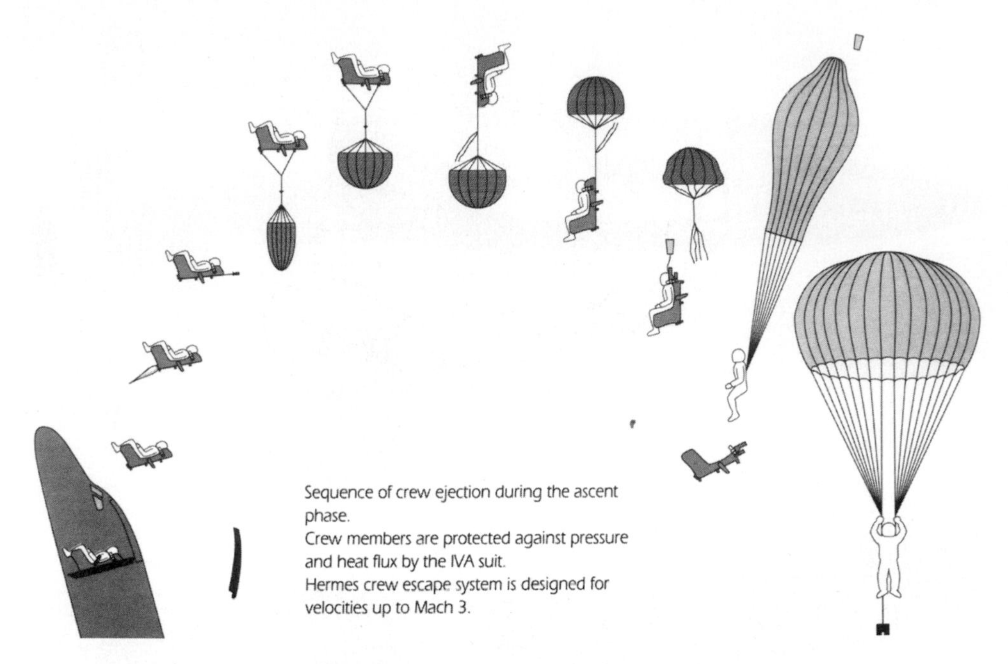

Ejection sequence from Hermes. Courtesy ESA.

Following the 1986 Challenger disaster it was decided to incorporate a completely separate crew compartment to rescue the entire crew in the event of a Challenger-type disaster. This added greatly to the launch weight and complexity of the design.

Two options were evaluated. First, the complete separation of the front part of Hermes to place it outside the discernible atmosphere allowing for a re-entry of the lifting body shape and suitable flotation gear after an ocean landing. The other method was an Apollo Command Module–type ballistic capsule fitting inside the Hermes payload area and available for emergency use. This, however, proved to be difficult to incorporate and, though the first option was prepared, both were costly in the form of mass added to the overall weight of the vehicle, and some considered perhaps over-designed with respect to foreseeable risks (Colrat *et al.*, 1988, pp. 269–274).

In March 1987 it was determined that the current provisions for crew safety were not sufficient enough to withstand an exploding Ariane 5. It was determined that the way forward was by an ejectable cabin but this would only be considered in the extreme situation phase. These studies evolved the closed cargo bay configuration of the Hermes towards the final design configuration. It was therefore necessary to amend the idea of an ejectable nose section, making it simpler to include just the crew compartment. Studies were begun on comparable ejectable cabins such as the U.S. B-1 or F-111 bombers. In the end the second solution of a lighter ejectable cabin as in the military aircraft—not the whole nose section—was the preferred option.

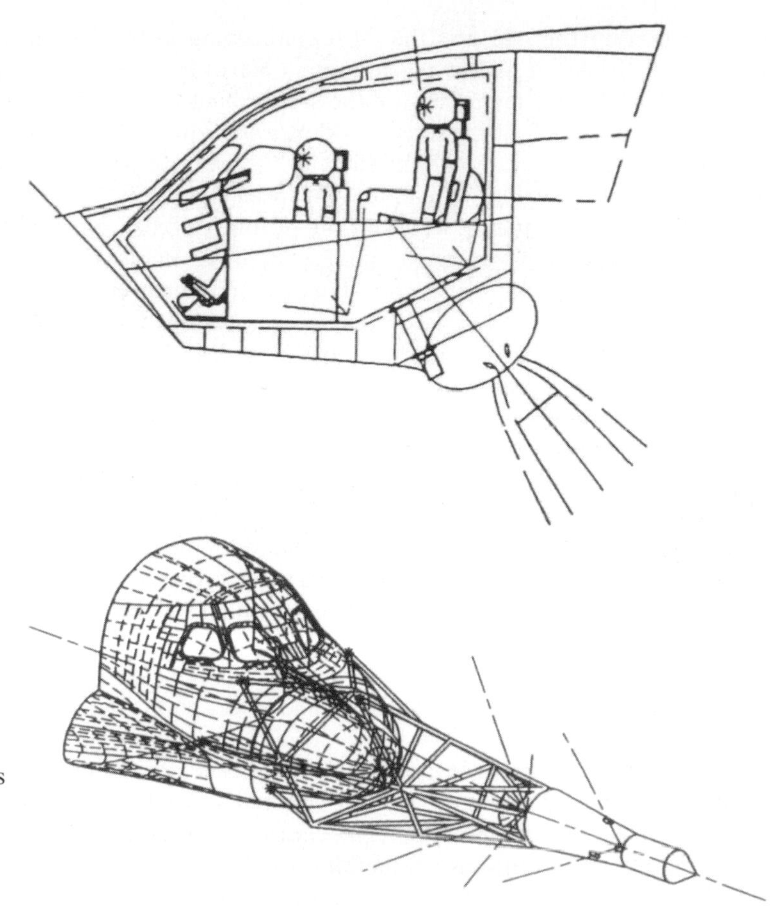

Capsule ejection system for Hermes (top). Escape tower suggestion for Hermes (bottom).

This meant that most of the unwanted mass was left with the main vehicle retaining the lighter configuration containing the crew, controls and associated support and recovery subsystems.

The Crew Escape Module (CEM) of Hermes was proposed as a 3,365 kg mass pressurized compartment where the crew would be stationed during the ascent and entry phase of the mission. Constructed of titanium the floor was specially designed to absorb impact loads, and under the floor a "dampening" device was to feature crushable structures or dampening bags. Recovery parachutes and stabilization debris (winglets and air brake flaps) were located in the rear, front and side areas. In the lower rear was the escape motor that pushed the compartment away from the vehicle with a so-called "champagne cork effect".

An overhead hatch provided access to and from the crew compartment either on the launch pad or during emergency operations, though it was impossible to conduct EVA or docking and transfer operations through it. These would be performed through other hatches in the rear of the vehicle.

Ejection on the pad was the most challenging in the design, but requirements meant that the cabin had to reach an area 600 m from the pad to prevent collision with the launch vehicle and ensure the safety of the crew. The nominal sequence of events for the ejection of the rescue module were firing of the booster rocket; severing the pyrotechnic and electrical umbilical to the main vehicle; separation of the cabin and ejection of the cabin with an inertial speed slope of 45°, the boosted phase with the angle of attack ending at about 40°; cabin stabilization and drogue incurring speed reduction; deployment of pilot parachute; deployment of the three main parachutes at 7 meters per second; dampened landing; ejection of parachutes; and recovery of crew.

There were studies of various abort altitudes of the CEM and these included maximum dynamic pressures; the 50 km altitude as the Ariane 5 engine burns out; and abort to orbit or immediate suborbital trajectory to emergency recovery. As well as conducting wind tunnel tests on the model, configuration studies were still conducted into the feasibility of an ejectable nose or even installing an escape tower system on the nose of Hermes to pull the crew compartment away from an impending explosion of the launcher. At the time, the development of the CEM was planned for 1988–1990, with a full test program operating between 1991 to 1995 to qualify the tests prior to the first manned flight. Crew training for the rescue systems was to take place at Trondheim, Norway.

However, by 1990 Hermes had grown too heavy and too expensive and a major redesign was required, making some elements non-reusable and significantly reducing the mass and cost. In late 1990 it was decided to replace the complicated and heavy crew ejection module with individual ejection seats and reduce the crew of five down to three (commander, pilot, mission specialist) similar to the core Orbiter crew operated by NASA (commander, pilot, mission specialist/flight engineer) used for ascent and entry operations on the Shuttle. The ejection seats were to be based on those planned for use on Buran.

Over the next two years the program suffered further indecisions and delays, gradually pushing the maiden launch well into the new millennium. Funding and technical issues dogged the program, and discussions about the future of the program which varied from considering cooperation with the Russians to abandoning the program altogether continued to plague it. During 1992 suggestions were made for ESA to study an assured crew return vehicle for the American Freedom Space Station program and focus more on the Columbia science laboratory for the Space Station, perhaps considering Hermes as the space station rescue vehicle. However, the Hermes spaceplane program slipped quietly into the background of other ESA commitments and was essentially abandoned in the mid-1990s, though the exact termination date is unclear. Hermes would not fly; Europeans would need to rely, at least for the foreseeable future, on launch systems provided by the United States and Russia. A decade after Hermes there was talk of European cooperation on a replacement of the Soyuz vehicle, and a new venture for European access to manned spaceflight capability was suggested, but this time looking at ballistic-type capsules that were partially reusable but certainly not along the lines of the spaceplane or Shuttle concept.

REFERENCES

Isaak P. Abramov and Å. Ingemar Skoog (2003). *Russian Spacesuits*. Springer/Praxis, Chichester, U.K.

J. Colrat, H. P. Nguyen and H. Hirsch (1988). Hermes Escape System. Paper presented at the *Int. Symp. on Europe in Space: Manned Space Systems, Strasbourg, France, April 25–29, 1988*, ESA-SP-277. ESA, Noordwijk, The Netherlands

Rob Godwin (ed.) (2000). *X-15 NASA Mission Reports*. Apogee Books, Burlington, Ontario.

Rex Hall and David Shayler (2001). *The Rocket Men*. Springer/Praxis, Chichester, U.K.

Rex Hall and David Shayler (2003). *Soyuz: A Universal Spacecraft*. Springer/Praxis, Chichester, U.K.

Brian Harvey (2003). *Europe's Space Programme: To Ariane and Beyond*. Springer/Praxis, Chichester, U.K.

Bart Hendrickx and Bert Vis (2007). *Energiya-Buran: The Soviet Space Shuttle*. Springer/Praxis, Chichester, U.K.

Robert S. Houston (1959). Pressurization and escape. *Development of the X-15 Research Aircraft 1954–1959*, 59WC-2184. Wright Air Development Center, Patterson AFB, OH.

J. D. Hunley (ed.) (1999). *Toward Mach 2: The Douglas D-55* Program*, NASA-SP-4222. NASA, Washington, D.C.

Dennis R. Jenkins (2001). *Space Shuttle: The History of the National Space Transportation System*, Third Edition, *The First 100 Missions*. Midland Publishing, Hinckley, U.K.

Jay Miller (2001). *The X-Planes: X-1 to X-45*. Midland Publishing, Hinckley, U.K.

MOL and Gemini B Office (various dates). *Crew Safety Briefing Objectives*. Manned Orbiting Laboratory and Gemini B Office, Los Angeles, CA.

NASA (2008). *NASA Ejection Seats: The X-15*. Available at The Ejection website *http://www.ejectionsite.com/x15seat.htm* (last accessed July 7, 2008).

Hilary A. Ray Jr. and Frederick T. Burns (1967). *Development and Qualification of Gemini Escape System*, NASA-TN-D-4031, June 1967. NASA Manned Spacecraft Center, Houston, TX.

David Shayler (2001). *Gemini: Steps to the Moon*. Springer/Praxis, Chichester, U.K.

Asif Siddiqi (2000). *Challenge to Apollo*, NASA-SP-2000-4408. NASA, Washington, D.C.

Space Shuttle Transportation System, Press Information, January. Rockwell International, Downey, CA.

Kenneth S. Thomas and Harold J. McMann (2006). *US Space Suits*. Springer/Praxis, Chichester, U.K.

7

Away from Earth

The long training program is behind them. After hours in simulations, the preparation to make contingency, emergency and launch escape procedures has become second-nature. Hopefully, all has gone well and the crew is finally in Earth orbit and looking forward to their planned mission. After all the expectations, excitement and challenges of finally leaving the launch pad and making a rapid and at times violent ride into space in less than 10 minutes, the view out of the window can be breathtaking. Orbiting the world in 90 minutes 16 times a day certainly makes the effort even sweeter. However, as every crewmember is aware just making it into space is by no means a guaranteed safe ride to the end of the mission. Sometimes a problem that developed on the way into orbit means the stay is very short and contingency or emergency procedures have to be followed to get down as quickly as possible. Perhaps the stay in orbit is amended to a contingency or shortened, abbreviated visit to space and a challenging ride home.

At any point a mission can go seriously wrong and the lives of the crew and the safety of the spacecraft are threatened. This ever-present danger has to be addressed as best possible by the mission planners, spacecraft designers and flight controllers as well as the flight crew themselves. Training and procedures, facilities and alternatives have to be envisaged for use at any time in a mission. In these cases the opportunity of securing the chance to allow a safe return to Earth is perhaps more difficult than actually getting off the planet in the first place. Providing for the unknown is sometimes a challenge too far. However, it does not prevent alternative mission plans, contingency procedures or methods of escape being available, if the crew have the time to implement them, of course.

THE APOLLO ERA

In the early days of human spaceflight, the missions were of short duration and planned with the contingency of returning the crew to Earth if a serious situation occurred by providing redundancy in the de-orbit capability, constant tracking and data evaluation of critical systems and providing, as was the case for Vostok, an orbit that resulted in a natural decay after ten days if the de-orbit burn failed. In Mercury and Gemini, missions progressed in the knowledge that decisions to proceed with the mission or abort it were continuously under consideration. These go/no-go decisions have been a feature of manned spaceflight since the first humans ventured beyond the atmosphere. Under a nominal situation the decision to proceed with the next step or stage of the flight is made in conjunction with the flight crew, the ground controllers, and support teams evaluating real-time data.

Apollo alternative missions

When the Americans moved on to Apollo, the complexity and range planned for the mission offered the opportunity to plan for alternative missions should the primary mission plan become unobtainable. These would provide the maximum accomplishments of planned objectives while adhering to mission constraints that relate to ground rules, the safety of the crew and the conditions of the trajectory being flown. These offered the chance to achieve at least some of the objectives of the mission in Earth orbit out to the Moon but still wherever possible retain the safety of the crew and integrity of the spacecraft. Of course, serious situations would immediately terminate the mission as soon as possible and try to return the crew to Earth. This situation was of course employed by the Apollo 13 crew after the in-flight explosion in the Service Module wrecked the chances of landing on the Moon, and seriously threatened their ability to return to Earth.

Apollo Earth orbital

For the first maiden flight of the Apollo CM in Earth orbit (Apollo 7, October 1968) there were several alternative mission plans available had the primary planned 11-day mission been unable to proceed as scheduled. These were categorized as one-day, two-day or three-day mission plans and were all dependent on whether the S-IVB third stage was still attached, which spacecraft systems were affected, the amount and performance capability of the Service Propulsion System (SPS) and the condition of the crew (NASA, 2000a).

One-day. There were four alternatives. The first two designated 1a and 1b would have seen a recovery in the mid-Pacific after only six orbits, the other two (1c and 1d) would terminate in the mid-Atlantic towards the end of the first flight day. In this scenario the Service Propulsion System would only have been used for the de-orbit burn. Unless the SPS was required to place the CSM in Earth orbit, profiles 1b and 1d would have followed the first day's activities of the primary mission flight plan.

Two-day. Alternative Mission 2 featured three variants. None allowed all the test objectives to be accomplished. However, a rendezvous and two additional firings of the SPS (one for de-orbit) or no rendezvous and four maneuvering burns (one for

MISSION PRIORITY	1	5	6	7	11	12	13	14	18	38	
DTO#	7.19	3.15	3.14	1.13	2.5	20.13	1.10	20.8	2.6	57.21	
ALTERNATE MISSION	RADIATOR TEST	SPS PERFOR-MANCE	SPS MINIMUM IMPULSE	GNCS ΔV CONTROL	SCS ΔV CONTROL	CSM-ACTIVE RENDEZVOUS	SEXTANT TRACKING	TRANSPOSI-TION AND DOCKING	GNCS/MTVC ΔV TAKE OVER	SLA DEPLOYMENT	COMMENTS
1A				◑	●						
1B				◑				●		●	
1C	○			◑	●						
1D	○			◑				●		●	
2A	●			●	●	●	●	●		●	
2B	●	●	◑	●	●		○	○	●		S-IVB RELATED TEST OBJECTIVES SATISFIED IF COI=0
2C	●		●	●	●				○		MANUAL TAKEOVER BURN MUST BE AT LEAST 35 SECONDS LONG
3A	●	●	◑	●	●	●	●	●	●	●	
3B	●	●	●	●	●				●		
3C	●		●	●	●				○		MANUAL TAKE OVER BURN MUST BE AT LEAST 35 SECONDS LONG
ALTERNATE RENDEZVOUS	●	●	●	●	●	●	●	●	●	●	

DETAILED TEST OBJECTIVE ACCOMPLISHMENT FOR THE APOLLO 7 ALTERNATE MISSIONS

● FULFILLED

◑ PARTIALLY FULFILLED

○ POSSIBLY FULFILLED

Apollo (7) alternative missions (Earth orbit).

de-orbit) could be performed. Option 2a would have been taken if the S-IVB was made safe and rendezvous would be completed as the primary objective, with recovery in the Atlantic on the 32nd orbit. Alternative 2b tested the ability of the Stabilization and Control System for a contingency orbit insertion (COI), with recovery in the West Atlantic on the 33rd orbit. Option 2c assumed the S-IVB was not available and that a COI burn was sufficient to test the stabilization control system, again recovery in the West Atlantic on the 33rd orbit would be completed.

Three-day. Again a three-option category that depended on the availability of the S-IVB and the ability of testing the SPS. The primary objectives of these missions would be testing of the SPS and the Stabilization and Control System. De-orbit would be on orbit 46 or 47 with recovery in the West Atlantic.

Alternative rendezvous plan. This was the only alternative rendezvous plan considered if there was a one-day delay, others delays would affect the adequacy of ground tracking of the flight or the uncertainty of drag on the S-IVB. Mission

Mission Control—round-the-clock support on the ground.

tests would have continued as planned. Alternative missions longer than three days would have been planned in real time, had the situation required it. As it was Apollo 7 performed its mission with "101 percent" success, qualifying the Apollo CSM for manned operations.

Apollo 9, testing the LM in Earth orbit

There were no fewer than seven alternative missions for the Apollo 9 (March 1969) tests of the manned LM in Earth orbit (NASA, 1999a).

Alternative Mission A. If a contingency orbit insertion became necessary (essentially an abort to Earth orbit) and a CSM-only mission had to be performed, the primary objective would be the SPS test burn program if the LM was unable to be extracted from the S-IVB or its descent stage was deemed unsafe.

Alternative Mission B. If the failure was in the SPS or affected the lifetime duration of the CSM, such as an electrical problem, and the crew became dependent on the LM a real-time evaluation of the mission would have to be planned. Crew activities would have been rescheduled accordingly to try to achieve the maximum

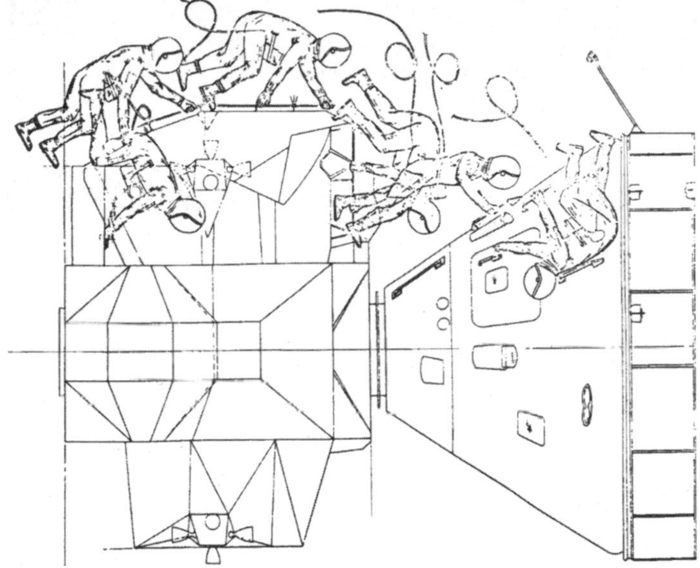

Apollo contingency EVA (top). Sequence of Apollo contingency EVA procedure (bottom).

number of mission objectives dependent on the situation. Recovery in the primary landing area was to be achieved by using the RCS on the Service Module.

Alternative Mission C. This was related to the integrity of the descent stage of the LM. Mission rules would have dictated jettisoning the descent stage if the LM was still to be used in a docked configuration. A planned EVA could still have been accomplished if the ascent stage was available and safe to use.

Alternative Mission D. CSM lifetime or LM electrical problems could have seen a change to the flight plan to rearrange time-critical objectives and retargeting the mission to the primary recovery area as soon as possible after the identified objectives under this category had been accomplished. Real-time decisions would be an overriding factor in planning for this type of mission.

Alternative Missions E, F, and G. A range of systems failures on the spacecraft would have been considered and alternative missions would have been planned depending on the extent of the failure and when and where it occurred. If problems occurred in the undocked LM mode, then the CSM would have been the active spacecraft to "rescue" the LM and its crew. The EVA planned on Apollo 9 was to have simulated the contingency operation of transferring an LM crew to the CM by EVA if the internal transfer tunnel was not available. This was partially accomplished on Apollo 9 and demonstrated the theoretical EVA transfer of a crew. This was also demonstrated by the Soviet Soyuz 5 cosmonaut in an EVA transfer to Soyuz 4 in January 1969.

Lunar distance

Apollo 8, first humans to the Moon

There were several alternative missions planned for Apollo 8 (December 1968) and the scope of such options was highly dependent upon exactly when the need became apparent to switch to an alternative plan. These were (NASA, 2000b):

Alternative Mission 1. Early shutdown of the S-IVB prevented nominal insertion into Earth orbit and the use of the SPS was required to place Apollo 8 in Earth orbit. If possible the use of the SPS, dependent on the available fuel, would have seen a high-apogee (7,200 km) Earth orbit mission (one of the early studies for the original Apollo 3 mission which the crew trained for) over ten days.

Alternative Mission 2. The S-IVB achieved Earth orbit, but not the full burn to provide sufficient velocity for the translunar insertion burn. Again the SPS would have been used to achieve a 7,400 km apogee orbit for the CSM to complete two or three of these revolutions around Earth and then use the SPS to lower the orbit for the rest of the 10-day mission.

Alternative Mission 3. If the S-IVB TLI burn cut off early, then depending on the apogee the SPS would have been used to complete a mission similar to Alternative Mission 2 and complete program landmark sightings. If the apogee ranged from between 1,853 km and 46,250 km then a maneuver would have been performed by the SPS to shift a later perigee over a network station where the crew would have performed a de-boost burn that would lower the apogee to about 740 km where the mission would have continued. The third alternative, 3c, would have been used if the resulting apogee was between 46,250 km and 111,000 km. Here the CSM would remain in the established trajectory and subsequently make a direct entry, the fuel aboard the SM would probably not have been enough to lower the orbit to 740 km and still have sufficient for de-orbit. The plan was to place Apollo 8 in a semi-synchronous period of about 12 hours allowing for two opportunities for a perigee de-orbit burn over the Pacific and one over the Atlantic. The nominal duration could have been flown with a direct entry from a high-ellipse apogee orbit fulfilling partial entry test objectives planned from the lunar distance. Option 3d would have seen an early TLI cut-off, but one that would have produced an apogee greater than 111,000 km. This would have allowed for a free-return trajectory to complete a circumlunar flyby (without entering orbit) with the SPS providing the required additional burns, and completing a direct entry.

Fortunately, Apollo 8 did not require any of these alternatives and became the first human spacecraft to enter orbit around the Moon. The historic mission of Borman, Lovell and Anders over Christmas 1968 is a milestone in the story of manned space exploration and set the stage for the successful lunar landing mission three flights later.

Apollo 10, LM in lunar orbit

With Apollo 10 (May 1969), the second manned mission to the Moon, there were alternative missions inserted into the flight plan to be available at different stages into the mission. There was also a series of go/no-go decisions available to the controllers and the flight crew (NASA, 2000c). The CSM could have flown alone in Earth orbit or in a high-elliptical high-apogee deep-space trajectory. The CSM/LM combination would also have been able to provide a range of alternative missions in Earth orbit or at semi-synchronous orbit.

There remained the opportunity to operate a free return trajectory, with the CSM/LM combination making a flyby of the Moon before heading home, or completing a CSM-only mission in lunar orbit or lunar flyby and a number of alternatives using the descent stage engine, ascent stage engine or both depending on the extent of the problem, timing of the mission and safety of the crew. Also planned were alternative rendezvous modes to recover the LM ascent stage in the event of an off-nominal situation.

Apollo, humans on the Moon

As the lunar mission progressed, so did confidence and experience in the systems and procedure, though there was constantly the awareness of an off-nominal situation, and planning of alternative missions for the Apollo landing flight continued.

Apollo 11 (July 1969) was the first to fly the full Apollo mission profile and in doing so rewrote the pages of history, though there were provisions to overcome as many unplanned incidents as possible in order to achieve this goal (NASA, 1999b). These aborts were also available to later lunar landing crews, with amendments to the specific flight profile, objectives and hardware.

From Earth orbit, a "return-to-Earth abort" could be initiated to bring the crew home at the earliest opportunity. Following the translunar insertion maneuver there were a couple of opportunities to abort the burn and return the crew to Earth. The "ten-minute abort" was available in the remote possibility of an immediate return to Earth during the short period of the TLI burn. If this had been the case then the TLI maneuver would have been terminated early and the crew perform a retrograde SPS burn calculated onboard the spacecraft. This would have to be completed just ten minutes after the cut-off of the S-IVB to ensure a safe CM entry. The timing from the initiation of the abort to getting the crew back in theory would be dependent on the length of the TLI burn prior to cut-off and this could vary from 20 minutes to five hours. The use of the SPS would also have to be considered for course corrections to refine the entry conditions. This was an extreme emergency abort scenario, with crew survival being the uppermost consideration. As a result of the speed, this abort would be chosen, but the landing point could not be confirmed at the time of the abort and due to the huge number of variables and real-time situations prevailing, an accurate landing point position could not be determined. The crew would probably have to draw upon some of the contingency and wilderness training if the landing came outside the primary or secondary landing zones.

A 90-minute abort was possible with the crew checking off-nominal situations during the TLI burn. If, during this time frame, it became apparent that the abort and crew return to Earth was necessary the abort would be initiated about 90 minutes after the cut-off of TLI burn. This time however a pre-selected landing area in the Atlantic, Pacific or Indian Oceans could be targeted to an area called the "recovery line". Following an SPS retrograde burn and mid-course correction the nominal entry conditions would hopefully see a nominal CM entry profile.

Translunar coast. Normally, the three-day journey between the Earth and the Moon was termed the translunar coast. If an abort had been necessary during this time then the profile could have been similar to that of the 90-minute abort above. Information was sent to the crew about the length of the SPS burn and the attitude of the spacecraft. The timing of the abort burn during the mission and the trajectory resulting from it would have dictated the location of the landing area. The rotation of the Earth meant that the preferred landing in the mid-Pacific could only be accomplished once every 24 hours. For this reason landing in the Atlantic or Indian Ocean would have been considered because of three landing opportunities every 24 hours.

As the spacecraft travels farther from the Earth towards the point in space where the Moon would be in its orbit at the end of the translunar coast, the capability for a return to Earth changes. As soon as the influence of the Moon's gravity equals that of the Earth, the lunar sphere of influence increases making a circumlunar abort a better choice than a direct return-to-Earth abort.

LOI insertion. If the burn to place the spacecraft into lunar orbit had to be terminated early, a resulting abort procedure was available that fell into three modes depending on when the aborted burn occurred. Any of these abort modes would have seen the CSM return to a mid-Pacific recovery zone.

- Mode I would have been used if the SPS engine cut off at any point from the moment of ignition until 1.5 minutes into the burn. In this case a posigrade Descent Propulsion System burn of the LM descent stage engine would have been initiated about 2 hours after cut-off which would have placed the spacecraft back on a return-to-Earth trajectory.
- Mode II would have been used between LOI ignitions lasting 1.5 minutes and 3 minutes. This was a two-stage burn. Initially, the DPS would be used to ensure the lunar orbital period would be reduced and that the spacecraft combination would not be sent on a catastrophic impact with the lunar surface. A second burn would place the spacecraft on a return-to-Earth trajectory.
- Mode III would have been used from the 3-minute mark until nominal shutdown. After that the spacecraft would have been inserted in a nominal lunar orbit where future actions could be determined. If an abort was called during this time a couple of lunar orbits would be completed before a posigrade DPS burn would be initiated to place the spacecraft in the required return-to-Earth trajectory.

Lunar orbit options. Aborting operations in lunar orbit essentially meant performing the trans-Earth insertion burn earlier than planned. When this abort mode was called would depend on the timing in the lunar orbital operations flight plan. Should an LM abort become necessary during the descent, ascent or orbital phase of its mission, then the lander could make the necessary burns to rendezvous with the orbiting CSM. Had an LM been unable to complete the rendezvous, then the CSM could maneuver to rescue a stranded LM in orbit. A nominal Pacific landing would be the result of end-of-mission targeting in this situation.

LM-powered descent. On the Apollo Lunar Module the Primary Guidance and Navigation System (PGNS, pronounced "Pings") abort program, or the Abort Guidance System (AGS, or "Agges") would control any abort during descent to the lunar surface. The status of the DPS and the PGNS is paramount in this phase. If both are functioning correctly then a crewmember could initiate an abort by depressing the "Abort" button on the display console inside the LM. This "DPS Abort" would have continued under the control of PGNS until orbit insertion of the spacecraft, loss of the descent engine by DPS failure or depletion of the available propellant. Should the descent engine cut off resulting in a less than 9 m/s gain in velocity then the descent and ascent stages were separated by manual command and the Reaction Control System was used to make it to orbital insertion. If the velocity gained is in excess of 9 m/s the "Abort Stage" button is pressed which separates explosively the descent and ascent stages and ignites the ascent engine which would have been used to place the ascent stage in the required orbit. Should the DPS fail the abort would have been by means of the Ascent Propulsion System initiated by pushing the "Abort Stage" button.

Should the PGNS fail then the abort would have been controlled by the AGS system. With an operational DPS, thrust level was controlled manually and steering controlled by the AGS. Had the descent propulsion system not been operational or gained a velocity of over 9 m/s the crew would have manually staged the vehicle and used the RCS to insert the LM into orbit. Should both the PGNS and AGS fail the manual abort techniques of using the horizon angle for reference could have been used by the crew.

Lunar surface stay time. Once the LM touched down on the lunar surface and the descent engines confirmed the shutdown, the question most astronauts wanted answered and their training dictated was if they were OK to stay for a while. Controllers on the ground would assess the situation and confirm the decision to stay and proceed with post-landing operations if the situation allowed. If not, then there were two preferred lift-off times which actually began at the time of powered descent initiation (igniting the descent stage engine in lunar orbit). The first had a 15-minute time slot from PDI until three minutes after touchdown in which to be achieved. The second was about 9.5 minutes after touchdown (21.5 minutes after PDI). If either of these abort options had been taken, the ascent stage would have been fired to take it and the two astronauts into a 16.6×55.5 km orbit which was acceptable for an active LM rendezvous with the CM. If this was performed at the touchdown plus 33-minute stage then an extra orbit and the so-called "CSM dwell orbit" would have been used to improve rendezvous conditions. If the second option of landing plus 9.5 minutes abort was called, then two orbits would have been added to the rendezvous operation.

There would have to have been real-time adjustment for any variations in the CSM orbit and as the time on the surface increases so the time to abort the stay changes. Optimum time was shortly after the CSM passed over the landing site once each revolution (about every two hours). Of course, abbreviating surface activities would depend on when and where the crew were on the surface; inside the LM would have been somewhat easier than from the surface during an EVA.

An immediate lift-off at any time could be performed within some time constraints and performance values. This contingency though available was considered highly unlikely. Nevertheless, with rendezvous phasing fairly low during some periods this may have had to be considered. This was such a low probability abort mission that there was no need to develop any highly detailed plans to cope with such a situation. This abort would have been under the control of PNGS if it was operational, back-up was via AGS, and a manual guidance system was also developed if both systems failed. This mode would have used the Flight Director Attitude Indicator for attitude reference or the lunar horizon could have been used instead.

LM-powered ascent. If a problem occurred during the short period of powered ascent there were three abort profiles available. A PGNS failure would have seen the AGS command the ascent. If the AGS failed the ascent, the RCS would have provided orbit insertion as long as of course there was sufficient propellant available at the time. If both the PGNS and AGN had failed the crews would have to manually initiate the abort mode with the RCS.

Trans-Earth injection. Should the SPS shut off early there were a number of alternatives for the crew to choose from, many derived from the LOI abort procedures, except in reverse and attempts to re-ignite the SPS would be the primary objective. If a cut-off was experienced in the first 1.5 minutes of TEI then Mode III LOI abort procedures would be followed, between 1.5 and 2 minutes then MODE II LOI abort applied, and if a shutdown occurred between 2 minutes to the end of the nominal burn then Mode I would apply and the 2-hour coast period would have been deleted.

Trans-Earth coast. On the way home the only real abort option was to use the SPS or RCS to increase or decrease the trajectory and change the longitude for the landing. Once the entry-minus-24-hour point had been reached there would be no further burns to change the landing point. This allowed the CM to maintain the optimum velocity and angle of flight path to ensure the safest entry possible.

Entry. If the Guidance Navigation and Control System (GNCS) failed during entry then a guided entry to the desired end-of-mission targeting point would not have been achieved. Should this have been experienced on one of the missions, the crew onboard the CM had at their disposal the Entry Monitor System (EMS) to assist in landing guidance. Though this would take the vehicle slightly away from the intended target landing area it would still be in the ocean. If both GNCS and the EMS had failed, then it would have been possible to fly a "constant *g*" deceleration entry which would have resulted in a landing farther uprange of the original intended target.

As with the earlier missions there were a number of alternatives Apollo 11 could have flown had problems developed in the mission that did not prevent an immediate abort situation, but did prevent progress to the next key point in the mission. For example, these included a 10-day CSM lone mission in low Earth orbit or a semi-synchronous orbit, if the third stage of the Saturn failed to operate correctly. There were three alternative missions for combining CSM and LM operations in low Earth orbit, or a semi-synchronous one either with the LM undocked or as a combination to further gather experience on deep-space activities and the consequences of flying the CSM and LM together or alone.

At the Moon the crew could have performed a fly-around and free return trajectory if they had been unable to enter lunar orbit, and photographed future landing sites. The CSM may have operated alone in lunar orbit, again conducting landmark photography and tracking activities for future mission planning. There were also alternatives that included flying the LM solo in lunar orbit if it had been unable to receive a go for landing and as an added back-up for a CSM with a communication failure. The ascent stage would have been retained to be used as a communication relay, possibly until approaching Earth when the modules need to be separated for CM entry. This was dependent on the level of the failure.

With the objective of landing on the moon achieved, the capabilities of Apollo could now be expanded to develop a more in-depth program of scientific and engineering objectives at the lunar distance, but retain the option for alternative missions should hardware or procedures fail at strategic points throughout the flights. Of course, the safety of the crew at all times remained paramount and having a range of abort or contingency options available in the event of a series incident provided redundancy and added safety as far as possible. This decision was clearly demonstrated during the Apollo 13 mission during April 1970.

Had Apollo 12 (November 1969) been unable to perform the TLI burn, then the LM would have been extracted and for the first 24 hours a photographic mission would have been completed (NASA, 1999c). On Flight Day 2 (FD 2) the LM would have been de-orbited and the CSM placed in an elliptical orbit to complete a photographic mission during FD 3 to FD 5. If the objective had been completed by a ground-elapsed time (GET) of 100 hours, on FD 5 the spacecraft would have been recovered.

If the crew had been unable to extract the LM from the S-IVB then they would have proceeded to lunar orbit for a photographic mission of landmark tracking, stereo and high-resolution photography over three days before returning to Earth. If the descent propulsion system on the LM was unable to support a safe landing then the commander and Lunar Module pilot would have returned to the CSM and transferred the necessary equipment before jettisoning the ascent stage and completing a CSM plan change. The alternative photographic and observation mission would have focused on eight potential future Apollo landing sites before the crew headed home. If a problem had been found with the LM prior to undocking then the descent propulsion subsystem, if available, would have been used to change the orbital plan, saving fuel on the SM and then jettisoned. The crew would then continue a site-landmark-and-feature-photography program. A further plan change by the CSM would allow additional photography farther westwards. In all, ten potential landing sites would have been targeted.

On Apollo 13 (April 1970) the alternative mission in Earth orbit would have seen the crew dock with the LM and transfer photographic equipment (NASA, 2000d). The LM would then be de-orbited into the Pacific and the CSM would have plan-changed to achieve an orbital inclination of 40° allowing photography in daylight of the continental United States on each pass for the duration of the mission.

If the nominal lunar orbit-only alternative mission had been flown the CSM/LM combination would have focused on photographic objectives and the targets would have been Censorious, Descartes and Davey Rille. To avoid obscuring some of the CM windows the LM would have been jettisoned prior to commencing photographic activities. If it was determined that a CSM alone would be entering lunar orbit then the hybrid transfer would have been deleted and the CSM placed back on a free-return trajectory. In this case two LOI burns would place the spacecraft in a 111 km circular orbit. This would have allowed photography of Censorious and Mosting C.

Of course, Apollo 13 suffered an in-flight explosion in an oxygen tank *en route* to the Moon which overtook any ideas of an alternative mission, other than the swiftest return of the crew to Earth. For Apollo 14 (January/February 1971), which flew to the original landing site of Apollo 13 (Fra Mauro), the alternative missions were essentially the same as for the aborted lunar mission (NASA, 2000e).

The final three Apollo missions flew under the "J" series with additional hardware, consumables, expanded scientific equipment and capabilities; they were termed the "super-science" Apollo missions. As a result the alternative missions for this flight were also amended to reflect their added capabilities on nominal missions.

Had the Apollo 15 (July–August 1971) mission been unable to leave Earth orbit, then a mission of 6.3 days was the proposed alternative (NASA, 2001). The LM would have been ejected and de-orbited over the Pacific and the subsatellite ejected from the Scientific Instrument Module (SIM) bay after the CSM was placed in a 1,299 × 213 km high apogee Earth orbit to maximize the life of the small satellite. Its objective would then have been to obtain data on Earth's magnetosphere using the satellite's gamma ray spectrometer. The CSM would then have been returned to a 444 × 211 km orbit with apogee over the continental United States. This would have allowed the Apollo 15 crew to complete a range of photographic tasks using the SIM bay cameras, and CMP Al Worden would have conducted an EVA to retrieve the camera cassettes on the final day of the alternative mission. In addition, the crew would have evaluated the use of the alpha particle spectrometer, mass spectrometer and laser altimeter to verify their operational use. It was also planned to use the X-ray fluorescent equipment for partial mapping of the Universe and obtain cosmic background data readings.

Should the mission make the translunar trajectory with the landing deemed impossible, then (with the LM attached) the trajectory would be maintained within the descent propulsion system capabilities should an insertion into lunar orbit not be performed but allow acceptable return-to-Earth options. With the LM descent stage engine as a back-up the SPS would retain the prime role of performing the burn from lunar orbit to begin the trip home. Around the Moon (with the LM still attached) the crew would have completed orbital science and photographic tasks at a high-inclination 111 km orbit over a 4-day period.

If the LM was not available then the CSM could have continued a translunar trajectory with the Service Module Reaction Control System capacity as an acceptable return-to-Earth option. If the SIM bay door did not jettison, in this case then LOI would not have been performed. If orbit was accomplished then again a high-inclination 111 km orbit would have seen scientific and photographic tasks taking place over a 4-day to 6-day period. Had the landing been aborted either prior to or after PDI then a CSM-only lunar orbital science mission would have been completed for about 6 days. In this case the LM would have rendezvoused, docked and transferred the crew before being jettisoned to continue the alternative mission.

Apollo 16 was the penultimate Apollo and was flown in April 1972 (NASA, 2002a). If the mission remained in Earth orbit then a 6.3-day mission would have been flown with the transfer of necessary equipment to the CM before jettisoning and de-orbiting the LM. SIM bay operation would be primarily over the United States and the X-ray fluorescence spectrometer would be operated to obtain additional galactic X-ray sources. Data on Earth gamma-ray albedo would be gathered using the gamma-ray spectrometer. It would also have been used for gamma-ray astronomy objectives. The subsatellite would have been released into a high-apogee orbit to increase its orbital life and gather data from the particle detectors. The other SIM bay experiments would also be operated on a non-interference basis gathering further engineering data for future applications. CMP Ken Mattingly would have completed an EVA to retrieve the SIM bay film cassettes near the end of the mission.

The lunar orbit alternative missions included one with an inoperable DPS similar to that available for Apollo 15. There an orbital scientific mission would have been accomplished over 6 days. If the DPS was deemed inoperable prior to an LOI maneuver then the LM would have jettisoned after redocking and crew transfer and the Service Module would have been used to circularize the orbit for scientific and photographic tasks also over a 6-day period. As with Apollo 15, if the LM on Apollo 16 was unable to complete the descent to orbit and return to the CSM then a 6-day orbital science mission would have been conducted.

The final mission of Apollo to the Moon was Apollo 17 in December 1972, closing out an historic and highly successful series of missions. The alternatives for this flight if the nominal mission had not been possible included the normal Earth orbital or lunar orbital options (NASA, 2002b).

The Earth orbital alternative missions would have lasted about 6.5 days and would see the transfer of useful equipment to the CM before jettisoning and de-orbiting the LM. An SPS burn would shift the orbital inclination to 45° enabling optimum conditions for maximizing the scientific return. The RCS would have provided a back-up de-orbit burn capability at all times. As for science the lunar sounder, far UV spectrometer and IR scanning radiometer in the SIM bay would have been used to gather data on the Earth's atmosphere and terrain and for astronomical observations. Using the mapping and panoramic cameras the crew would have photographed selected Earth targets, and the film from the experiments would have been retrieved by CMP Ron Evans towards the end of the mission.

The operations in lunar orbits were again dependent on an operable DPS on the LM, on a non-operable system, on whether the CSM operated alone prior to DOI or following it and on any aborted DOI maneuver. Again orbital science and photography would be the priority in such cases, though of course there were no future flights to gather information on potential landing sites. The J series would have gathered information to expand our knowledge of the lunar surface from orbit and photo-document as much of the surface as possible.

Support on the ground and in space

Apart from Apollo 13, all the landing missions achieved their objectives and returned safely to Earth. Even with Apollo 13 the crew were recovered safely after a harrowing few days following the explosion that aborted their landing objective. The success of these missions depended on a number of other factors that supplemented the redundancy of the hardware and flight profile, the availability of alternatives and abort mission options.

Lunar landing sites

The original landing sites on the visible side of the Moon were chosen after an intense two years of discussion and debate. From 30 original sites the opportunities for each mission were narrowed down initially for smoothness, approach, propellant requirements and expected slope of the terrain in the approach path and landing area to ensure a safe launch of the LM's ascent stage. In addition, should a Saturn V launch be delayed then the sites could be recycled to allow for the delay. Initially, each site allowed use of a free-return trajectory looping around the Moon should a major problem occur that required an early return to Earth. As the missions grew in number, confidence increased and early rules and plans were expanded to include landing sites whose criteria were the opposite of those followed for the earlier landing sites. The new sites had more craters and boulders, there were hills and rougher terrain to fly over and explore, and more propellant would be required. To reach these sites a departure from the free-return trajectory to a hybrid trajectory was required on the way to the Moon.

Free return or hybrid

Even before Apollo was committed by President John F. Kennedy to place American astronauts on the Moon before the end of 1969, the safest method of doing so was understood. This was called free return, where lunar gravity is used to effectively steer spacecraft rather than using rocket fuel alone, which would have added a greater mass to the vehicle, probably preventing the spacecraft from being lifted off the Earth, resulting in a more complicated assembly of spacecraft components in Earth orbit, adding to the time to master these techniques and increasing the risk of multiple components failing to work together once in orbit.

Flying a figure-of-8 trajectory the spacecraft departed Earth orbit and out towards the vicinity of where the Moon would be in three days time as it travels in its own orbit around the Earth. Using the velocity of the spacecraft and the Moon as well as the gravitational force of the Moon a spacecraft could loop behind the Moon and continue back towards Earth without entering orbit, and any refinements to the trajectory could be accomplished by the smaller Reaction Control System onboard the spacecraft, if the main propulsion system experienced some type of failure. A variation of this approach was flown by the Soviet unmanned Zond spacecraft and could have seen the first cosmonauts loop around the Moon prior to the American Apollo 8 mission in December 1968 had the Soviet program and hardware progressed as smoothly as the American Apollo system. The gravitational influences of the Sun and the major planets in the solar system also needed to be considered. It was also important to define exactly how the flight plan worked to allow landing on the Moon. As the flights continued and confidence grew in the system and hardware so did refinements to this basic trajectory to allow visits to more scientifically interesting sites.

The missions of Apollo 8, Apollo 10 and Apollo 11 followed the free-return trajectory. For the H series of missions following the first lunar landing (Apollo 12, Apollo 13 and Apollo 14) they amended their flight path to the Moon by a small burn of onboard engines, after confirming the trajectory was safe to continue. This could be returned to the free-return trajectory by a small burn of the spacecraft engines, either the SPS engine on the Service Module or the larger Descent Engine (or smaller

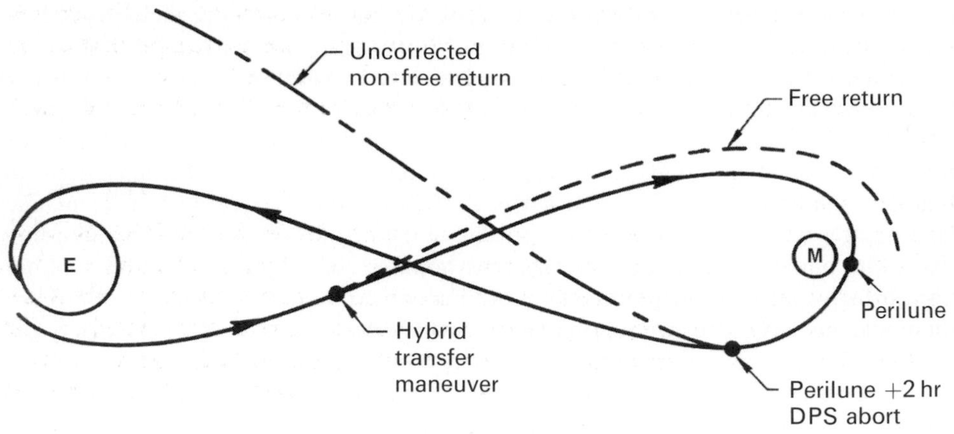

Apollo lunar mission flight paths to the Moon.

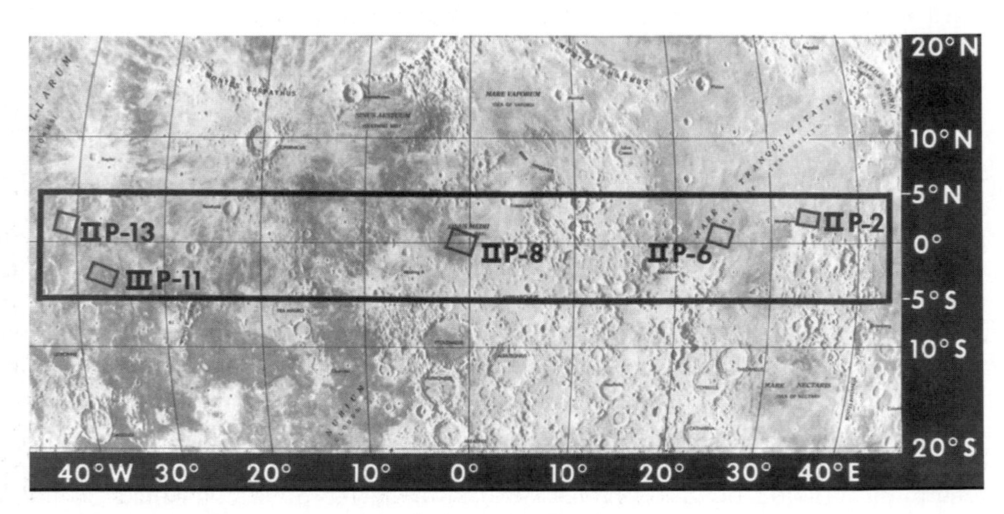

Apollo lunar landing zone.

Ascent Engine) on the Lunar Module. This approach was called the hybrid trajectory and was less "safe" for a return to Earth than the free-return option. For the J missions, right from the start of their trajectory towards the Moon they used the hybrid option and would only have gone back to a free-return trajectory by using the major onboard engine systems. Luckily for those missions, this was not required. However, on Apollo 13 it was the availability of the still-docked LM that allowed the crew to steer the spacecraft and provide the required correction burns to return them to the Earth after the loss of the SPS on the Service Module on FD 3.

Throughout the mission, planning refinements to the vector and required engine burns in the near future were part of a system of data uplinks from Mission Control to the crew. The initial information contained the primary and preferred set of data, later information referred to back-up or contingency action in the event of an abort or loss of communications. This of course was brought to the attention of the general public during the Apollo 13 mission when changing data were frequently being uplinked to the crew and handwritten down by the astronauts to allow them to proceed with their off-nominal actions. This was later included in the dramatization of the Apollo 13 events in the 1995 feature film. Another "safety" element of these uplinked data, called pre-advisory data or PAD, was the information required to bring the crew home in the event of a loss of communications with Earth. Mission Control provided the constant role of support to the crew flying in space, as well as monitoring aspects of the flight beyond the capabilities of the crew. Their support in mission safety is explained in more detail from p. 280.

In his book in this series author W. David Woods provided an excellent explanation of how the technology on Apollo worked and how each hurdle in the flight was approached and overcome (Woods, 2008). The details of the Apollo 13 mission have already been covered by this author in a previous book in this series (Shayler, 2000, pp. 277–307).

No alternatives

Safety on the lunar surface was another added concern. The hardware, procedures, alternatives, back-ups and technology can only be as good as those who design and develop the theory, test the practical and support the operations. Failure on Apollo was never far from success; fortunately, there was much more success than failure throughout the scope of the program. There were of course some things that simply could not be backed up or made with a redundant system. The Ascent Engine of the LM simply had to work, as the smaller RCS did not have the power to do the job. If the lone astronaut in the Command Module could not come down and rescue his two colleagues, his would be a long lonely trip home, as eloquently explained by Apollo 11 astronaut Mike Collins in his 1974 autobiography (Collins, 1974) as he recollected his thoughts whilst awaiting confirmation of the engine burn of LM Eagle bringing back to orbit his colleagues Armstrong and Aldrin.

Collins writes of feeling like a nervous bride, and all of his experience in 17 years of flying including making his first spaceflight three years before in Gemini 10 had never made him sweat out a flight like that of the launch of the LM from the Moon. Ever since he had been assigned to the mission his secret fear was having to leave his colleagues behind and come home alone, and as the seconds ticked toward that defining point he knew he would not commit suicide but make the decision and come home alone if they crashed or failed to leave the surface "I will be a marked man for life and I know it . . . almost better not to have the option . . . one little hiccup and they are dead men" (Collins, 1974, p. 412).

Holding his breath seemed to have helped as the ascent stage of Eagle soared into orbit carrying Armstrong and Aldrin with it. Indeed, the technology, procedures, skill of the astronauts and controllers, worked each time Apollo flew with men aboard, even Apollo 13 was deemed a successful failure in that the crew returned home safely, though the primary goal of the mission to land on the Moon was not achieved. If anything, Apollo 13 vindicated the years of planned procedures and development of contingencies, some of which, such as using the LM descent engine to "steer" the docked CSM, had been trialed on the Apollo 9 Earth orbital mission the year before. The fact that the old rule books had to rewritten during Apollo 13 added to the drama and excitement, and helped create the image of a can-do NASA, an image that would come back to haunt the agency years later in the Shuttle era.

Apollo's Buddy System

On the top of each astronaut's Portable Life Support System (PLSS, pronounced *pliss*) was an Oxygen Purge System (OPS) that provided a contingency supply of 45 minutes of gaseous oxygen in two small bottles, allowing a back-up emergency supply to the primary life support system and an opportunity to return quickly to the LM and hook up with the spacecraft system should a problem develop in an EVA on the surface.

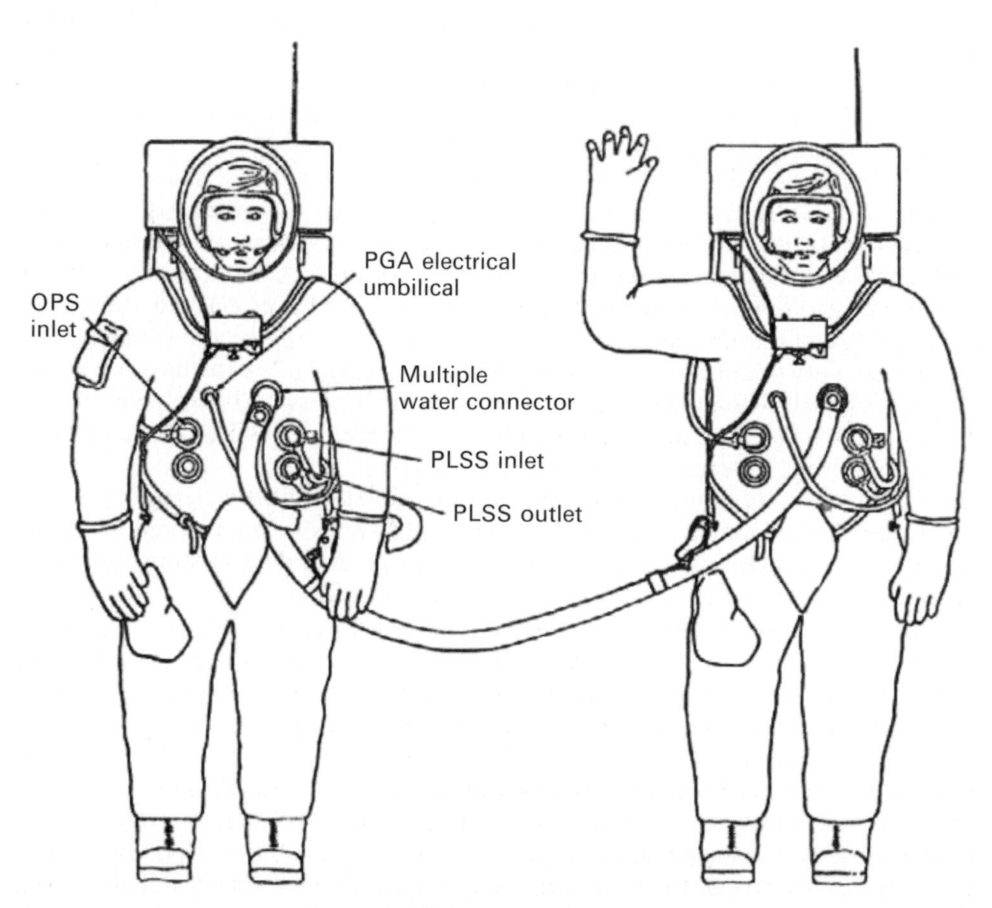

Apollo Buddy System.

As lunar excursions increased in complexity so did the need to provide an adequate method of supporting an emergency in one of the life support systems the astronaut needed to survive on the surface outside of the LM. This was devised from Apollo 14 onward and termed the "Buddy" System.

The Buddy Secondary Life Support System (BSLSS) was a connecting hose which allowed one crewmember with a failed PLSS to share coolant water from a second astronaut's PLSS to allow a return to the LM as soon as possible. This would lessen the load on the OPS. Had a complete failure in one PLSS been experienced, then the BSLSS would route breathing and pressurization oxygen from the OPS while the circulating water in the Liquid Cooling Garment would have removed metabolic heat. The BSLSS was stored on the Modularized Equipment Transporter (MET), a two-wheeled "rickshaw" cart used to carry tools and equipment during the second EVA which featured a traverse farther away from the LM than had been accomplished on Apollo 11, Apollo 12 or the first EVA on Apollo 14. From Apollo 15 this system was stowed on the Lunar Roving Vehicle.

When we return to the Moon the expanded surface exploration will have to include suitable rescue and contingency procedures and systems, like the Apollo Buddy System, to provide redundancy for surface exploration crews in the event of a suit failure during EVA far from the lander or base site.

Contingency Moonwalks

As the crew never really knew if their stay on the Moon was to be full term as planned or terminated due to an unforeseen problem, there was a desire in mission planning to include the gathering of a sample of lunar surface material as early in the first EVA as possible to ensure at least a small sample was brought back to Earth should the surface EVA be terminated early.

In the event of only one Extravehicular Mobility Unit (EMU) being serviceable, then contingency EVA operations were in place to allow at least one-person short EVAs on the surface if the second astronaut was unable to leave the LM, and remain attached to the environmental control system of the spacecraft supplied by an umbilical. Also, should a problem occur in some of the LM systems where constant monitoring was required, an amended contingency one-person EVA may still have been possible, though restricted in content and distance depending on the problem encountered at the time.

Mission rules dictated a one-person EVA could be almost full term or of minimum time, amending the surface tasks as required. The EVA astronaut would have remained in sight of the second astronaut in the LM crew compartment as much as possible. The structure of the EVA meant that each astronaut prioritizes tasks and roles so that in the event of a contingency EVA being required the most important tasks would be assigned to one astronaut with secondary tasks assigned to the second astronaut. The commander would probably have conducted the contingency EVA, to ensure the maximum opportunity to complete the revised primary objectives, while either the commander or Lunar Module pilot could have conducted a second or even third contingency EVA (Shayler, 2002, pp. 294–295 and 299–300).

GRAND PLANS AND DESIGN STUDIES

Throughout the years there have been a wealth of ideas and proposals for recovering an astronaut from space and these have been incorporated not just in stories of science fiction but also in formal design studies and proposals for operational space activities (Burgaust, 1974; Wade, n.d.). Many have never left the drawing board while others have been incorporated, or proposed, into past or current programs. Still others were formally proposed and planned for but were never used due to budget restrictions or loss of the hardware they were supposed to support. A summary of the major design proposals for the more personal and individual "space rescues" during the 1960s and early 1970s are listed below (Greensite, 1970).

Space lifeboats. The U.S. company Bloom & Quillinan proposed either one-person or three-person lifeboats covering the range of multi-crewed spacecraft of the 1960s which provided short-duration capsules, which could separate from the main spacecraft but which would not be capable of re-entry. Presumably, they would be retrieved by a second "rescue craft". The capsules contained emergency food, oxygen, water, a solar array and batteries for independent power, onboard lighting, signaling equipment and for long-duration designs a waste collection subsystem.

The one-person design had a lifetime of 14 days, a length of about 2.1 meters and maximum diameter of 1.05 m with a 266 kg mass. A long-duration variation of this design was also proposed, this time for use on future Mars or Venus expeditions. The lifetime of this model was given as 365 days, though it has been pointed out that the sanity of the occupant after this time in what was essentially a space coffin could be questioned! The three-person design measured 5.2 m, had a diameter of 1.6 m with a mass of 716 kg. On the long-duration design the length was increased to 32 m; this design featured regeneration of contaminated air, attitude control, propulsion and navigation equipment offering some capability of maneuverability.

A second company, ASD, proposed a five-person lifeboat which was separable but not capable of re-entry; it had only a short-duration lifetime and its mass was 1,422 kg.

Crew shelters. ASD also put forward a design for a ten-person emergency shelter capsule which was not capable of separating from the main spacecraft, and thus was not capable of re-entry. Instead, the crew would be able to isolate themselves from the rest of the "space station", awaiting rescue from another vehicle using a dedicated docking and transfer facility built into the shelter. This design measured 3.3 m, had a diameter of 4.5 m and a mass of 3,012 kg. Rescue from these shelters had to be achieved within the 14-day lifespan of the facility. Some designs of the Grumman LM featured crew "shelters" for expanded exploration of the Moon in what was called the Apollo Applications (lunar) program. These could support crews on the surface for up to 14 days but also could shelter a stranded crew on the surface awaiting rescue from a second LM.

Bailout re-entry

Some of the most exotic and challenging proposals were for personal re-entry systems, in which a crewmember wears (or rides in) a personal re-entry system allowing passage through Earth's atmosphere and thence a parachute or paraglide landing to safety. The image of a lone astronaut passing through the fireball of re-entry is not new—all the Mercury and Vostok crewmembers did it, and so have Soyuz crewmembers and one Shenzhou occupant. It's not the re-entry alone that's the problem, it's what you would be riding in that would push your belief and trust in the technology far beyond that of "normal" spacecraft re-entry techniques.

So-called "satellite life rafts" proposed a counterpart to a life raft at sea. The seaborne life raft would lack power and controls and simply drift hopefully towards the shore but always at the mercy of prevailing currents. In space the "life raft" would provide insulation for the occupant as the device hurtles towards the Earth, sometimes not under its own power but always under the influence of gravity!

Space life rafts would have been located on the outer walls of large spacecraft, the shield protruding outwards in the vacuum of space, its entry hatch open facing the living or working quarters and clamped securely with airtight seals to the main spacecraft. When the "abandon ship" signal was raised, the astronaut floated as quickly as possible into the life raft, sealed the door and pressed the emergency ejection control mechanism or explosively separated the now-sealed life raft from the main spacecraft. Using compressed air the lone occupant would activate systems to orientate and de-orbit the unit by onboard propulsion systems (backed up by the compressed air orientation system), then determine the orientation of the craft before re-entry, where the space life raft would be protected (probably by ablative coatings), after which parachute landing would be accomplished with appropriate cushioning devices for land or water recovery and survival packages for an extended stay while awaiting recovery. Additional transponders and signaling equipment, some activated at separation from the spacecraft, would assist the search and rescue teams in locating the downed astronaut. These vehicles could hold up to three or more crewmembers as required. Presumably several of these escape vehicles would be located in strategic places around a larger spacecraft or space station. In much the same way, Soyuz spacecraft today are available to ISS crewmembers to provide emergency escape from the main space vehicle.

Variants of this type of rescue system included

Rib-Stiffened Expandable Escape System. North American Aviation proposed an articulated rib structure that would be deployed from a storage canister to a mechanical, rigid aero-shell shape, which was pressurized inside to provide a shirt-sleeve environment for up to three crewmembers. Measuring 3 m long and 4 m in diameter its mass was 660 kg; it was designed to be implemented from a space station.

Emergency Earth Orbital Escape Device. This Lockheed design was for three astronauts inside a capsule that was based on the USAF Discovery unmanned

Airmat escape concept
(Goodyear)

- Two-man
- Suits required
- Inflatable
- Ejection seat
- 518 kg
- New technology requirements
  —flexible heat shield
  —material

MOSES escape concept
Manned Orbiting Shuttle Escape System
(GE)

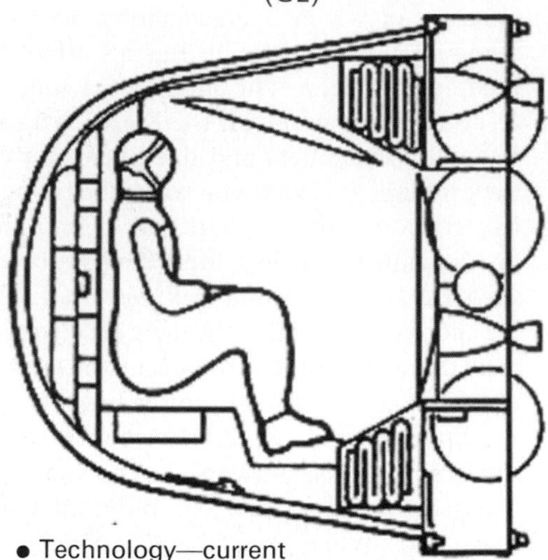

- Technology—current
- One-, two- or four-man concepts
- 1,810, 2,880 and 5,110 lb, respectively
- Escape suits required
- Ballistic entry
- Parachute recovery
- Proven system for satellite recovery vehicles

ENCAP escape concept

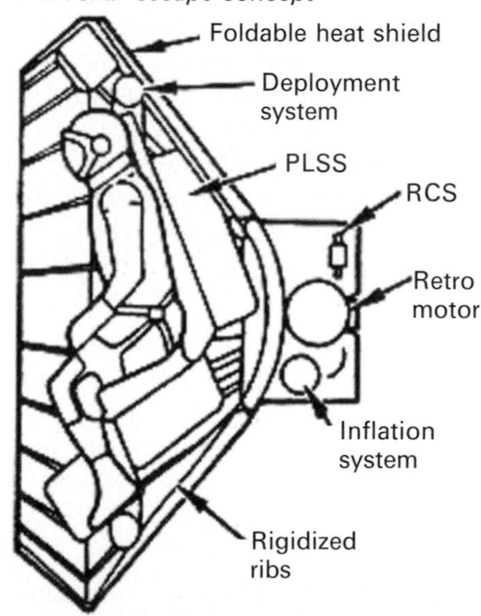

- One-man
- Suit
- EVA
- Mech rigid
- 24 kg
- New technology requirements
  —mechanical deployment mechanism
  —foldable heat shield

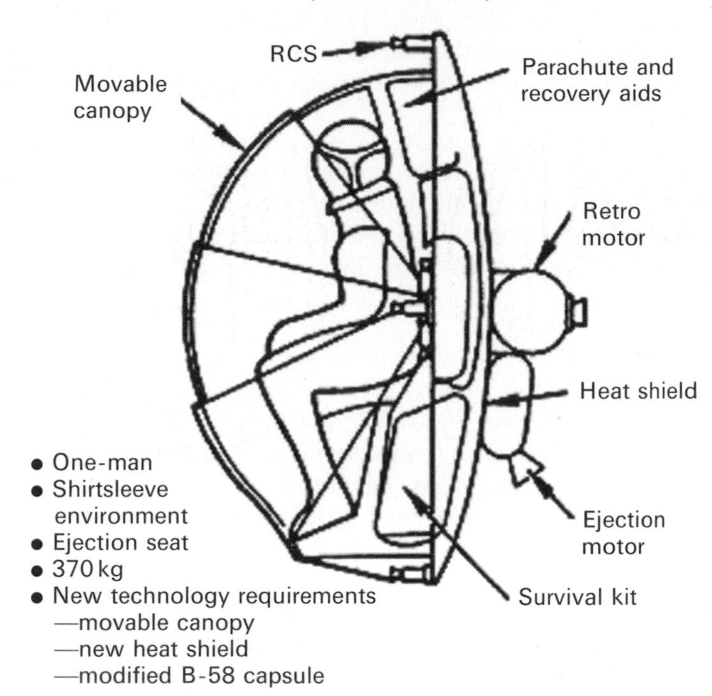

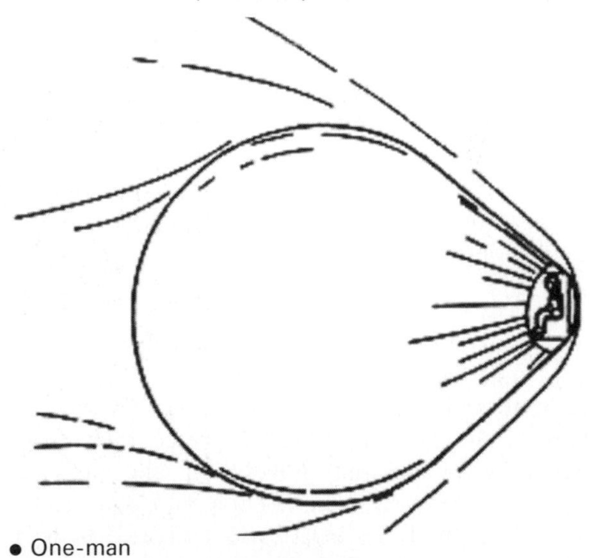

Proposals for escape from orbit. Courtesy Mark Wade Encyclopaedia Astronautix

retrievable satellite capsule concept used in the early 1960s supporting the military space program. Measuring 3.70 m with a maximum diameter of 2.90 m and a mass of 1,240 kg, it used solid-propellant rockets for de-orbit, and its "new technology" heat shield material made the claim that the three-person vehicle would be lighter than the one-person Mercury spacecraft flown between 1961 and 1963.

Spherical Heat Shield. This was a Rockwell concept which could return two crewmembers from a space station in a pressurized environment. Its length was 1.50 m, and it had a maximum diameter of 2.20 m and a mass of 445 kg. Its shape was hemispherical similar to a Vostok Sharik design cut in half.

GE Life Raft. General Electric proposed a three-person, rigid, unpressurized life raft. This required the crew to wear full-pressure suits (IVA rescue suits similar to Gemini or Soyuz Sokol suits—not EVA suit designs). The aero-shell concept had new non-ablative material with a foam core design and measured 1.80 m long, 3.00 m in diameter and had a 480 kg mass. Integral RCS cold-gas retrofire capability was manually aligned for firing by a head-up display unit. Each crewmember had a personal parachute for independent landing after re-entry.

SAVER. This was a single-person escape concept utilizing a personal ejection seat (similar to those used on Gemini or early Shuttle OFT missions). North American Rockwell proposed a nose cape behind the seat to absorb most re-entry heating, then a large inflatable balloon (10.00 m in diameter) would be deployed from the seat and by changing the size of the balloon the re-entry loads and drag effects could be varied.

Airmat. Goodyear came up with this design featuring an inflatable cocoon around two space-suited astronauts still on ejection seats. This required development of a new and flexible heat shield, with a design length of 1.5 m, diameter of 2.20 m and mass of 1,140 kg. Goodyear was thinking of a two-person ejection system from a space station that would then be enclosed in the protective re-entry cocoon and land by personnel recovery parachute.

ENCAP. As the name implies this design was for an encapsulated concept, based on a foldable heat shield. A space-suited astronaut would exit the spacecraft and a gas-powered system would unfold the ribbed structure of the heat shield around the rear of the seat structure. Integrated retro-rockets and reaction control systems would allow personal orientation and individual parachutes would allow for personnel recovery after re-entry and seat separation (similar to the Gemini and Vostok method). Its length was estimated to be 1.40 m with a maximum diameter of 2.40 m and mass of 266 kg.

Three of the more advanced studies were

Emergency Global Rescue Escape and Survival System (EGRESS). Developed by Martin Marietta and based on real designs of an encapsulated ejection seat concept developed for the B-58 bomber by Stanley Aviation in the 1960s, EGRESS featured a 2.0 m long, 1.00 m diameter, 370 kg design with a dish-shaped heat shield. It incorporated an attitude control system, guidance unit, environmental control systems, UHF communications, retro-rockets and a drag stabilization system. A manual selector switch would choose the appropriate system operation when the unit was activated, from on the pad through ascent, orbital operation, re-entry and landing. The sequence started with the upper, center and lower sections of the frontal door lowered to the down position from the normally stowed position in the unit's hood. This effectively would have sealed the occupant, who now would have initiated ejection a fraction of a second after the spacecraft outer hatch was explosively separated. A short 13*g* acceleration would have been experienced; however, deceleration drag could be as much as 22*g*. If ejection was to be in orbit the ejection rocket was isolated and only small jets would have separated the unit prior to the de-orbit burn. An attitude control system would have orientated the unit, and the astronaut using an onboard timer would have prepared to initiate the retro-fire burn. The onboard life support system could have sustained a crewmember for 1.5 orbits giving an opportunity to choose the optimum time for re-entry and emergency landing. The recovery parachute would have automatically deployed after re-entry and a crushable struts and stabilization system would have softened the landing. Should the landing have been on water the design was such that flotation bags would ensure it turned on its back allowing it to float on water like a huge enclosed saucer with the astronaut inside.

MOOSE. This was the most widespread personnel rescue system design of the early 1960s. The Man Out Of Space Easiest rescue system was, as its name implies, a design to get an astronaut out of orbit quickly. Newspapers of the day emblazoned headlines about astronauts "jumping" overboard and falling to Earth in a protective cocoon. Proposed by General Electric, it measured 0.87 m in length with a maximum diameter of 1.8 m and a mass of 215 kg. The way the system worked was a space-suited crewmember would strap MOOSE to his back, open the hatch and push himself into free space, basically "jumping" overboard. A tug on a ripcord would inflate a foldable heat shield and between that and the astronaut a quantity of form-fitting polyuthathine would be injected to fill the cavity between the shield and the astronaut's back. A hand-held orientation device and retro-rocket vehicle would provide de-orbit and orientation capability so that the astronaut would "fall" back into the atmosphere, the heat shield and foam protecting him on the way. After re-entry a ripcord on the chest pack could deploy a personnel recovery parachute and separate him from the MOOSE device. This design even got as far as tests by GE to prove the concept worked. Foam tests recorded a temperature of 100°C at the core but there was no heat transfer to the subject. Ablative material was tested and provided further confidence in the system. Apparently, test subjects suggested a little oil be added to the mixture of foam to allow a smooth separation after entry, and in one test in Massachusetts a test subject jumped from 6 meters and survived the landing in a river, with the system and occupant safely floating downstream. It was pointed out at the time though there would be a slight difference in falling 500 km from orbit than 6 m from a bridge. The design was intended for the X-20 Dyna Soar and certain military space stations such as the Manned Orbiting Laboratory; however, the program was canceled prior to flight operations and MOOSE was abandoned.

Paracone. Douglas developed this concept which was similar to EGRESS and featured an gas-inflated structure shaped like a cone with a large spherical nose. Paracone actually formed part of the astronaut's seat and included an ejection device to separate from the spacecraft, orientation motors, re-entry motors, and a landing and recovery package. Following ejection from the vehicle the astronaut would deploy the Paracone retro-rocket stashed on a strut above his seat and orientate himself face-forward to fire the retro-rocket, thus slowing his velocity and allowing re-entry. He would then re-orientate himself with his back toward the re-entry burn and deploy the lightweight re-entry shell from the seat using gas pressure, which also operated the RCS. This was a very light and simple design for use in land landing only—there was no parachute! Paracone was to be made from Rene-41 fabric with a Teflon coating. As ballistic re-entry would impart no more than 9.6*g* on the occupant, the pressure suit could handle the expected heat reduction loads. Landing velocity was absorbed by the crushable structure of the nose cap. Basic tests on this system were made in the wind tunnel along with full-size impact tests and free-fall model drop tests. They were encouraging enough to demonstrate the basic feasibility of the design. Manual re-entry options were chosen in this design giving a maximum orbital life of three hours (two orbits). It was 3.0 m long, had a maximum diameter of 2.0 m with a deployable Paracone of 7.6 m across its base and a mass of 227 kg. This system could also support off-the-pad, ascent or post-entry ejection as well as emergency recovery. Its woven fabric design, made of nickel-based alloys, provided its own braking system and was often described as the astronaut's "flying carpet" escape system.

All of these personnel escape systems relied on the development of new technologies and the invention of new materials and procedures. Other designs—both American and Russian—featured the use of proven technology such as variants of Gemini, Apollo or Soyuz capsule-type spacecraft, and were incorporated into Shuttle or Space Station concepts (see below).

SPACE SHUTTLE CONTINGENCY OPERATIONS

A 1972 report from North American Rockwell focused on safety in Earth orbit that featured designs of the Space Shuttle then being developed, and proposed large space stations. This was the time of discussions about international docking missions

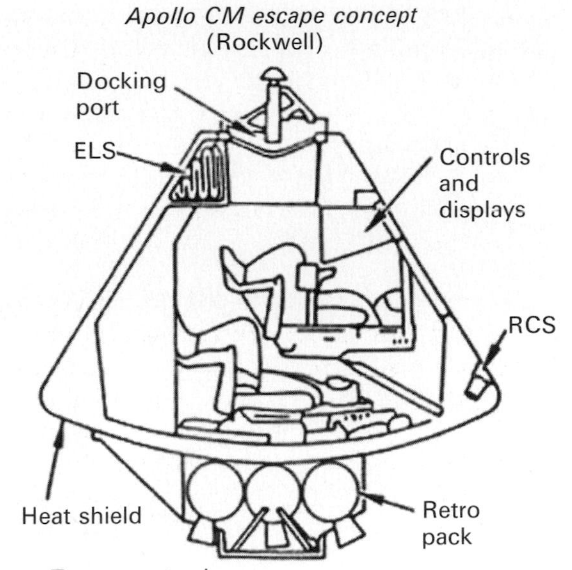

CM-type escape capsule for Shuttle. Courtesy Mark Wade, Encyclopaedia Astronautix.

between the American and Soviets with the view of developing an international docking and space rescue system, the forerunner of which became the Apollo Soyuz Test Project. These featured either escape from the Shuttle vehicle, or using the Orbiter as a rescue vehicle (Jenkins, 2001, pp. 156–159). The study, which had lasted for 12 months focused on five areas of safe operation of future manned spacecraft and fell into the areas of flying hazardous payloads, docking the Shuttle to the Space Station or other vehicles (such as a second Shuttle or future Russian spacecraft), the provisions for onboard survivability, the effects of a tumbling spacecraft, and options for escape and rescue.

Focus of study

The objective of this study was more about crew safety than to prevent damage or the loss of the spacecraft; only the orbital phase of the flight profile was evaluated. Earlier studies by North American Rockwell had also indicated that the Shuttle was very sensitive to uncontained explosions in the payload bay as small as 0.05 kg of TNT. To put this in perspective a hand grenade was equivalent to as little as 0.01 kg of TNT. In the aftermath of the Challenger explosion a decision was made to remove the liquid-fueled Centaur upper stage from deploying payloads from the Shuttle. A fully fueled Centaur would carry normally the equivalent of 2,720 kg of TNT!

Structural failures of large payload stages and upper stages could also prove fatal to the integrity of the Orbiter and safety of the flight crew. Recommendations

Landing the Shuttle (STS-3) at White Sands, New Mexico, a backup landing site to KSC and Edwards AFB.

for handling such hazardous payloads included the early opening and late closure of the Orbiter payload bay doors, dumping liquid propellant prior to return to Earth, strengthening the Orbiter structure, human-rating hazardous payloads, and incorporating the jettison capability of certain payloads to reduce the landing mass.

Onboard survivability focused on evaluating the projected movement of astronauts around the Orbiter during nominal operations, potential escape routes and how safe the crew compartment was to in-orbit failures. Quick-don survival pressures suits without the need for pre-breathing were one option. It was also suggested that the crew compartment of the Shuttle should feature two separate pressurized compartments, divided by a bulkhead that was smoke-resistant and flame-resistant. It was further suggested that one of the "compartments" could indeed be the proposed airlock of the Orbiter, to accommodate all crewmembers who would not require pressure suits. At this time the nose of the Orbiter provided access to a rescue craft via a docking and transfer facility, or via a ground access hatch. In any event the design of the Orbiter required some of the flight crew to be suited up and to remain on the flight deck to control the vehicle in as many scenarios as possible.

A tumbling Orbiter could spell the loss of both crew and vehicle, as it would be unable to orientate itself for entry and make it almost impossible to be docked to a second vehicle.

To provide a rescue capability, studies were directed at providing the Shuttle with an escape means for the whole crew and provide a safe return to Earth. The capacity to be rescued by a second vehicle was also reviewed as well as providing means to

Simulated emergency evacuation from Shuttle.

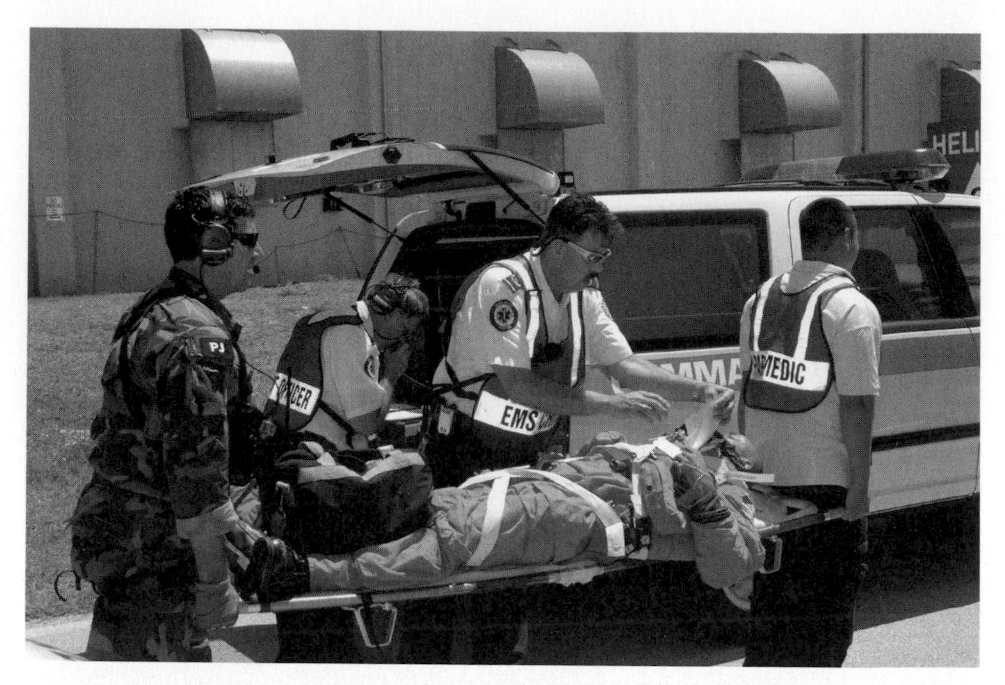

Simulated recovery of injured Shuttle crewmember.

Helio evacuation simulation.

survive a major incident. In some cases the Shuttle acted as a 'rescue' craft, but of course there should be a requirement to "rescue" a stricken Orbiter itself. Thirty years after this study was published the feasibility of launching a rescue Orbiter to the Columbia crew was also reviewed, but only after the STS-107 crew had been lost. One of the more favored designs in the 1972 study was for a modified eight-person Apollo CM–type vehicle located in the forward end of the payload bay, and connected by internal transfer tunnels to the mid-deck. It was also proposed to place a "Spacelab" single module–type facility in the bay to provide a safe haven for a crew in trouble. The difficulty here would be providing separate life support means when the Shuttle Orbiter normally supplies that service. The Spacelab was also integral at that time to the Shuttle and not a free-flying unit, therefore attitude control and thermal control would be via the Orbiter.

As with most studies of space rescue systems, they were found to be dependent on the availability of the crew to access the system and have the time to isolate or jettison the source of the problem. As events have clearly demonstrated, on Mir and the Shuttle that time was not always available to the crew, and they may not all be in the right place to put the rescue or safety procedures into action effectively.

The 1972 study suggested making some of the residual Apollo hardware available for a rescue capability on the early Shuttle missions, while other Orbiters were still being constructed. A rescue Apollo CM on top of a Saturn 1B or Titan IIIC launch vehicle was an idea mooted from Skylab and suggested for the 1974–1978 time period when Shuttle test flights were expected to fly. Studies also included reviewing the early 1960s' designs for the personnel rescue systems detailed above (Airmat, Paracone, EGRESS and ENCAP as well as lifting body designs).

The study was made at the time the Shuttle was finally authorized and a flight rate of one Shuttle mission a week would occur from the Cape in Florida and Vandenberg in California, each mounting one mission a fortnight. This meant there would be sufficient Orbiters available to sustain that rate. It did not suggest a dedicated Shuttle rescue vehicle but more an adaptation kit available to be installed on the next Orbiter in the launch-processing cycle to support the rescue mission. This was the method employed for Skylab. There was not a dedicated Apollo Saturn 1B vehicle during orbital operations for Skylab, but the next launcher and spacecraft were made available as soon as possible to rescue the resident Skylab crew, who would remain onboard the Space Station if they could not use their own Apollo CSM to come home.

As grand as these plans might be, by 2008 the reality of the limits of the program, funding restrictions and technical hurdles meant that the Shuttle had rarely sustained a flight rate of 8 missions a year, never mind 52 missions a year. Therefore, it is difficult to imagine a second Shuttle being readied quickly enough to mount a rescue mission, though this was almost implemented in part for missions after STS-107.

Contingency Shuttle crew support (CSCS)

Following STS-107, NASA instigated a study into the feasibility of rescuing a stranded Shuttle crew at the International Space Station, when it was determined

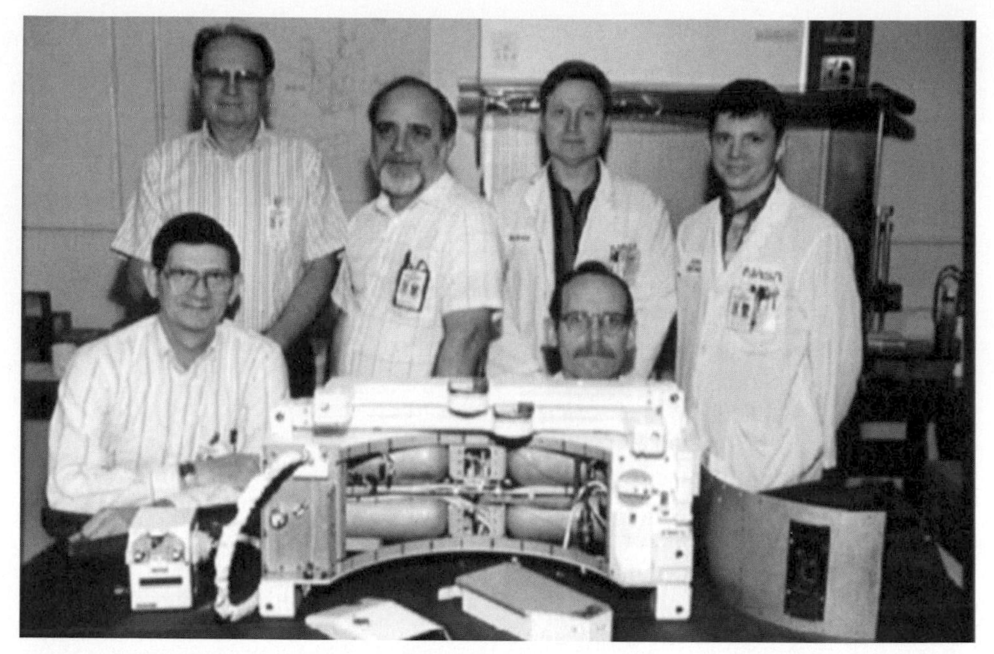

Integral but often overlooked ground support and development teams. SAFER Development Team (STS-64) 1993. (Front, left to right) Russell Flack and Bob Lowe. (Rear, left to right) Jack Humphreys, Chuck Deason, Bill Wood and James Brown.

Evaluating the rescue of a simulated stranded crewmember during EVA.

from launch imaging on the ground, from the vehicle (cameras on the SRB, ET and Orbiter), from an inspection by the resident ISS crew prior to docking and evaluation of data downlinked to the ground that the condition of the Orbiter for entry and landing was not safe. It should be possible to coordinate matters such that a second Shuttle was advanced enough in its processing to rescue the stranded crew and that the stranded crew should be able to live onboard the ISS until the arrival of the rescue Shuttle, with ISS, in effect, becoming a safe haven for the stranded crew.

Contingency Shuttle flights for the two return-to-flight missions (STS-114 and STS-121) were designated STS-300 and STS-301 for planning and processing purposes. The astronauts assigned to those rescue missions should they be required were drawn from the next two flight crews in training. These were STS-121 (STS-300) and STS-115 (STS-301) the specific assignments being

STS-300	Steve Lindsey Mark Kelly Mike Fossum Piers Sellers	Commander and back-up RMS operator Pilot and primary RMS operator MS1 and EV2 MS2/FE and EV1
STS-301	Brent Jett Chris Ferguson Joe Tanner Dan Burbank	Commander Pilot and back-up RMS operator MS1, EV1 and primary RMS operator MS2/FE and EV2

Post-flight investigations and incorporation of recommendations not only from the Challenger and Columbia accidents but also other missions have commonly been incorporated in the Shuttle Program and provided amendments to the flight rules, procedures and operations to improve safety and security not only during dynamic periods of flight but also in orbital flight operations.

SAFER EVAS

One Shuttle-based orbital rescue system was elaborated on in Chapter 1: that of the Shuttle rescue sphere. A second rescue capability was also developed initially for Shuttle EVA operations but was extended to ISS EVA operations. This is the SAFER manned maneuvering unit.

This was a small low-cost maneuvering device to be attached to the life support system of the Shuttle EMU and available for use in the case of an emergency situation. There are no inbuilt back-up systems as the SAFER itself was only ever intended as a back-up system. Full description of this unit, which was more simplified than the larger Martin Marietta Manned Maneuvering Unit was described in the companion volume *Walking in Space* (Shayler, 2004, pp. 74–76).

This unit is now standard equipment on both Russian-based and American-based EVAs from ISS. During the time a Shuttle is docked to ISS normal Shuttle EVA contingency rules, where the Orbiter can maneuver to "capture" a stranded astronaut, are prohibited by the time it takes to undock and separate from the station, clear the structure and pursue the drifting astronaut. The SAFER units therefore offer the redundancy to return to the station should a tether become disengaged or a problem occur with a pressure garment and allow a quicker, more direct route than the nominal hand-over-hand or RMS-supported EVA operation.

During Shuttle EVA operations it's always been the policy to train at least two astronauts in EVA techniques to complete together both planned and contingency EVAs. Unplanned contingency excursions could include the opening or closing operation of payload bay doors, separation or lowering of deployed payloads, alignment of communications equipment, closing of drain, vent or ground access doors or inspections of the surface of the Orbiter's skin. Normally, EVAs accomplished in the payload bay are in sight of other astronauts supporting the EVA by photo-documentation, operating the RMS or coordinating the events. During the early orbital flight tests single-person EVAs were possible for contingency purposes but fortunately were never called upon.

Early EVA experiments on Voskhod and Gemini crewmembers were tethered to the spacecraft and demonstrated not only the feasibility of performing work in a pressure suit outside a spacecraft but also the limitations and difficulties encountered in such work. Shuttle-based EVAs have expanded the knowledge and database of repair and servicing tasks, but the real expansion of EVA activities in Earth orbit have been at the various space stations orbited since 1971, and these offer other challenges for the safety of participating crewmembers.

SPACE STATION EMERGENCIES

The call for a multi-crewed space station offered designers the opportunity to develop a multi-crew re-entry and rescue vehicle in the event of an emergency aboard such space complexes. These rescue systems featured both on-orbit survival and Earth launch rescue systems, some of which were incorporated in the operational space station concept, others like the personal rescue systems proposed in the early years of the program hardly left the drawing boards. Design studies for space rescue systems similar to the lifting body concept have been prototypes since the 1960s, and were featured in the 1968 film *Marooned* and have continued to be studied right up to recent proposals for implementation during expanded crew operations aboard the International Space Station.

From FIRST to X-38

Using lifting body spaceplane concepts for rescue and recovery has been ongoing for almost 50 years. In the early 1960s Aerojet looked at the Rogelio wing concept suggested for recovering Gemini spacecraft for its own space rescue profile called FIRST. Since then a number of lifting body concepts have been proposed, and one by one dropped from planning mainly for financial reasons.

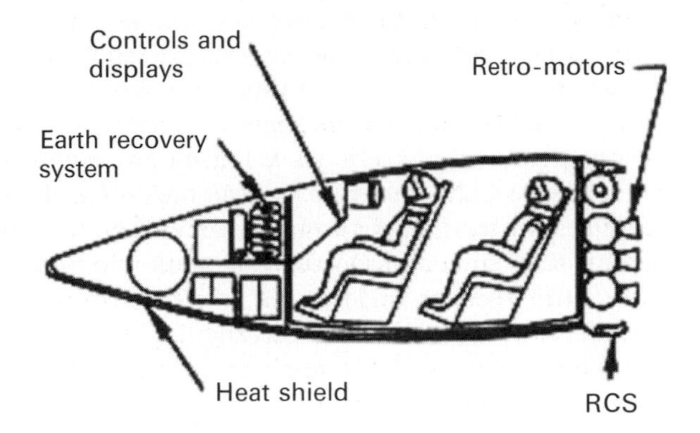

- Three-man
- Shirtsleeve environment
- 1,850 kg
- New technology requirements
  —new heat shield
  —re-entry technique
  —high-speed pilot techniques

Proposal for winged re-entry vehicles for crew rescue from orbit. © Mark Wade, Encyclopaedia Astronautix

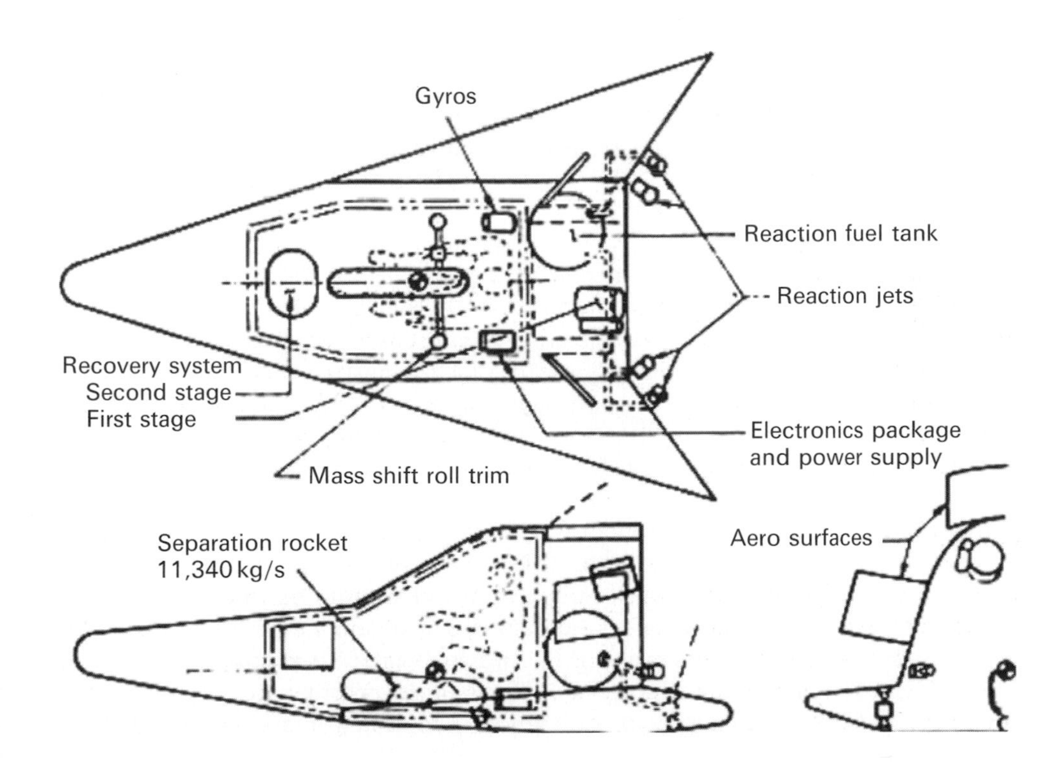

FIRST—fabrication of inflatable re-entry structures for test. This was a 1960 design to be stowed in a cylindrical package and attached to the external surface of the space station. In an emergency the astronaut would float into the cylindrical, coffin-sized central compartment, seal the hatch and blow the unit free of the station. A paraglider, made from ultra-fine filament, super-alloy, wire fabric impregnated with an elastomeric silicone matrix material would be deployed and inflated. The occupant could then use a gas stabilization and control system to orientate the unit for retro-fire and be required to keep the unit at the correct attitude during re-entry. It was estimated the unit would re-enter the atmosphere 26 minutes after firing the retro-rockets from a proposed space station altitude of 600 km. The lift-to-drag ratio of 0.5 and re-entry angle of 70° would allow partial control, either automatic or manual, with *g*-levels not exceeding 2.0. It was 6.5 m long and had a maximum diameter of 0.71 m and a mass of 407 kg. The maximum span of the device was given as 2.75 m. The lift-to-drag ratio once subsonic afforded a 345 km range to reach a suitable landing point with the pilot maneuvering the device to any available flat area landing at about 55 km/h, but if a flare was performed just prior to landing then the horizontal speed would be reduced to 9 km/h. In the event of a parasail failure after re-entry the occupant could separate the capsule and from below 9,000 m the unit would be blown open allowing descent by personnel recovery parachute.

LREE—lifting re-entry. This was a 1960 design for a three-person lifting re-entry capsule proposed by General Electric. The design featured a 5.17 m long vehicle with a span of 3.96 m and a mass of 1,303 kg featuring ablation surfaces and recovery parachutes.

REES—Re-Entry Escape System. An alternative-design, small lifting body–re-entry capsule design by ASC in 1963 for one occupant. It was 6.82 m long, with a span of 4.57 m and a mass of 1,171 kg.

SOREEV. This was a larger version of the parasail escape system capable of rescuing up to six astronauts and proposed by Aerojet. Its length was 8.45 m, its span was 13.0 m and its mass was 3,514 kg. It afforded the crew a 24-hour supply of oxygen but required landing within a day of leaving the space station (or an advanced spacecraft).

LBEC—lifting body escape concept. This was a Northrop-designed three-person lifting body that built on that company's previous work on the HL-10 and M2-F3 lifting bodies. The crew compartment was to be pressurized affording a shirtsleeve environment and the pilot had to be proficient in the retro-fire, re-entry and landing phases of the flight profile. A quick exit was possible without putting on a pressure garment first, but on the downside more than one astronaut would have to be given training on this vehicle in the event of the primary "pilot" becoming injured or ill preventing operation in an emergency. This vehicle, which used a parasail for recovery, was 5.80 m long and had a span of 3.50 m and a mass of 1,950 kg. It

Tests of the X-38 Crew Rescue Vehicle development vehicle.

used solid rocket motors for propulsion in the de-orbit burn. A set of RCS jets provided attitude control in orbit.

ACRV—Assured Crew Return Vehicle. This was NASA's answer to providing emergency recovery from the Space Station at times when the Shuttle would not be docked to it. In some circumstances a crew would be able to use the station as a safe haven awaiting rescue from a Shuttle vehicle, but as the Challenger incident revealed this could not always be relied on. Prior to cooperation with the Russians, an alternative design had to be developed to provide sufficient crew rescue capability on the Space Station leading up to the Freedom configuration of the late 1980s. Various designs were studied ranging from a Station Crew Return Alternative Module (SCRAM) which was a six-person quick-return module that would have posed high-g loads on the occupants and cost in excess of $600 million. Then there was MOSES (Manned Orbital Shuttle Escape System), designed for the Space Shuttle based on experience from USAF Discovery recoverable military satellites; it required the use of pressure suits and could accommodate between one and four crewmembers (730 kg–2,320 kg mass). This proven hardware design by General Electric was adapted for the Space Station. The other designs considered for the station ACRV even featured refurbished and unused Apollo CMs from the Apollo Program as well as the HL-20 and X-38 lifting body vehicles.

HL-20. Not exactly the budget version at about $2 billion, this NASA vehicle out of the Langley Research Center was based very loosely on the Soviet BOR-4 spaceplane design used to support the development of the Buran space shuttle program. It was designed to carry a crew of eight and some designs envisaged a personal launch system, a sort of mini-shuttle being launched into orbit by a modified Titan IV rocket. However, its primary purpose was as a potential space station recovery vehicle. It entailed a significant amount of research and tests including one that evaluated ten test volunteers in mock-ups in both horizontal and vertical positions. This clearly demonstrated the advantage for rapid entry and seating in an emergency situation, though wearing partial-pressure suits restricted movement

Soyuz, a space station crew transport and rescue vehicle for over three decades.

somewhat. The design life of this vehicle would have been three days in Earth orbit independent of the station. Its length was 8.93 m, its span 7.16 m and its mass 10,884 kg. Though funding for this concept became excessive resulting in its cancellation, it did provide useful development information on such concepts and for the X-38 design which followed it. Two flight crew and eight persons could ride in the vehicle which featured a piloted landing and rollout on an airfield, offering a more comfortable landing than previous designs. The option of developing a launch capability offered both ground-based and in-orbit storage capabilities, but it was the high cost that finally sealed its fate.

X-38. The more recent spaceplane/lifting body concept for a space station crew rescue and bailout situation was a NASA program manufactured by Orbital Sciences that was based on earlier lifting body technology and designs. The design featured indefinite orbital storage capability (4,000 days has been quoted) and an independent design life of only 9 hours. Capable of carrying a crew of six, it was to be left attached to the station as a Crew Return Vehicle until required, and then use a cold nitrogen gas for attitude control. With a 1,300 km cross-range capability it had landing opportunities every two or three orbits of Earth (maximum six orbits). It should be pointed out that this design did not feature onboard crew controls as it was a totally automated landing system at an airfield. Following re-entry, drag chutes would be deployed followed by a steerable ram-air parafoil. It was 8.69 m long with a span of 4.42 m and a mass of 8,163 kg. Development since 1995 was in response to initial doubts about the availability and reliability of the Russian Soyuz spacecraft to provide adequate CRV capabilities at ISS. During most of the 1990s, tests on the design configuration were completed with models of a joint NASA/U.S. Army design (known as the Wedge) to validate the concept for the X-38 demonstrator vehicle. In total 36 flights were completed in Phase I of the flight tests featuring four models of the Space Wedge, 45 flights in Phase II and 34 flights in Phase III, between June 1992 and 1996, and the experience gained from these tests was applied to the X-38 design. Using existing technology and material, development costs were much lower than developing a brand new vehicle. Two-year atmospheric drop tests on three test vehicles were to be completed. Unmanned captive tests commenced in July 1997. The first drop test of the X-38 occurred on March 12, 1998 and was a success. It was planned to deploy an unpiloted test vehicle from the Shuttle Columbia during 2000 and program it for an automated landing on Earth, but this was constantly delayed and after a second drop test in 1999 the program was canceled due to budget restrictions in 2002 with only two test flights completed (Miller, 2001, pp. 378–383).

Capsule rescue options

Over the years consideration was given to adapting proven capsule designs to space station crew rescue vehicles, and in the event this was how the option for crew escape evolved.

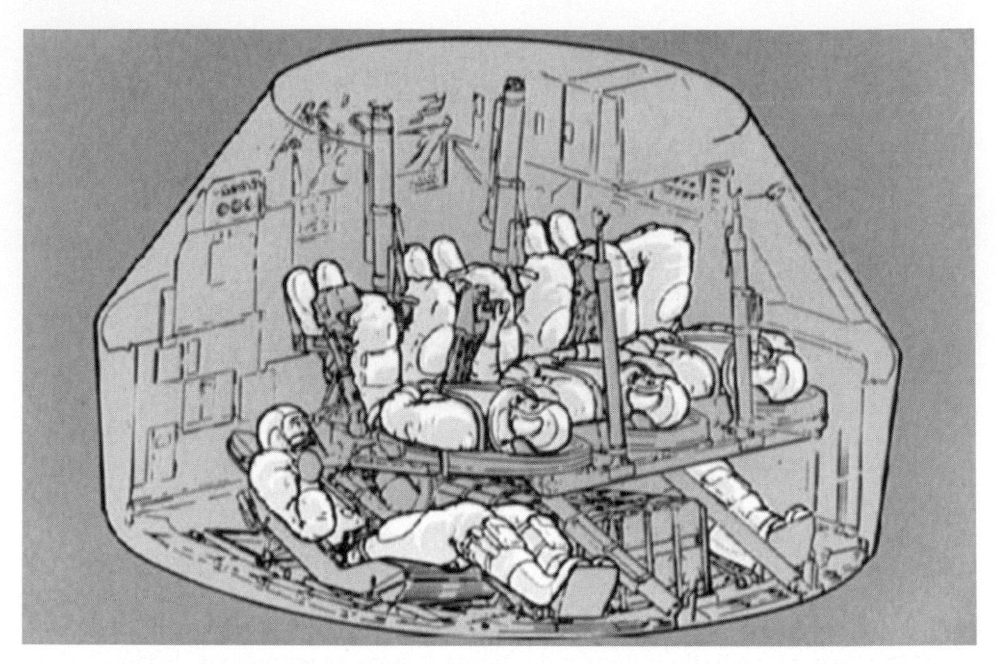

Cutaway of the Skylab Rescue CM configuration featuring five crew seats.

From early studies of the Apollo Application Program to the operational Skylab series, the flight crew were to be launched to and recovered from the Orbital Workshop by the Apollo Command and Service Module, which would have remained attached during periods of between 30 and 60 days of orbital operations at the station. During studies of the USAF Manned Orbiting Laboratory, two-person USAF crews would have ridden to and from orbit in a modified Gemini spacecraft which again would have remained attached to the laboratory for the 30-day duration missions. The use of the Soviet Soyuz for Salyut and Almaz space station missions between 1971 and 1986 and for Mir operations from 1986 through 2000 was also a clear demonstration of how (by rotating the Soyuz ferry craft) crews could be exchanged or the orbital life of the mission could be extended, offering a fresh vehicle every three to six months and providing a constant ability to leave the station quickly if an emergency situation occurred. The Soviets addressed the problem of making each cosmonaut safe—everyone had different body dimensions—by making the personnel seat liner interchangeable between spacecraft.

From these studies, rescue vehicles were proposed based on Gemini, Apollo and Soyuz designs.

McDonnell Douglas had suggested a larger Gemini for launch on a Titan IIIC capable of carrying five crew and potentially capable of recovering stranded astronauts from space stations in orbit. The Gemini re-entry module would have been extended to 3.05 m in diameter to provide a passenger compartment for three rescued astronauts and provide the baseline for the larger proposed Big

Gemini design and for potential rescue capability for a stranded crew on the Moon. Prohibitive development costs and the desire to move onto Apollo resulted in these designs not progressing any further than the drawing board (Shayler, 2001a).

During the Skylab program, NASA developed a kit to outfit an Apollo CM so that it could be launched with a two-person flight crew and dock to the second side-docking port of the workshop to rescue three resident but stranded astronauts and thus return to Earth with five astronauts. This option was almost called upon during the second manned occupation of the station when the resident crews, docked to the CSM, discovered leaks in the quads on the Service Module. When the problem was worked around by other methods the need for the rescue mission diminished and was not launched though the option was made available for the rest of the Skylab program. The system relied on the next vehicle being available under normal (or accelerated) launch processing. Thus, Skylab 2 (first manned mission) had the SL3 vehicle as its rescue craft, Skylab 3 would have used the vehicle intended for SL4 as its rescue craft, and likewise Skylab 4 relied on the Saturn 1B and CSM being prepared for ASTP as its rescue craft (Shayler, 2001b).

The idea of adding seats to the Apollo CSM was not new. In the 1960s studies were conducted on a third-generation Apollo CSM (Block III) to support the space station and extend the lunar surface program under Apollo Applications. There was even a study by Grumman for a rescue version of the LM used in Earth orbit with an attached Remote Manipulator System. Again these studies did not progress further than paper proposals (Shayler, 2002).

The six-person Apollo CM escape concept was revitalized for pioneering studies into Space Shuttle rescue capabilities by Rockwell in the early 1970s and featured an additional solid-propellant rocket pack strapped to the aft heat shield (similar to the Mercury retro-pack design). It was this idea that was again revisited when studying an assured crew rescue vehicle for the Space Station.

When cooperative talks led to Russia joining the ISS program in the early 1990s, it became clear that the most suitable and available vehicle for crew rescue (at least in the early stages of the ISS program) was the venerable Soyuz which had been performing a similar role for two decades in the Soviet national space station program. Some amendments had to be made to the then-operational Soyuz TM variant and resulted in the TMA version accommodating larger framed American astronauts in training for the early residence crews. This was chosen over the *in situ* design for a larger Soyuz planned for Mir 2 called Zarya. Studies in 1995 evaluated an ISS recovery craft based on Zarya with a solid retro-rocket motor and cold-gas thrusters with the potential of five-year storage at the station. A crew of eight could have been recovered. The Zarya-based space "lifeboat" had a mass of 12,500 kg, was 7.20 m long and had a maximum diameter of 3.70 m. It was to be delivered to the Space Station by the Shuttle, but its independent lifetime was only 24 hours. In 1996 these designs were rejected (along with that of Hermes) in favor of developing the Soyuz TMA design in the short term and (for a while, at least) the X-38 in the long term. Though Soyuz TMA is currently the ISS CRV, studies are continuing in Russia into using the Zarya design as a follow-on to Soyuz and in cooperation with the

European Space Agency as an independent crew space vehicle to ISS (and other stations) once the Shuttle is retired in 2010 (Hall and Shayler, 2003).

SAFETY ON THE SPACE STATION

In evaluating crew safety on the Space Station, consideration has to be given to the long-term design and functional lifetime of components and subsystems. There are some elements and components that can be exchanged, replaced and upgraded but there remain significant major components that simply are unable to be exchanged or removed from the complex. This was learned on the early space stations (especially Mir) where it became impractical, or just impossible to dispose of outdated or unwanted equipment prior to the arrival of the Space Shuttle due to the limitations of payload return of Soyuz.

When designing ISS such criteria as suitable certification guidelines and limits had to be defined and adhered to. The complexity of ISS and its international scope made this task even more complex. Initially, the elimination of hazards was by design or operation, any hazard that remained had to be minimized by safety factors, containment or isolation. Priority was given to preventing hazardous items being incorporated in the design and the provision of redundancy, back-up and work-around systems. Safety devices and a network of caution and warning systems was also to be developed along with a workable maintenance and repair program and provision of support work tools and spares.

Problem areas

Aboard a long-duration space station, potential explosive mixtures are a major risk and are required to be located away from habitation areas as far as possible. These include high-pressure fluids, volatile gases and critical fluids. Cabin pressure design should feature leakage before rupture and incorporate safety features to prevent both over-pressurization and under-pressurization. Additional safety factors are required on all valves, regulators and other pressurized components, and pressure shells should be made available for inspection, maintenance and repair wherever possible.

In addition, the use of toxic materials, hazardous fluids, explosive devices and contaminants should be kept to a minimum and methods to isolate the crew or other components from hazards should be thoroughly tested, especially where there are drains, vents and exhausts.

Crew health is paramount and any toxic material or contaminants must have adequate health and safety protection equipment and, if required, maintenance or servicing procedures. There must also be adequate provision for fire control, protection from excessive exposure to radiation, variations in temperature and micro-organisms.

Safe haven

When the Space Station Freedom was first proposed in 1984, NASA recognized the requirement for a safe haven for the crew pending rescue from an alternative vehicle if one could not be provided on station. In March 1984 this provision stated:

> "The Space Station shall provide area(s) to which the crew may retreat in the event of the development of hazardous or life-threatening conditions. The safe haven area shall be accessible at all times from anywhere within the station, isolatable from the hazardous condition, and shall contain emergency equipment including fire suppression, life support, communications, medical supplies and provision to maintain the crew over a period to be determined. The Space Station shall provide the capability to transfer crewmembers from the safe haven(s) to the rescue vehicle in the worst case i.e. through an unpressurized module" (NASA, 1984).

In the almost 25 years since this document was written, there have been countless changes to the Space Station Program though Freedom and ISS have always kept this goal in sight but achieving it has been elusive. Delays in the program, reductions in budgets, the loss of Columbia, the decision to retire the Shuttle and operational difficulties have seen the Soyuz remain as the only operational ISS safe haven and crew rescue vehicle. Although the design of the station prevents the crew assembling in one dedicated location with all the provisions identified above, it still provides access to an escape vehicle that bypasses the hazardous area.

During the fire and collision on Mir in 1997, it was found that the path to a second Soyuz was blocked by the hazardous gas and that the updated trajectory analysis information required for the re-entry of the Soyuz was the same for both vehicles, as it had not been a requirement to separate both Soyuz vehicles and perform entry profiles at the same time. Therefore, both vehicles had similar entry datapoints, a situation now amended for ISS where different de-orbit data are provided for each vehicle. As the station grows to its present size (2008), coupled with the plan to move to six-person permanent crewing from 2009, the provision of at least two constantly available Soyuz TMA vehicles remains. How this develops from 2016, which will mark the end of U.S. priority on ISS, remains to be seen.

There have been provisions for on-orbit puncture repair kits for some time. Studies and demonstrations of on-orbit repair to the Shuttle thermal protection system have continued since the return to flight of STS-114, and a system called KERMIt (Kit for External Repair of Module Impacts) has been proposed. The theory underlying Shuttle tile repairs in orbit has been around since the early 1980s and it is clear that EVA will play a part in future maintenance and repair as the station gets older. Indeed, it might come to the point where some elements of ISS will have to be closed and isolated and therefore no longer habitable, as was the case with the Spektr module following the collision of Progress with Mir in 1997.

HELLO, HOUSTON

It is the nature of manned spaceflight that the primary focus of attention is to the flight crew and the mission being flown. Often forgotten and overlooked are the hundreds of team members within NASA, support contractors and establishments across not only the U.S. but internationally that make each mission a success or support this effort by endeavoring to bring an accident or emergency situation to a happy and successful conclusion. At the forefront of this ground-based effort is of course Mission Control, whether in Houston (Texas) or Moscow (Russia).

Often controllers on the ground have far more data available to them than the astronauts or cosmonauts in space. The training, stress and dedication inherent in a flight controller's role are as high as those experienced by those who fly the mission. Essentially, flight controllers are the unseen crewmembers who "fly" every mission on the ground. The story of Mission Control is one waiting to be told in full, but there are some works that provide an excellent overview of the development, operations and characters in the MCC room at Houston in Texas (Liebergot and Harland, 2006; Kraft, 2001; Kranz, 2000; Murray and Cox, 1989).

Role of Mission Control

It is a fact that Mission Control does not actually *control* the spacecraft—it's the onboard systems and the crew that does that. The role of MCC, whether in Houston or Moscow or more recently Beijing, is one of analysis, whether that be of the onboard systems or the trajectory, the means of verbal communications from the crew, visual information by TV/video-link or data supplied by telemetry and tracking. The crew in space works closely with the "crew" on the ground; they have completed joint training and simulation sessions together covering both nominal stages of the flight but also off-nominal situations including abort situations, contingency and serious malfunctions, devised and thrown into the training session by simulation engineers, who not only have a somewhat strange sense of humor but also at times a warped one, designed to challenge some of the most experienced and hardened flight controllers or astronauts.

There is a lot of the spacecraft and its myriad of systems that the crew simply cannot see or be expected to monitor every second of the flight. Onboard and ground-based computer networks can handle this wealth of data and the team on the ground can quickly analyze what they see and hopefully act accordingly. The commander in the spacecraft can take the next step he or she feels necessary to control the spacecraft, both safely and efficiently; this is based not only on the current flight plan but also updates and advice from the ground. In the case of a serious problem the commander and the rest of the crew are expected to (and traditionally do) work closely with the controllers on the ground to overcome the problem, reach a safe conclusion or terminate the mission. The commander can of course make his or her own decision but must be ultimately responsible for their actions and answerable for the consequences if the decision conflicts with advice from the ground. This could result in an examination of the actions taken in light of real-time conditions, suggestions from the ground and any evidence retrieved from the mission. It could result in the appropriate course of action being taken that may result in the commander never flying again. Indeed, poor crewmember performance can also reflect on their future space career.

The role of Mission Control has evolved along with the program and hardware and is based on aviation models both civilian and military. Mission Control is responsible for incorporating into the flight plan multiple opportunities to proceed with the flight, change it or terminate it at that point. These have been termed go or no-go decisions. The latest information and updates are fed to the crew wherever possible to provide them with the most accurate and up-to-date information available from the ground so that a joint decision can be made. The new sequence of events are then interpreted as the current flight plan, and though the commander may choose to change or delay a certain step if they think it is better to do so, their decision, if correct, would be applauded; if not, they would have to explain to "the powers that be" why such a step was taken.

The balance of a strong, experienced and confident team on the ground, headed by flight directors who often come from the ranks of flight controllers, added to that of an experienced, confident and strong team in space, commanded by a veteran crewmember, is usually a winning combination. In support of the Mission Control teams operating the rotating shifts of flight controllers, there are teams of "back-room" support staff covering crucial situations, systems and hardware, some of whom are representatives from leading contractors, all of which in turn is supported by a countrywide network of aerospace academics and managerial teams. With the ever-increasing focus on the ISS being *international* the safety and success of each resident crew is based on the constant 24/7 support of an international team of controllers, engineers, managers, theorists and technicians across the globe.

In the event of a serious problem or situation, the data recorded on ground equipment at the time, leading up to and shortly after the incident must be protected for later analysis. Termed "protecting the data" these incidents—which include Apollo 13, Challenger and Columbia—form a database for subsequent investigation and enquiry.

CONTINGENCY ACTION PLANS

In addition to planning a hopefully nominal mission, there have to be plans laid and prepared for when things go wrong. Contingency action plans are available for every Shuttle mission and are a good way to explain how the ground responds to a serious problem that falls outside that of nominal mission operations such as the rescue and recovery of a crew from space (NASA, n.d.).

Who is responsible?

For flight preparation, launch ascent and post-landing operations the responsibility of contingency plans rests with the Launch Integration Manager at KSC. If during an ascent a stable orbit is not achieved then the Launch Integration Manager continues lead responsibility right up to landing. Once the Shuttle has entered a safe orbit any contingency responsibility lies with the Space Shuttle Program Integration Manager right up to landing.

The decision to proceed with the flight and its continuity are handled by a Mission Management Team made up of representatives from engineering, systems integration, the Space Flight Operations Contract Office, Shuttle Safety Office and the JSC Director of Flight Crew Operations, mission operations, and space and life sciences. This team is convened a few days prior to launch and remains until the safe landing of the Orbiter. The chair of this team reports directly to the Shuttle Program Manager. This team is tasked with resolving outstanding problems outside the responsibility of the launch directors or flight directors. During activities prior to launch the chair for the MMT is taken by the Launch Integration Manager at KSC, and during the flight it's chaired by the Shuttle Program Integration Manager at JSC.

During launch (or landing) preparations this team meets at crucial times to review the current situation and address any open issues in a series of go/no-go decisions. During the flight the mission is also evaluated from a Mission Evaluation Room, managed by Vehicle Engineering Office personnel, who report any situation, anomaly or engineering analysis to the MMT.

Should a contingency action be called, then the Manager of the Space Shuttle Program determines when the MMT no longer has operational oversight of the plan that has been determined in light of the contingency situation.

NASA has established criteria for contingency which are summarized below and cover damage to property, facilities and equipment (and their estimated value) as well as personal injury or death, and the resulting level of investigation.

Close call. No equipment property damage greater than $1,000 and no injuries to personnel or interruption of productive work. Any investigation is in accordance with its potential for a higher class of mishap. Should the event involve more than one field center an investigation board may be recommended. This is an event that has the potential for higher severity for any of the following types of mishap.

Mission failure. A mishap of such great severity that it prevents achievement of the primary mission objectives (such as Apollo 13). Here an investigation board is required and a Type A or Type B mishap investigation process is followed (see below).

Incident. Damage equal to or greater than $1,000 but less than $25,000, with personal injury less than a Type C mishap but requiring more than basic first-aid assistance. The investigation and/or analyses are the same as for a Type C mishap.

Type C. Damage equal to or greater than $25,000 but less than $1,000,000, occupational injury or illness resulting in the loss of a working day. The Deputy Associate Administrator appoints an investigator or investigation.

Type B. A mishap equal to or greater than $250,000 but less than $1,000,000. The permanent disability of one or more persons or hospitalization of three or more persons. The Associate Administrator Office of Space Flight (AA-OSF) or Deputy Associate Administrator would appoint an investigation board and this board would then investigate the mishap.

Type A. Incidents resulting in death (such as the Apollo 1, Challenger and Columbia incidents) and property damage greater than $1,000,000. The AA-OSF appoints an investigation board or the NASA Administrator chooses to appoint an investigation board and that board investigates the mishap.

Responsibilities

The Manager of the Space Shuttle Program is responsible in accordance with the OSF Contingency Plan to ensure that each center has contingency response actions; that the program is able to manage the required actions to minimize losses and preserve evidence in the event of any contingency; and that the program is prepared to manage the required contingency situation prior to handing over to the formal investigation board as required.

At KSC the Launch Integration Manager is responsible for the management of contingency activities after a suspected launch or an EOM landing contingency has been called. He or she then appoints the chair of the Mishap Investigation Team (MIT) and activates the team as necessary with approval from the AA-OSF.

Should the incident be on-orbit then the Space Shuttle Program Integration Manager handles this responsibility up to landing when it then becomes the responsibility of the Launch Integration Manager once again.

Should the failure, accident or incident involve any of the hardware or facilities under the Shuttle Program Elements (OV, SRB, ET, payload facility, etc.), it is the responsibility of the respective element manager to take appropriate action to prevent further injury, to secure the scene and records as required and inform higher management. These areas include projects managed at Marshall Space Flight Center, EVA, payload processing, Shuttle processing, integration, vehicle engineering and Flight Crew Operations Directorate issues.

Should the incident, accident or mishap occur during mission operations then the on-duty flight director would be the person to put contingency procedures into effect. Here all flight control and support personnel would be required to participate and log the status of each individual piece of equipment prior to and after the time of the potential contingency. The Flight Director Mission Log would be used and completed as soon as possible by coordinating logs from other personnel and forwarding them to the Mishap Investigation Board as required. At the time of the incident/accident a mishap is declared and each member of personnel would verify that their logs are up to date and institute a "hands off" policy with regard to switches, push buttons, indicated knobs and other controls to protect the security of information while observing continued flight safety rules. The MCC would remain in support and active until released by the AA-OSF.

Team players

Two weeks before the proposed launch a list of qualified personnel is drawn up who remain on alert for contingency scene action, if required. These teams, which include appropriate medical evacuation members and Orbiter recovery crews, are

Mishap Investigation Team. The MIT are responsible for immediate travel to the contingency site to gather first-hand information, witness statements, and material valuable to the investigation board. Any turnaround/salvage team cannot commence their operations until directed to do so by the MIT team.

Rapid Response Team. A group of personnel to support contingency landing operations at TAL sites including recovery of the Orbiter; the team arrive at the site within 18 hours of the contingency landing. Should a landing occur at a TAL site another trained member can be transferred to the landing site from other TAL locations, the team is then known as the Augmented Landing Site Rapid Response team. Should a landing occur at an alternative or contingency site a Non-Augmented Landing Site Rapid Response team will be implemented from the TAL site or KSC.

Crew Recovery Team. In the event of a non-continental U.S. landing then a KC-135 would be dispatched from Ellington Field by JSC FCOD to the location of the flight crew with the aim of returning them to the U.S. as quickly as possible. If necessary, any injured crewmember would be evacuated by appropriate means to the nearest U.S. military base.

Summary

With the crew either safely awaiting the end of their normal mission or preparing for an emergency or contingency landing, then support for the landing, wherever it may be, has to be instigated and organized so that any termination of the mission earlier than planned can receive such support.

During the period the crew are away from Earth their wellbeing and safety has to be monitored and guaranteed as far as possible by making the flight plan flexible, by monitoring onboard systems and by actions taken by the flight crew themselves and those at mission control and the supporting facilities.

The challenge that faces any crew at the end of a mission is returning safely back to Earth.

RETURN TO THE MOON, TO MARS AND BEYOND?

Discussions on a return to the Moon and human spaceflight to Mars have been ongoing for the past three decades, ever since we first traveled across the lunar distance. Follow-on programs are the subject of debate, planning, discussions, budget fluctuations, the whims of politicians and acceptance by the public. As the Shuttle

Program winds down and ISS operations continue, the advent of an expanded Chinese manned space program and opportunities for "tourist" spaceflight operations highlights once again the need for safety, rescue and recovery from the lunar distance: our next step in space.

It is far too early to determine exactly what procedures will be in place to ensure the safety of a crewmember at the lunar distance, whether in lunar orbit or on the surface, be it for short or long duration. There, the chance of a rapid return to Earth is more difficult than from Earth orbit bases of the future. It is not, however, as difficult as supporting abort or rescue scenarios from the distances of Mars, the asteroids or beyond. The opportunities for a relatively rapid return to Earth similar to Apollo (3 days) are not so easily achieved on a six-month to nine-month mission towards the Red Planet. In this case, it may be easier, safer and more suitable to abort a mission to the Martian orbit or surface and await rescue than try to bring the crew home in a crippled spacecraft. Earth support at these distances is compounded by the time it takes to send and receive signals; therefore, it is conceivable that the majority of mission control capability will be on the Mars spacecraft itself than on the Earth.

The next few years will see the appearance of the Orion CEV, the Ares launch vehicles, their development and associated procedures to sustain the crew in space. Depending on the accident, mishap or contingency this will become more apparent as the new hardware is tested and eventually flies.

REFERENCES

Erik Burgaust (1974). *Rescue in Space: Lifeboats for Astronauts and Cosmonauts.* G. P. Putnam Sons, New York.

Michael Collins (1974). *Carrying the Fire.* Farrar, Straus, Giroux, New York.

Arthur L. Greensite (1970). Emergency escape systems. *Analysis and Design of Space Vehicle Flight Control Systems.* Spartan Books, New York.

Rex D. Hall and David J. Shayler (2003). *Soyuz: A Universal Spacecraft.* Springer/Praxis, Chichester, U.K.

D. R. Jenkins (2001). *Space Shuttle.* Midland Publishing, Hinckley, U.K.

Chris Kraft (2001). *Flight: My Life in Mission Control.* Plume Books, New York

Gene Kranz (2000). *Failure Is Not an Option.* Simon & Schuster, New York.

Sy Liebergot and David M. Harland (2006). *Apollo EECOM: Journey of a Lifetime*, Second Edition. Apogee Books, Burlington, Ontario.

Jay Miller (2001). *The X-planes.* Midland Publishing, Hinckley, U.K.

Charles Murray and Catherine Bly Cox (1989). *Apollo: Race to the Moon.* Simon & Schuster, New York.

NASA (n.d.). *Space Shuttle Program Contingency Action Plan*, NSTS 07700, Volume VIII, Revision E. NASA, Washington, D.C.

NASA (1984). *Space Station Program Description Document Book, 3: System Requirements and Characteristics.* NASA, Washington, D.C. [prepared for the Space Station Task Force in March 1984].

NASA (1999a). *Apollo 9 Press Kit, 23 February 1969*, Release No. 69-29, Apollo 9 NASA Mission Reports. Apogee Books, Burlington, Ontario.

NASA (1999b). *Apollo 11 Press Kit, July 6, 1969*, Release No. 69-83K, Apollo 11 NASA Mission Reports, Volume 1. Apogee Books, Burlington, Ontario.

NASA (1999c). *Apollo 12 Mission Operations Reports, November 5, 1969*, Release No. M-932-69-12, Apollo 12 NASA Mission Reports, Volume 1. Apogee Books, Burlington, Ontario.

NASA (2000a). *Apollo 7 Press Kit, October 6, 1968*, Release No. 68-168, Apollo 7 NASA Mission Reports. Apogee Books, Burlington, Ontario.

NASA (2000b). *Apollo 8 Press Kit, December 15, 1968*, Release No. 68-208, Apollo 8 NASA Mission Reports, Second Edition. Apogee Books, Burlington, Ontario.

NASA (2000c). *Apollo 10 Press Kit, May 7, 1969*, Release No. 69-68, Apollo 10 NASA Mission Reports, Second Edition. Apogee Books, Burlington, Ontario.

NASA (2000d). *Apollo 13 Press Kit, April 2, 1970*, Release No. 70-50K, Apollo 13 NASA Mission Reports. Apogee Books, Burlington, Ontario.

NASA (2000e). *Apollo 14 Mission Operations Report, January 22, 1971*, Release No. M-933-71-14, Apollo 14 NASA Mission Reports. Apogee Books, Burlington, Ontario.

NASA (2001). *Apollo 15 Mission Operations Report, July 17, 1971*, Release No. M-933-71-15, Apollo 15 NASA Mission Reports, Volume 1. Apogee Books, Burlington, Ontario.

NASA (2002a). *Apollo 16 Mission Operations Report, April 3, 1972*, Release No. M-933-72-16, Apollo 16 NASA Mission Reports, Volume 1. Apogee Books, Burlington, Ontario.

NASA (2002b). *Apollo 17 Mission Operations Report, November 28, 1972*, Release No. M-933-72-17, Apollo 17 NASA Mission Reports, Volume 1. Apogee Books, Burlington, Ontario.

David J. Shayler (2000). The Apollo 13 explosion, 1970. *Accidents and Disasters in Manned Spaceflight*. Springer/Praxis, Chichester, U.K.

David J. Shayler (2001a). *Gemini: Steps to the Moon*. Springer/Praxis, Chichester, U.K.

David J. Shayler (2001b). *Skylab: America's Space Station*. Springer/Praxis, Chichester, U.K.

David J. Shayler (2002). *Apollo: The Lost and Forgotten Missions*. Springer/Praxis, Chichester, U.K.

David J. Shayler (2004). *Walking in Space*. Springer/Praxis, Chichester, U.K.

Mark Wade's website. Available at *http://www.astronautix.com* [for further details of the varied space rescue concepts over the past 50 years].

W. David Woods (2008). *How Apollo Flew to the Moon*. Springer/Praxis, Chichester, U.K.

8

Return to Earth

Over the past five decades it has been demonstrated several times that getting back on Earth is just as difficult and dangerous as leaving our planet in the first place. Developing the technologies for protecting the spacecraft and its human cargo from the vacuum of space, through the increasingly denser layers of our atmosphere and finally for a gentle landing on Earth is one of the more challenging aspects of spacecraft design.

The added complexity of human survival in both nominal and emergency situations and providing the ability to recover, or rescue a returning space crew adds not only to design considerations but also the expense of each mission. Being a planet with two elements on which a landing can occur has meant a division of recovery techniques over the years from either the ocean or on the land. If that was not complicated enough, consider the weather conditions, the terrain, distances, access, logistics and communications—all of which have to be incorporated in the planning of the mission to terminate the "flight". Add to this the inevitable, political consequences of unforeseen landings in hostile states and whether the hardware needs to be used again—a mammoth task for those who do not make the actual flight into space. For those who do, the ride home can be equally as thrilling, frightening and at times as deadly as the short rocket ride they made to leave Earth in the first place, or as challenging as learning to live and work off our planet.

THE LANDING SCOREBOARD

Humans first ventured off planet Earth in April 1961 and, at the time of writing (July 2008) some 47 years later, we have recorded 260 missions into space of which two (Mercury 3 and 4) were suborbital flights in 1961, and two (Soyuz 18-1 in 1975, Challenger STS-51L in 1986) were failed launches. In addition there have been 13 X-15 "astro-flights" between 1962 and 1968, surpassing the 80 km air/space decision and three "astro-flights" of SpaceShipOne in 2004 claiming the "X-Prize". Finally, there was one pad abort of a Soyuz spacecraft in 1983.

Of these all but one of the X-15 "astro-flights" completed a land landing and rollout; the other astro-flight resulted in the loss of the pilot and break-up of the vehicle in flight. The three SpaceShipOne "astro-flights" also returned to a land landing and rollout.

Both suborbital flights under the Mercury Program were recovered from the Atlantic Ocean and instigated a trend for U.S. astronauts to be recovered from the Ocean (Atlantic or Pacific) during all of the Mercury-Gemini and Apollo/Skylab/ASTP missions flown between 1961 and 1975.

Since 1981 all American astronauts have intended to complete their mission on the land, either in the Space Shuttle, with 121 landings out of 123 launches (the failed landings being the launch accident of Challenger in 1986 and Columbia during its landing phase in 2003) or on the Russian Soyuz spacecraft.

The Soyuz has been returning cosmonauts from space since 1967 and will continue to do so for some years yet. Of the 98 manned Soyuz launches, all have returned to Earth (territory in the Soviet Union from 1961 to 1991 or Kazakhstan since 1991) like Vostok/Voskhod before them, though two (Soyuz 1 in 1967 and Soyuz 11 in 1971) have resulted in the loss of the crew during the closing stages of their missions. There have been several hard landings throughout the Soyuz program, and one (Soyuz 23 in 1976) actually splashed down in a frozen lake. The 1983 Soyuz launch pad abort also resulted in the safe landing of the crew under emergency conditions.

Finally, there have been two manned Shenzhou launches with landings on land in China. With a third mission planned for 2008, continued use of a parachute recovery system in Chinese territory is the plan for future Shenzhou manned operations. With the certification of Orion over the next decade it will be a return to parachute landing on U.S. territory or in the ocean for American astronauts after the retirement of the Shuttle in 2010 unless the ride home is in a Soyuz from ISS.

GET-ME-DOWN SUITS

Early "spacesuits" developed from high-altitude military pressure garments were essentially back-ups to the life support system in the event of a failure in the primary supply of oxygen and pressure, with the ability to sustain the crewmember until a suitable opportunity arose to de-orbit the spacecraft and bring home the crewmembers safely. These became categorized, like the military pressure garments, as "get-me-down suits". These then were the objectives of intravehicular activity (IVA, activity inside the vehicle) pressure garments worn during Vostok, Mercury and Gemini.

The Gemini EVA suits had added protection for the wearer and the Voskhod 2 cosmonauts wore pressure garments in support of their EVA activities. The crew of Voskhod 1 did not have room to wear pressure garments and it was decided that, for crews from Soyuz 1 on, pressure garments would not be required except for EVA.

Coming home on the Shuttle middeck.

It was this decision that tragically cost the lives of the Soyuz 11 cosmonauts during the re-entry sequence in June 1971 when the cabin air was lost due to a faulty valve and they died before they could close off the leak. From 1973 Soyuz crewmembers have worn a Sokol pressure garment for all dynamic phases of their flight.

For Apollo (and Skylab/ASTP) operations the crew wore full-pressure garments for dynamic phases of the mission that had added protection for EVA. After the Shuttle Orbital Flight Tests, Shuttle crews wore clamshell-like helmets only for launch and entry and not pressure garments. Following the loss of Challenger in 1986 escape suits for all Shuttle crewmembers were reintroduced into the program (they are described on p. 329). Had Buran flown with a crew a variant of the Soyuz Sokol-type suits would have been worn that was suitable for orbital emergency operations as well as launch and entry escape.

Soyuz landing recovery team.

In June 2008 a contract was awarded to Oceaneering International Inc. of Houston for the design, development and production of a new spacesuit to support the Constellation Program. The contract runs through September 2014 and it is hoped will provide a suitable pressure garment to support the first manned flights on the Orion crew capsule from 2015, eventually supporting flights to ISS and the planned return to the Moon. Configuration 1 is a pressure garment designed to support launch and landing operations and contingency IVA activities; it would also be used in the event of launch aborts and loss of pressure in the vehicle, as well as for a contingency EVA should the need arise. This is similar to the Russian Sokol suit described below. Configuration 2 is designed to be used on the surface of the Moon and will be reconfigured by the astronaut from Configuration 1 with elements dedicated for surface explorations.

SOKOL K (*KOSMICHESKY*, SPACE FALCON) PRESSURE SUIT

The longest serving suit of this type is that worn on Soyuz; it was introduced in 1973 as a direct result of the loss of the Soyuz 11 cosmonauts two years before. Developed at Zvezda Plant 918 it is lightweight, personally tailored to each cosmonaut and comparable with the Kazbek shock-absorbing seat liners inside the Soyuz Descent Module.

Its primary aim is to sustain the cosmonaut in the event of a cabin depressurization during launch, in orbit or during entry and has received one major upgrade

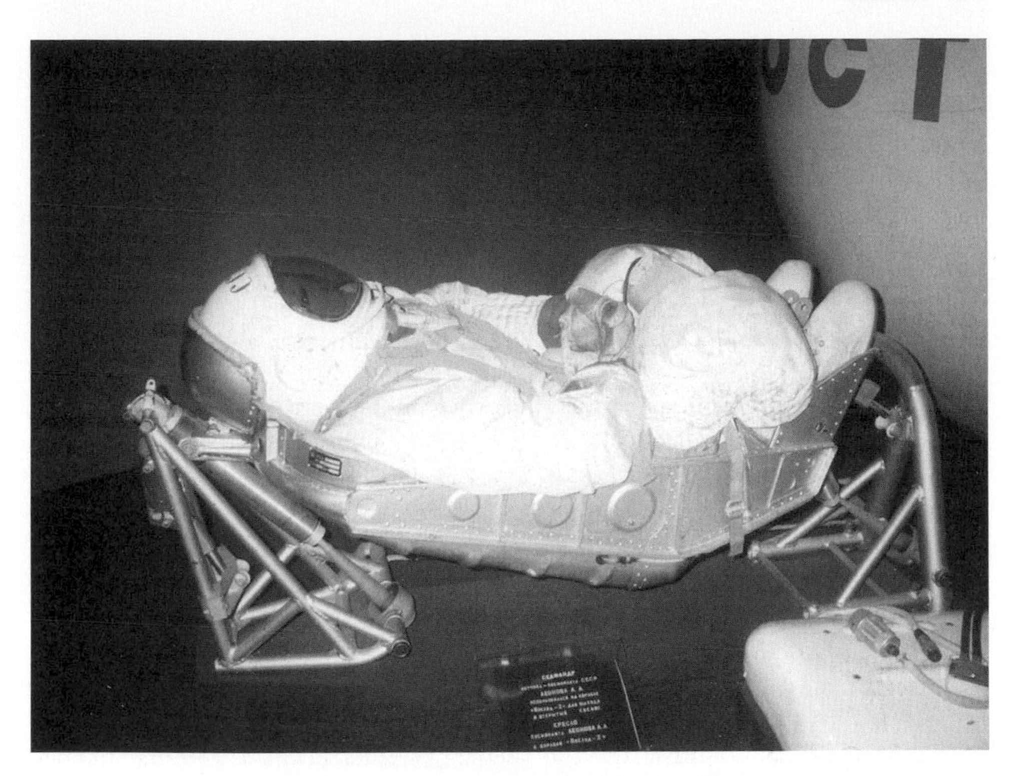

Soyuz pressure suit and cosmonaut couch.

(KV-2) to support Soyuz T operations, Soyuz TM and TMA variants of the main spacecraft. Sokol K or KV-2 operate at 41 kPa in full-pressure mode and have a 26.7 kPa capability in back-up mode. The Sokol K could support a cosmonaut in the suit in a pressurized cabin for up to 30 hours or in an unpressurized cabin for two hours. Weighing about 10 kg it is a quick-don suit and can be put on in just a few minutes, reaching the operational pressure condition in just 30 seconds. Though intended solely for IVA, it does have the capability and facilities to support an emergency EVA supported by a small portable backpack. Additional protective layers have to be added which are normally stored in the Soyuz. It was first tested on Soyuz 12 in September 1973 and though some difficulties were encountered in the tests these were soon overcome and the Sokol suit became a standard item of survival and rescue equipment on all manned Soyuz vehicles since that time, and it is expected to be a focus element of future Soyuz flights for some time. The original Sokol K included 66 test and training models and 89 flight models during its operational lifetime of 1971 to 1981. It was worn by cosmonauts flying on Soyuz 12 (1973) through Soyuz 40 (1981). Following a 6-year test program KV-2 was introduced in 1980 for the Soyuz T-2 mission and has been worn by all crews since that date. The number of test and training models is in excess of 60 and the number of flight models to date has exceeded 220 (Hall and Shayler, 2003; Abramov and Skoog, 2003).

PARACHUTES

The idea of a deployable canopy designed to trap air inside to slow down the descent of a payload suspended below is centuries old and has been an element of safety for pilots of balloons, gliders, aircraft and microlights for well over a century.

The dictionary describes a parachute as "a rectangular or umbrella-shaped apparatus allowing a person or heavy object attached to it to descend slowly from height." It was this capability that attracted the designers of the first human spacecraft: it was far easier to develop a parachute landing and recovery system than a prepared winged vehicle that could be flown to a landing site and refurbished for relaunch.

Though runway landings by the American Shuttle were to feature during the three decades from 1981, it has still been the parachute that has brought home the space explorers of many nations for the past five decades. Parachutes were not only used in the nominal landing of a space crew, but also available (where carried) for any emergency and contingency landings of space crews.

Vostok and Voskhod

Following retro-fire and descent through the upper atmosphere, the cosmonaut entry/exit hatch was jettisoned at 7,000 m and the catapult ejection seat initiated to eject the lone cosmonaut out of the Descent Module for descent by separate parachute system. The Vostok Descent Module, now unoccupied, descended under its own parachute system. At 4,000 m the parachute compartment hatch was jettisoned and a drogue parachute was deployed pulling out the main chute at 2,500 m. Meanwhile, the cosmonaut would separate from the ejection seat at 4,000 m to descend under a personnel recovery parachute system landing close to the empty Descent Module. Full details of the development of the Vostok/Voskhod landing system and technique can be found in Hall and Shayler (2001) and Siddiqi (2000).

Often witnessed by local people working in farms across the landing zone in Kazakhstan the cosmonaut would await the landing support team who would perform initial medical tests on him/her before being airlifted back to Moscow for more intensive post-flight debrief and evaluation. The Soviets had trained a special group of doctors to parachute in if an emergency condition occurred. They could then treat the injured cosmonaut while the main recovery teams made their way to the landing site. A fleet of helicopters and support personnel as well as engineers arrived at the site, some focusing on the cosmonaut and others on the landed Descent Module. This format has over nearly five decades been expanded by the Russians and now includes a specialist unit for the recovery of crews from space, though its basic format remains the same as in the early days.

Parachute training and ejection tests prepared the cosmonauts for the events at the end of their Vostok mission, and the hardware had been extensively tested to support ejection not only at the end of the flight but also during the launch of the vehicle into space. The most extensive test program was during the Volga high-

Though Vostok cosmonauts expected landing on land they had to train for possible water landings.

altitude balloon program, where experienced parachutists ejected or jumped from stratospheric balloons or aircraft to evaluate Vostok hardware and procedures. Such a test on November 1, 1962 resulted in the death of one of the most experienced Soviet parachute testers, Colonel Pyotr Dolgov. Tests of the parachute and ejection system were also tested on unmanned Vostok missions (Korabl-Sputnik) which included biological specimens, animals and manikins.

Interestingly, under IAF rules a pilot had to take off and land in the same vehicle to qualify for a new aeronautical record. Of course, the Vostok cosmonauts did not do this; it was not until 1978 that the initial reports of Gagarin staying in his Vostok from launch to landing were revealed as being not exactly true. He like the other

Recovery teams are soon on the scene for a Vostok landing.

five cosmonauts who flew Vostok ejected as planned and landed separate from his spacecraft.

For the upgraded Vostok called "Voskhod", the ejection seats were removed; therefore, the two crews who flew the missions had to land while inside the Descent Module. To cushion the invariably hard landing a three-part parachute system (a drogue, a braking canopy and two primary canopies) would be deployed. A solid-propellant braking rocket located in the parachute harness would be activated by a 1.18 m probe extending below the capsule. The firing of the rocket reduced the landing velocity from about 8–9.75 m/s to a more comfortable 0.15 m/s with added support coming from the cosmonauts' spring-loaded suspension Elbrus couches. If anything had gone wrong with the soft-landing rocket, or indeed the parachute system on Voskhod, there would have been hope for the survival of the cosmonauts. However, they could not have exited their spacecraft to descend by personnel recovery parachute as the Vostok cosmonauts had available to them.

Mercury

Development of the recovery system for America's pioneering manned spacecraft program focused on a parachute landing in the ocean and a deployable "bag" suspended below the capsule attached to the heat shield used to cushion and stabilize the spacecraft in the water. During 1959 and 1960 the Rogello wing was considered for recovery of the Mercury capsules, but though its study was continued at NASA for some years (and led to serious consideration for Gemini) it was not adopted for the American manned spacecraft of the 1961–1975 era.

The Recovery Section of the Mercury capsule, the uppermost section at the apex of the design, featured the parachute stowage assembly. For both suborbital and orbital missions one of two barostats would sense the pressure of the atmosphere and either could trigger activation of the landing system at 6,400 m. The 1.8 m diameter ribbon-type drogue parachute was deployed, for deceleration and stabilization purposes, followed by the 19.2 m diameter main parachute or the ring sail–designed parachute deployed at 3,050 m by means of jettisoning the antenna canister and the drogue chute.

Another pair of barostats signaled separation of the drogue and deployment of the main chute, which was initially deployed in reefed condition to 12% of the maximum diameter for four seconds which would minimize the opening shock, then to full deployment. Should the main have failed then the reserve would be manually deployed by the astronaut. Either main or reserve parachutes would provide a sink rate of 9 m/s at sea level. The recovery compartment also contained an antenna, a flashing beacon light and communication gear to aid in the recovery of the astronauts.

Twelve seconds after deployment of the main parachute the landing bag was deployed. This was a rubberized cloth assembly approximately 1.2 m long; it was folded and contained between the spacecraft heat shield and pressurized compartment. Once mechanical latches were released it dropped down to its full length. For a water landing the bag, with air trapped inside, attenuated the landing impact from

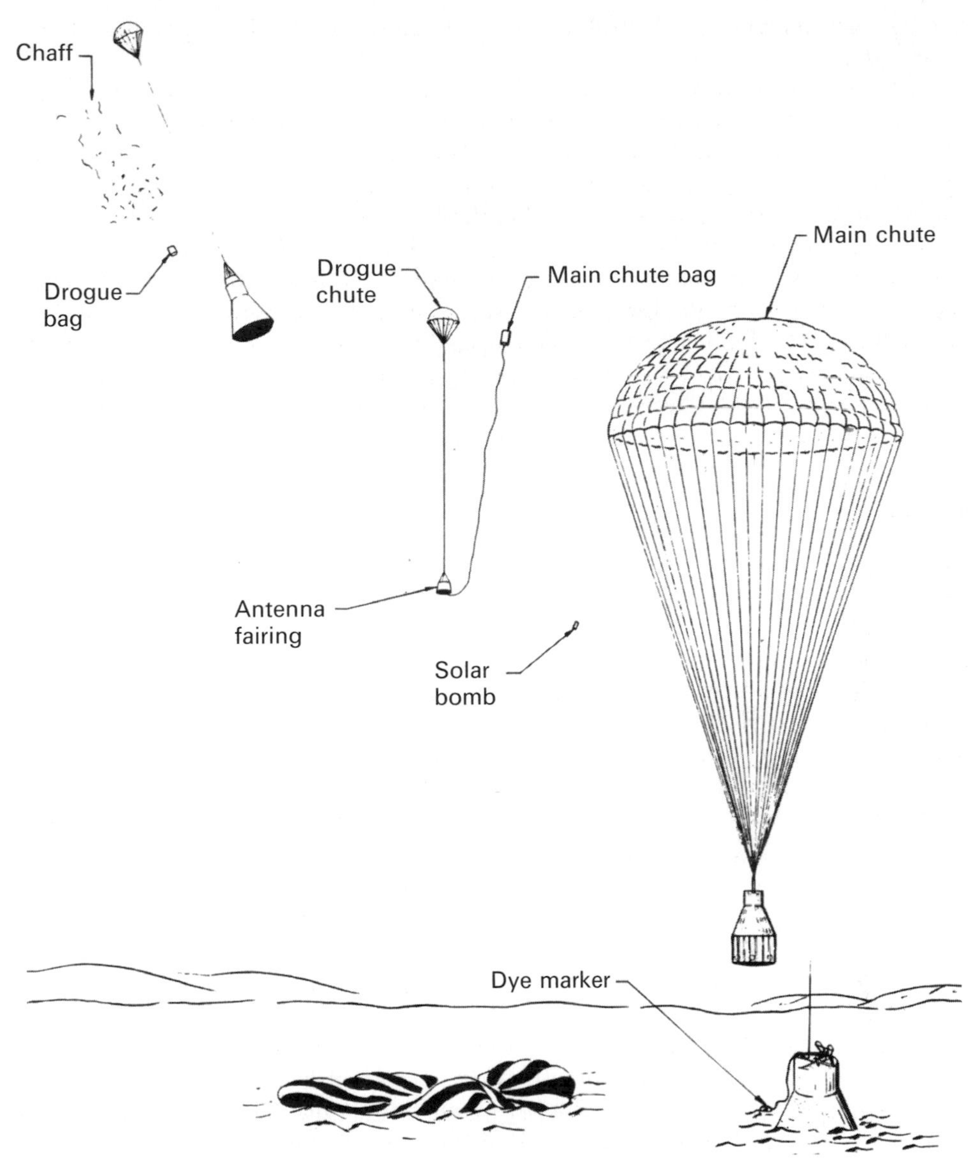

Mercury parachute landing sequence.

about 45g to 15g; it then filled with water providing an "anchor" to assist in flotation and stability while awaiting recovery (NASA, 1999; Catchpole, 2001).

Development and testing

Testing of the parachute descent system commenced before NASA came into existence. In August 1958 the then Space Task Group had requested a series of parachute

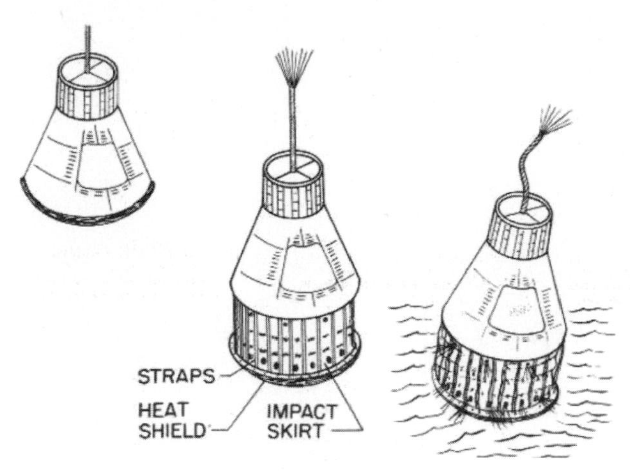

Mercury landing bag deployment.

drop tests designed to qualify the "Earth Landing System" for use in the proposed piloted satellite vehicle. Less than a week later, negotiations with the Pioneer Parachute Company focused around the development and prediction of a test program parachute. This was to be smaller than the production model at 14.1 m as the proposed test vehicle would be lighter than the actual flight vehicles. Approval came for the drop test program the following month.

This was to be a three-phase program. Phase 1 would be from a helicopter and involved an active parachute canister attached to a 208-liter oildrum filled with concrete (which would act as ballast) to simulate the Mercury capsule. During the tests, pyrotechnics were fired which not only opened the canister but also deployed the parachute.

Phase 2 involved a single drop of a boilerplate capsule out of the rear door of a C-130 transport aircraft on September 29, 1958. This was a test of the deployment system to be used during Phase 3. The system was attached to a pallet and when the aft aircraft door was opened a parachute pulled out the pallet and the boilerplate which then separated and descended under its own parachute system. Phase 3 consisted of five separation tests of this system between November 25, 1958 and March 24, 1959. The sequence was filmed by a T-38 chase crew and drops ranged from 1,524 m to 7,010 m and produced varied results that initiated changes to the design of the ELS (as these tests were intended to do prior to certification for manned flight).

Instability during the first two drops required the addition of a high-altitude drogue parachute, to be deployed first, in order to help stabilize the falling spacecraft prior to deployment of the main parachute. Initially, the Mercury heat shield was jettisoned, but on the third drop test the simulated shield impacted the spacecraft and on landing it floated on its side and could not be uprighted, and on the next test the simulated shield struck the parachute container and closed the release circuit which resulted in the jettisoning of the system partway through its full deployment cycle. The result was a tumbling spacecraft hitting the water at such a velocity that it was

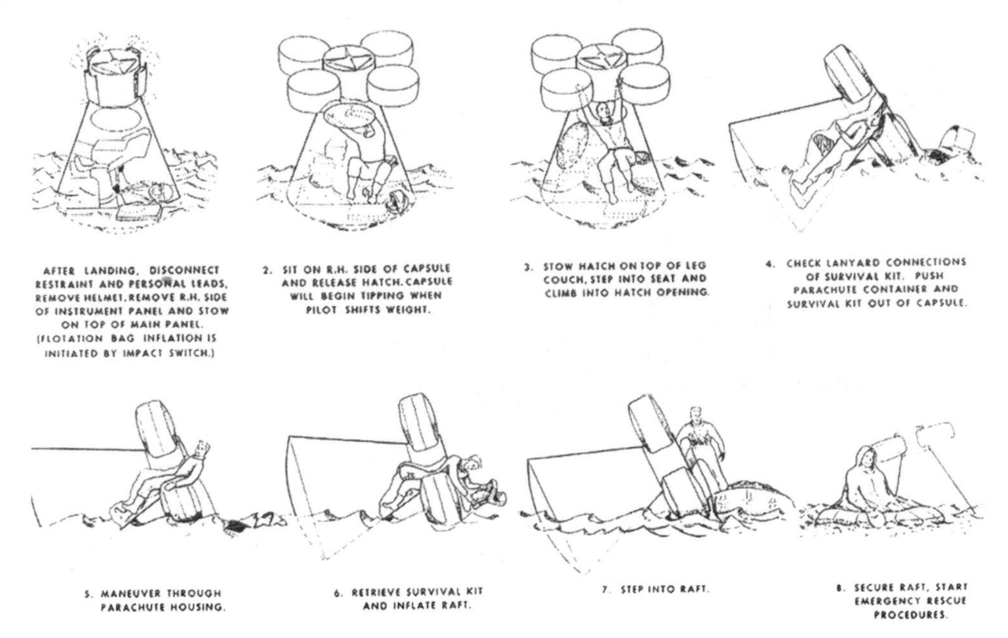

Mercury astronaut top-hatch exit procedures from Project Mercury indoctrination (May 1959).

destroyed. This resulted in changes to the plan to jettison the heat shield and adopt the original idea of dropping it 1.2 m to deploy the landing bag to cushion both the impact and retain the spacecraft in the upright position. However, when tested on the fifth drop the parachute constantly opened and closed ("squidding") and never fully deployed, again resulting in a hard impact with the water and a destroyed spacecraft.

To solve the problem, the parachute was changed to a 2 m diameter fist ribbon-type drogue and a 19.2 m diameter ring sail–type main parachute. Testing was carried out between April and July 1959, resulting in ten test drops. In addition, the series of drop tests were used to qualify the ELS for use on the Big Joe launcher that carried the first Mercury spacecraft on the Atlas launch vehicle. It was found that the original fist ribbon-type drogue was unsafe; it was replaced by a 2 m diameter conical drogue design instead.

There were 15 drops to test the new drogue deployment system, and 56 drops from helicopters and aircraft qualified the main parachute system up to altitudes of 9,144 m; helicopters were used to simulate altitudes to qualify off-the-pad aborts. Finally, a three-phase drop test program over water was completed in 1960 which featured boilerplates dropped with landing skirts deployed, a series of land landings, high wind speeds (up to 10 knots) and a chance that the attached heat shield would hit the spacecraft upon water contact. The testing was completed by two drops without landing bags to simulate a failed deployment; in this case no damage was sustained to either boilerplate spacecraft used in these tests. The qualification program was completed during Project Reef, a 20-drop test of the heavier one-day spacecraft with a 19.2 m ring sail–type parachute.

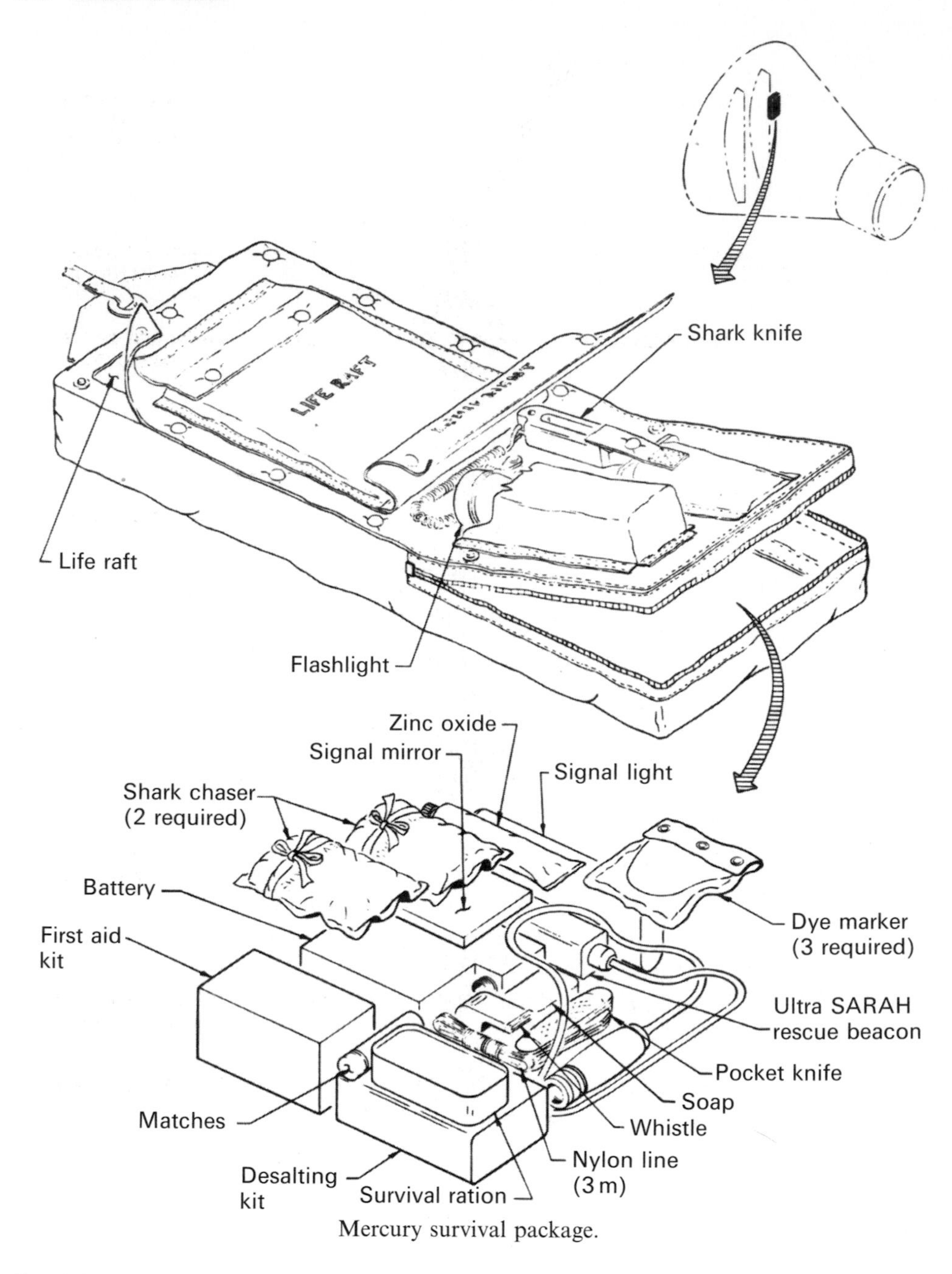

Mercury survival package.

Gemini

Tests on the ejection seat and personnel recovery parachute system had been completed under the launch abort program, and qualified the system for nominal ejection during descent, should the need arise. In the event, no Gemini astronaut required

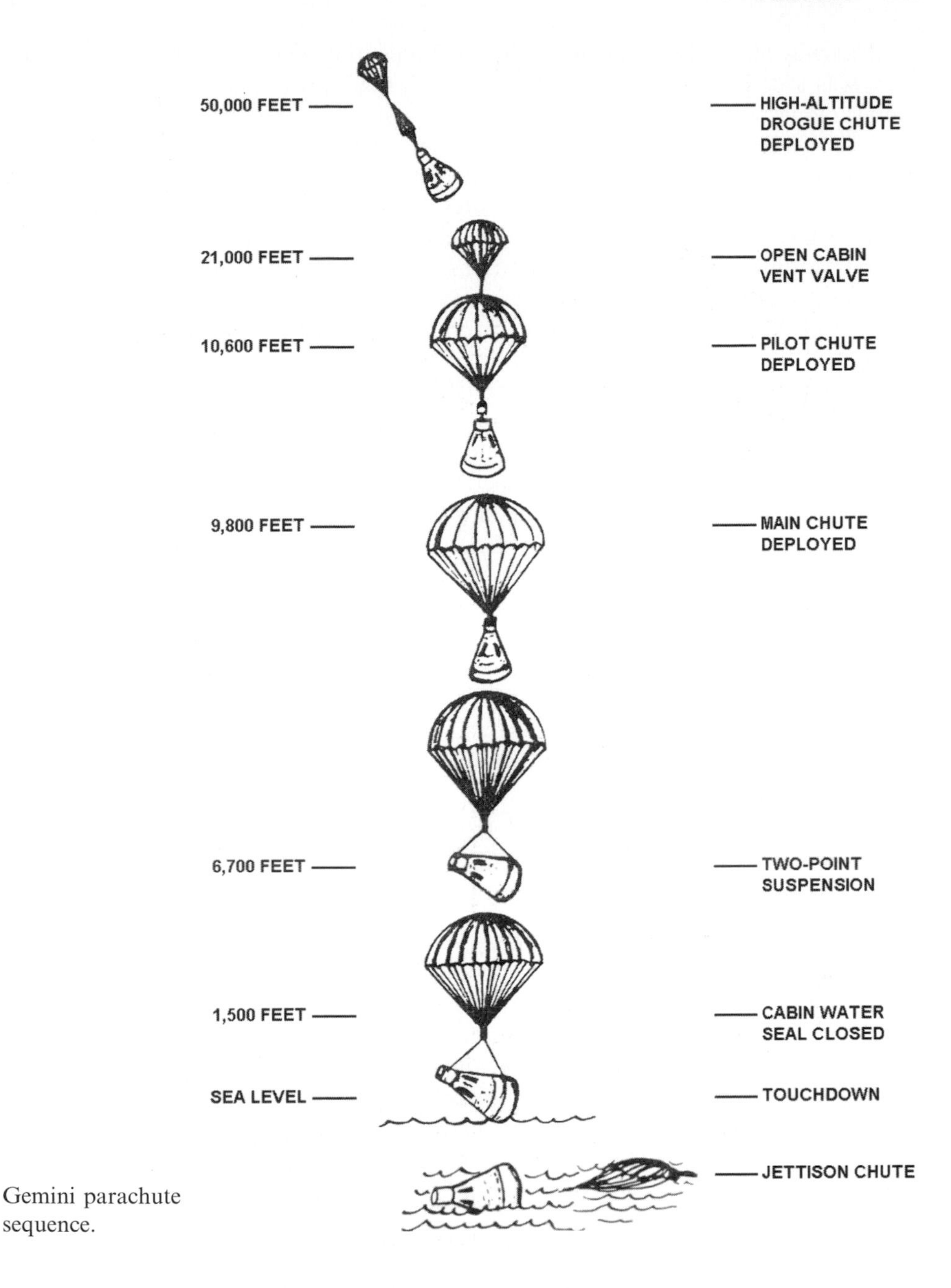

Gemini parachute sequence.

ejection from the spacecraft during descent. The spacecraft parachute system and the means of evaluating emergency recovery techniques required a separate test program, but also qualified it for both the NASA Gemini mission and the proposed USAF MOL program (which were not flown manned).

Mercury Mark II featured a nominal ocean landing under parachute, but in 1961 there was a desire to try to bring the two-man spacecraft back for a land landing, sliding along the surface on skids. With the assistance of a Rogello-type parasail wing, the development continued for four years until a series of difficulties and delays finally convinced NASA to abandon the program in August 1964 and commit Gemini to ocean landing under parachutes. Testing continued independently and, despite some interest from the USAF in using the system for controlled pinpoint runway landings (with military applications in MOL), the system was only tested on mock-up spacecraft of the Gemini design (Shayler, 2001).

The Gemini parachute recovery featured the deployment of a 3.3 m drogue at 38.6 km altitude; then, once stabilized two miles above the ocean, an 6.4 m pilot was deployed which also separated the rendezvous and recovery section of the spacecraft nose. An 25.60 m diameter ring sail–type parachute was deployed at 2.9 km altitude supported by a two-legged support structure located between the two pilot stations. It would then drop to a two-point suspension configuration, pitching the spacecraft 35° down from the horizontal offering a more comfortable seating position for the crew as they hit the ocean at about 32.8 km/h. At splashdown, parachutes were ejected and recovery aids deployed. The crew could open the hatches, deploy splash curtains and await rescue, when flotation devices would be used or life rafts, if required (which no astronaut did), and the spacecraft could be hoisted with the crew aboard on to the primary recovery capsule if they desired.

If the drogue had failed, an alternative deployment sequence would have been manually activated by the astronauts at an altitude of 3,230 m by depressing a switch to manually initiate four guillotines, freeing the drogue parachute from the re-entry module; five seconds later the pilot parachute pack would have been ejected and deployed, and a nominal landing sequence would have followed.

Development and testing

Initially, Mercury Mark II was to be recovered by two Mercury-type main parachutes, but the idea of a paraglider land landing appealed and development focused on that system as a primary objective for Mercury Mark II and then for early Gemini studies for a while. However, in March 1962 NASA directed North American Aviation to design and develop an emergency parachute recovery system for both half-scale and full-scale flight test vehicles under the paraglider program. This was subcontracted to Northrup Corporation's Radioplane Division in Van Nuys, California. The Parachute Landing System was to be used for the first unmanned test flight and employed a 25.6 m diameter ring sail design (Vince, 1966).

In May 1962 tests were commenced to qualify the emergency parachute system; the first two tests were successful but the third failed to deploy the main chute. The problem was resolved and a fourth test was enough to qualify the system for the half-scale test program.

That August, tests were commenced to qualify the emergency system for the full-scale test vehicle supporting the paraglider. Again these were not so successful with one of three chutes lost on the second test, the loss of two chutes on the third test

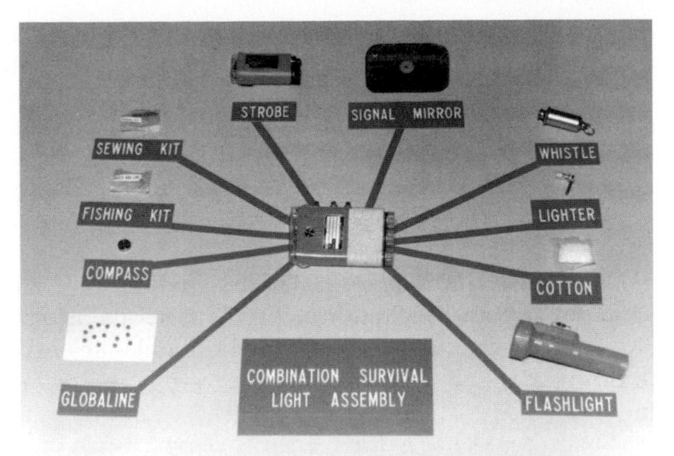

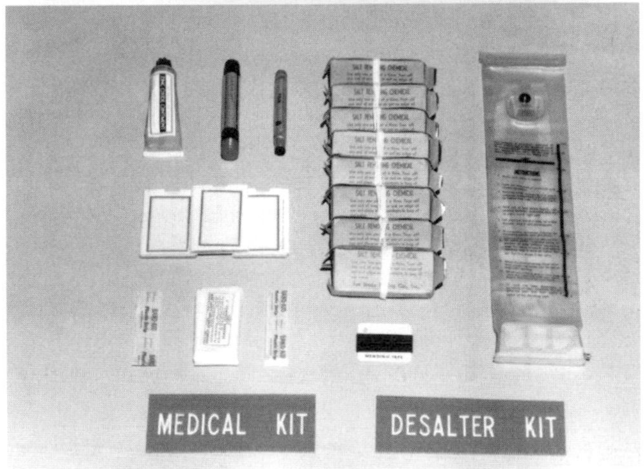

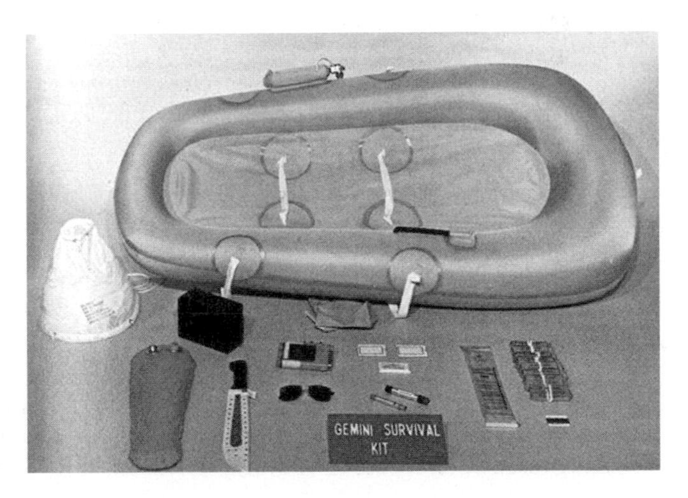

Gemini survival equipment.

(both sustaining minor damage) and the complete destruction of the boilerplate on the fourth test when all parachutes failed.

In February 1963 a series of 20 drop tests were completed on qualifying the parachute recovery system using a 5.5 m drogue parachute to determine the rate of descent of the dummy rendezvous and recovery section (four drops) and the 27.3 m diameter ring sail–type parachute using Boilerplate No. 1. Full deployment commenced on Drop 7 and, despite some small problems, the series was successful allowing qualification drop tests to commence in April 1963.

Problems with the paraglider development saw the use of the system delayed until the seventh then tenth of twelve missions before being finally deleted from Gemini altogether. Meanwhile, qualification testing of the parachute recovery system commenced in May 1963 but was delayed by problems in tucking in the edges of the canopy, added control tapes solving the problem. Tests resumed in August 1963 over the Salton Sea Range in California. The first five had been land impact tests, while the sixth and seventh were water impact tests. Studies were also completed on a parasail landing system (over a paraglider) and ring sail designs, but limited time and funding prevented further studies, and in August 1963 the idea of a parasail for land landing on Gemini was abandoned. Problems with stabilization saw the introduction of a drogue into the system in September 1963. This was qualified during a three-stage program. Phase I would develop the technique of pilot parachute deployment by the stabilization chute. Phase II was a drop test program that would complete the development series, while Phase III would qualify the system. In September 1963, 16 Group 1 and Group 2 astronauts commenced water and land parachute training to support potential low-level aborts under 21,336 m. Each astronaut was towed by a 7.3 m diameter parasail to 12.2 m and then released to glide down and land.

Tests on the Gemini spacecraft recovery parachute had to focus not only on horizontal water landings but also vertical ones if the two-point suspension system failed. Since a landing on land could not be ruled out tests were also completed in the California desert to obtain data on vertical land impact landings.

In all there were 89 tests over a 2.5-year development and qualification program. Drogue tests were completed at the Department of Defense Joint Parachute Test Facility at El Centro, California (15 drops); pilot parachute tests included 27 test drops (2 on water); and for the main parachute system there were 32 tests of the system. There were also a series of system qualification tests which comprised 3 tests for the unmanned flight program and then 10 drops in support of the manned flights.

Apollo

The Apollo parachute recovery system featured three main parachutes, two drogue chutes, a forward heat shield separation augmentation parachute and related electronic, mechanical and pyrotechnic systems to perform the operations. The recovery sequence was an automated system instigated by the closure of barometric pressure switches or manual initiation of time delay relays by the crew. The objective was to effect a safe and stable water recovery of the crew after nominal launch or abort

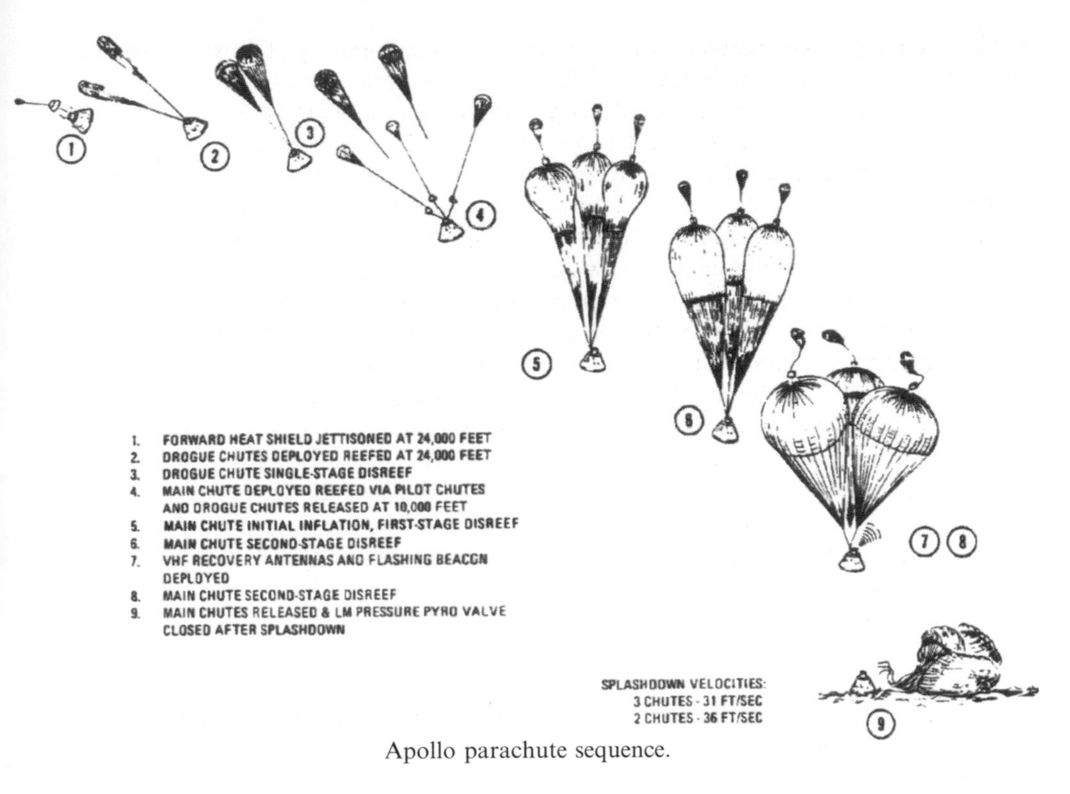

Apollo parachute sequence.

situations. Several development programs were highlighted by a constant battle with the overall mass of the system.

In January 1962 the original specifications were released for the Apollo Command Module to have an integrated parachute recovery system. The development and supply of this system was expected to take about 13 months. In reality, due to ongoing weight problems, various program delays and changes to the system, especially after the Apollo 1 pad fire of January 1967, the system was not qualified for manned use until July 1968, three months before the first Apollo crew flew into space. The same system, with amendments, flew on a total of 11 manned Apollo missions, three Skylab missions and one ASTP mission and only recorded the failure of one main chute to fully deploy (on Apollo 15 due to fuel coming into contact with the parachute fabric riser). It was successful in the safe recovery of all astronauts during the 1968–1975 period. The unmanned missions along with an extended ground test and airborne program contributed to the development of the final manned system.

The deployment sequence on a nominal landing is commenced at 7,315 m with the separation of the forward heat shield. Immediately, a small 2.2 m diameter parachute was mortar-deployed from the forward heat shield which creates a force to pull the unwanted heat shield away from the CM to prevent collision. After a delay of 1.6 seconds from the ejection of the forward heat shield two 5 m diameter conical ribbon drogue chutes were mortar-deployed. They remained reefed for the next ten seconds, then were fully deployed and remained attached until the CM had fallen to about 3,352 m. Then, the drogues were jettisoned and three 2.2 m diameter ring sail–type pilot parachutes were deployed, also by mortar. These supplied the required force to pull the main parachute pack from its stowage location. These 25.4 m diameter main parachutes were reefed in two stages to a fully open position, the first occurring 6 seconds after extraction, the second 10 seconds later at around 2,743 m. In the meantime, a VHF recovery antenna and flashing beacon was employed to aid in recovery location. Upon impact with the water the main parachutes were released allowing the vehicle to float on the surface. Impact velocity with all three parachutes in good condition was 9.7 m/s, with only two parachutes (as in the case of Apollo 15) impact velocity increased to 11 m/s.

In the event of an abort where the maximum altitude would have been below that allowing the opening of the baro-switches, a time sequence was initiated. The sequence in this situation was that the ELS was armed 14 seconds after the abort was initiated by the escape system. At the same time as arming the ELS, the launch escape tower, the booster protective cover and the docking ring were jettisoned, 0.4 seconds later the forward heat shield was separated. Two seconds later the drogues were deployed followed by the main parachute at 28 seconds (post-abort).

Development

It was recognized early in the development of the Apollo Program that to achieve the desired goal of landing men on the Moon by 1969 significant changes would have to be incorporated in the Command Module to support lunar distance missions over that of the original design. In consideration of the work already completed, the cost and time available, it was decided to continue with the original CM configuration (then called Block I) which would be used in unmanned and manned Earth orbital verification tests, and at the same time develop an upgraded version (Block II) for the lunar distance flight supporting the early lunar landing to achieve the program goal. There was also a design studied on a possible third version (Block III) designed to support extended exploration of the Moon, after the initial landings, and extensive Earth orbital scientific research missions under what was then termed Apollo Applications (later to be known as Skylab) using Apollo hardware for missions beyond its original intent. The Block III vehicle was never developed though upgrades to the Block II allowed additional scientific goals to be accomplished on the last three Apollo lunar missions (J series), and Block II spacecraft were also used to support the Skylab and ASTP missions that followed the lunar landing program.

Early planning for a lunar mission in December 1959 at NASA Langley Research Center included studies of parachute research for landing back on Earth. Over the next three years the design of what became Apollo was debated between ballistic and lift-over drag designs, eventually resulting in the decision to award North American Aviation the contract to build the ballistic capsule, the Apollo Command Module and the attached Service Module in November 1961 six months after President John F. Kennedy challenged NASA to place a man on the Moon by the end of that decade. The following year Grumman won the contract for the development of the lunar

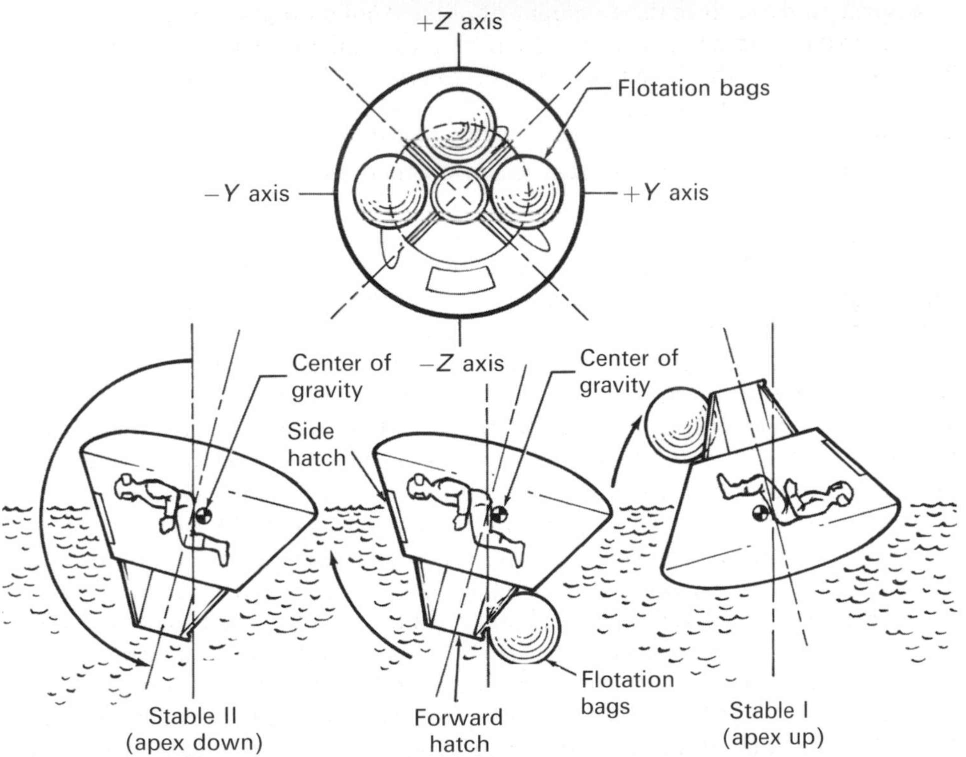

Two-parachute landing (Apollo 15) (top). Apollo self-righting system (bottom).

Apollo flotation collar and self-righting balloons.

landing module after it was decided to use lunar orbital rendezvous as the method to achieve the manned landing. Shortly after NAA received the contract it subcontracted the parachute and Earth landing system for Apollo to Northrop Corporation's Radioplane (later Ventura) Division. The preliminary plan for Apollo had been to include both a water-landing and a land-landing capability. In reviewing alternative concepts for landing (including rotating wings, paraglider, and parachutes) it was decided that three parachutes were the best option, especially since there were development problems with the parasail intended for Gemini.

On May 16, 1962 North American Aviation was told to proceed with the parachute recovery proposal, and the other systems were set aside for future review. There was strong support for parachutes and a water landing, especially after the successful recovery of Mercury astronauts John Glenn and Scott Carpenter. The idea of land landing for Apollo and for alternative methods to parachutes was not totally removed from Apollo long-term planning, but their prospect began to look very slim. The land-landing studies were not totally abandoned, however, as one of the first facilities built at NAA Downey specially for Apollo was the 46 m high Impact Test Facility in early 1962. Often likened to a giant playground swing it was designed to hold and drop a CM onto various services from height to test the impact velocities and the effects on the internal structure including astronaut couches and shock-absorbing systems. At one end of the facility was a pool of water, at the other end was a "beach" of sand gravel or boulders that could be banked or pitted with material to evaluate the different environments. By 1964 the reports from North American Aviation engineering studies concluded that land impact problems were so severe that they should be abandoned as the primary landing mode for Apollo. That sealed the fate for land landing in Apollo missions. The spacecraft would land under three parachutes and in the ocean for recovery by the Navy as Mercury had been, and like Gemini was being planned to do (Brooks *et al.*, 1979).

With the mode of landing decided, it was time to put theory to test and, though the systems retaining or deploying the systems from either a Block I or Block II CM changed slightly, the main parachute design remained the same for the two variants of vehicle. Great emphasis was placed on providing a CM recovery system that would perform under extreme and severe flight conditions. At the same time, the overall weight, the volume required to stow the gear and its integration into the complete system continued to increase. To overcome this, new manufacturing and assembling techniques were combined. To ensure it remained within the weight and volume assigned to the ELS and still perform as advertised was a major challenge to designers. Weight issues plagued Apollo for some years, but other challenges that had to be addressed in the development of the ELS included the size of the parachutes with the changing load limits of the CM, incorporating recommendations following the Apollo 1 pad fire, interfaces of the main parachute cluster, abrasions of the parachute risers, increasing the strength of the main canopies, the high-density issues involved in packing the parachutes and making sure they deployed correctly and sequentially, recontact of the forward heat shield and the burning of parachutes and/or risers from propellant dumps (as in the case of Apollo 15).

Testing

The tests and development program complemented each other by incorporating additional information from flight tests into the main design, which was then retested and evaluated to see if it remained within guidelines (mainly concerning weight and performance). A laboratory test program was completed on Block I to evaluate materials under conditions of temperature, vacuum, structure and abrasion, humidity and acceleration, physical strength and immersion. Most laboratory testing on Block II focused on changes between Block I and Block II. This included the redesigned main parachute deployment bag, the main parachute steel cable riser and the redesigned pilot parachute mortar.

Aerial drop tests were completed at the Air Force/Navy Joint Parachute Test Facility, El Centro, California. This phase of testing used an instrumented cylindrical test vehicle (ICTV) often referred to as a bomb drop vehicle, when the integration of a CM was not required, a Parachute Test Vehicle (PTV) and the Boiler Plate (BP) Test Vehicle. The PTV simulated the major features of the upper deck of a CM and was a simple cone shape below deck. The BP was a more accurate representation of the actual CM and included several spacecraft subsystems or features.

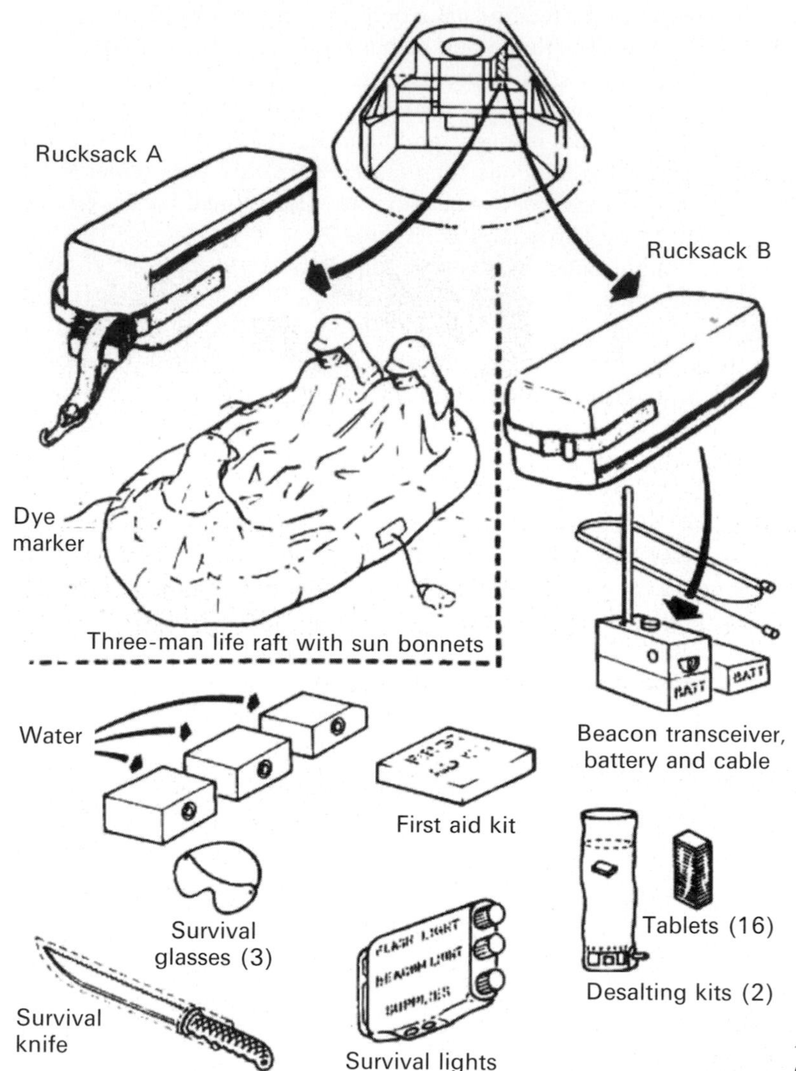

Apollo survival equipment.

The aerial drop tests were classified as either development or qualification tests with the later drops divided into individual tests depending on specific objectives. Initially single main parachutes were used for Block I to evaluate design concepts before moving on to multiple tests for Block II.

Between May 1965 and February 1966, 12 aerial drop tests were made to qualify the Block I ELS and rate it suitable for manned use (though no crew flew in a Block I spacecraft). From October 1966 to January 1967, total system drop tests were completed on what was then thought to be the final ELS for Block II. However, the increased capacity of Block II chutes resulted in further developmental and design

Apollo helicopter and diver support.

Primary recovery vessel approaches Apollo.

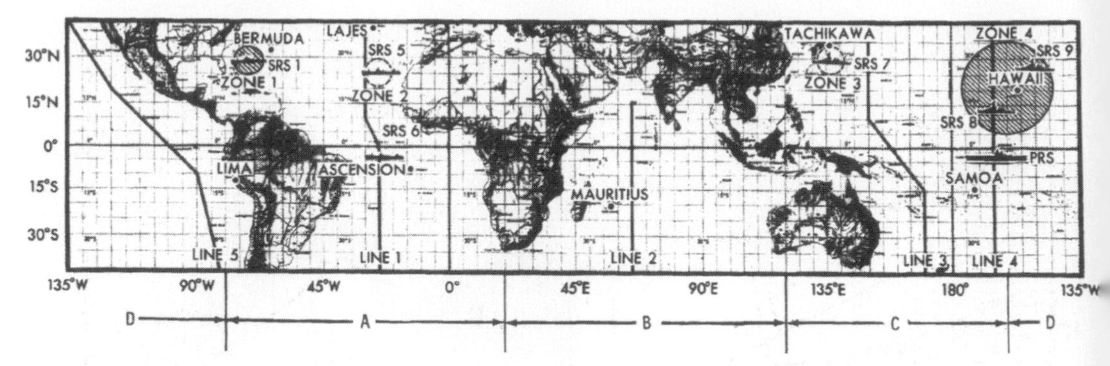

Apollo 8 landing zones. Recovery force deployment: recovery zones and contingency areas (A–D). Two HC-130s were stationed at each staging base.

verification tests between July 1967 and July 1968, in what evolved into a six-test program. From this a series of seven qualification drop tests were completed between April and July 1968. The completion of this program fully qualified the Apollo ELS for manned spaceflights. Valuable experience was also gained from operational use of the ELS in unmanned Saturn launch vehicles and recovery of boilerplate and operational CMs.

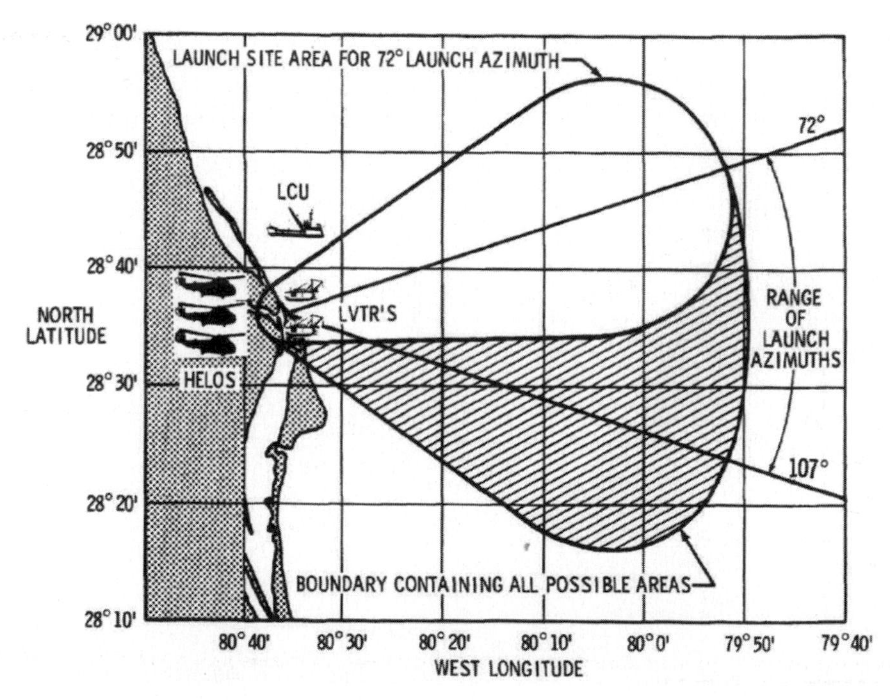

Apollo launch pad abort recovery forces. Launch site area and recommended force deployment.

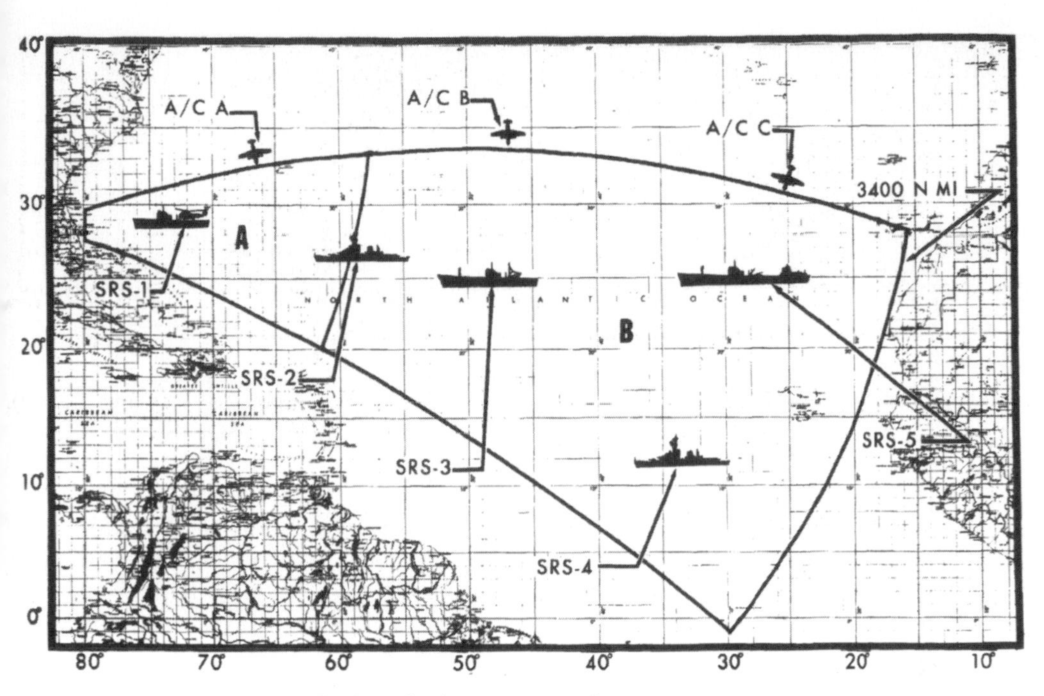

Apollo launch phase recovery forces coverage.

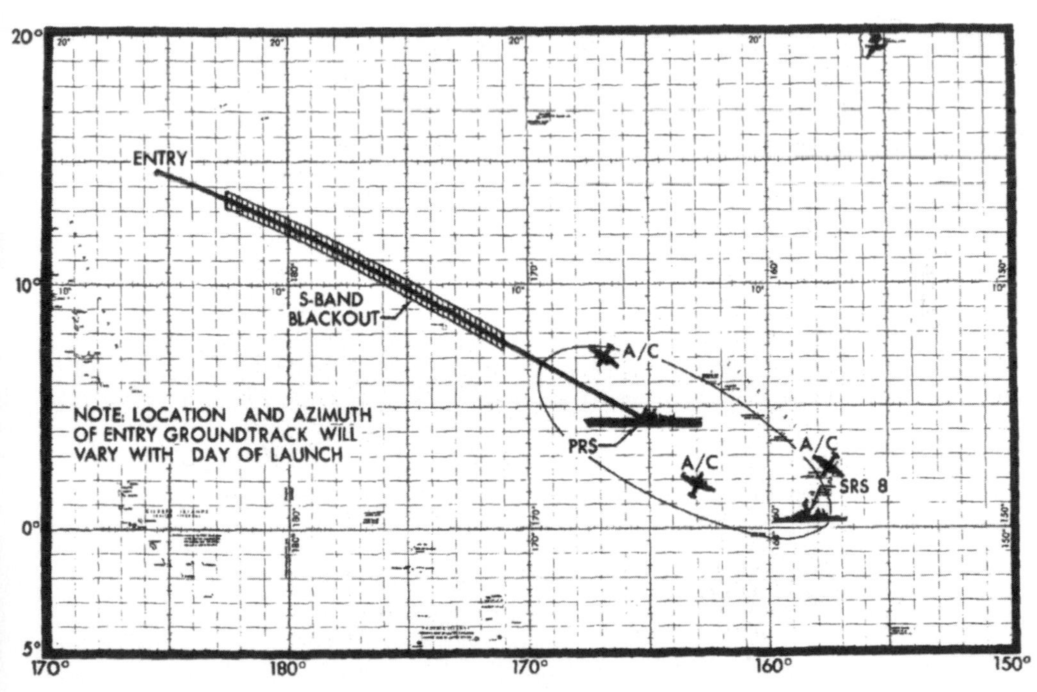

Apollo primary landing zone support.

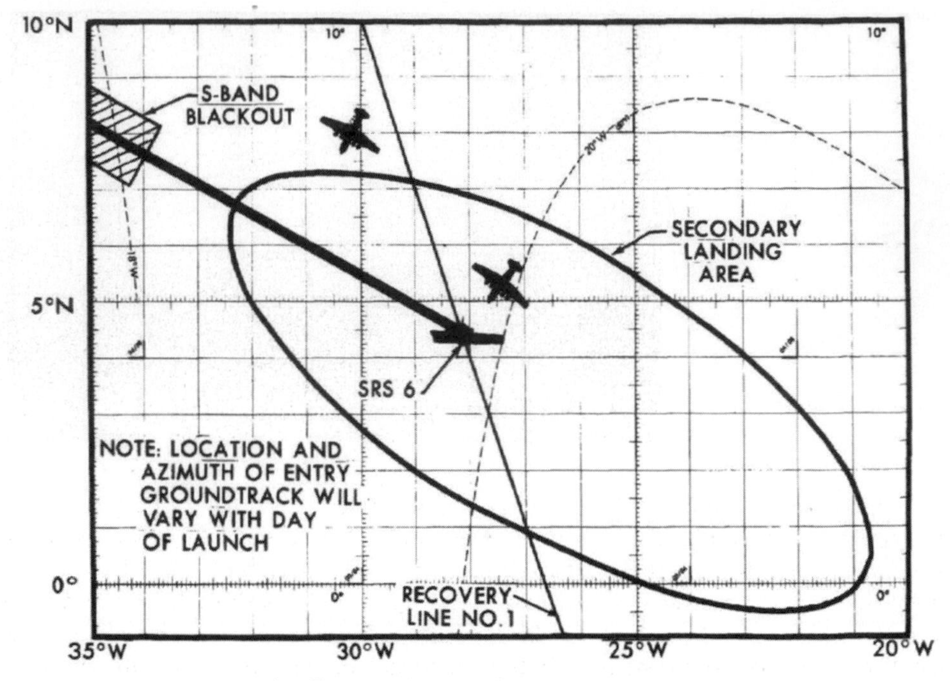

Apollo secondary landing zone support.

Soyuz

Parachute descent from Soyuz has been operational now for over 40 years and has proven to be highly reliable. There has been only one failure of the system in nearly 100 manned Soyuz missions, though some hard landings have been recorded. A two-parachute system is used, each in its own container. At about 10 km altitude the pilot and drogue chutes are deployed followed by the main chute at 5 km. The reserve parachute could be deployed between 3 km and 6 km, should it be required. At about 3 km from landing the thermal shield is jettisoned and at 1.5 m from the ground a gamma-ray altimeter commands the firing of four solid-propellant rocket motors on the base of the Descent Module which are designed to reduce the landing speed to 2 m to 3 m/s. In a worst-case situation in which the back-up parachute had been used and the solid rocket motors failed to fire, a landing of 4 m/s to 9 m/s would have to be endured.

Should a landing occur on foreign soil, then assistance directions are marked in Russian and in English on the base of the DM (as are several hazard-warning markings).

Development

The evaluation of the parachutes used on Soyuz goes back to 1961 when in a cooperative venture between OKB-1, the M.M. Gromov Flight Research Institute

Soyuz parachute recovery.

Plant 918, Plant 81 and the Parachute Landing Service's research and experimentation institute. In these early studies the engineers came up with a system similar to Voskhod in which a solid rocket braking system was installed in the base of the primary canopy. This was designed to reduce the landing speed to 8.5 m/s. Had this failed then a 10 m/s descent would have been recorded. A braking engine was not installed on the back-up system probably due to lack of available mass and volume in the storage compartment, instead it relied solely on the canopy to break the impact velocity. Initial mass model drop tests on a 7K Soyuz Descent Module occurred at the Flight Research Institute. After modification during 1963 and 1964 to improve the reliability of the back-up system when used in conjunction with the primary system, a new design was evolved from 1965. The parachute's area was increased from 574 m^2 to 1,000 m^2, a landing speed of no more than 6.5 m/s was adopted and the braking engines were relocated to the base of the module underneath the detachable heat shield. This allowed them to be used in either primary or back-up roles.

Extracting the crew from Soyuz after a long mission.

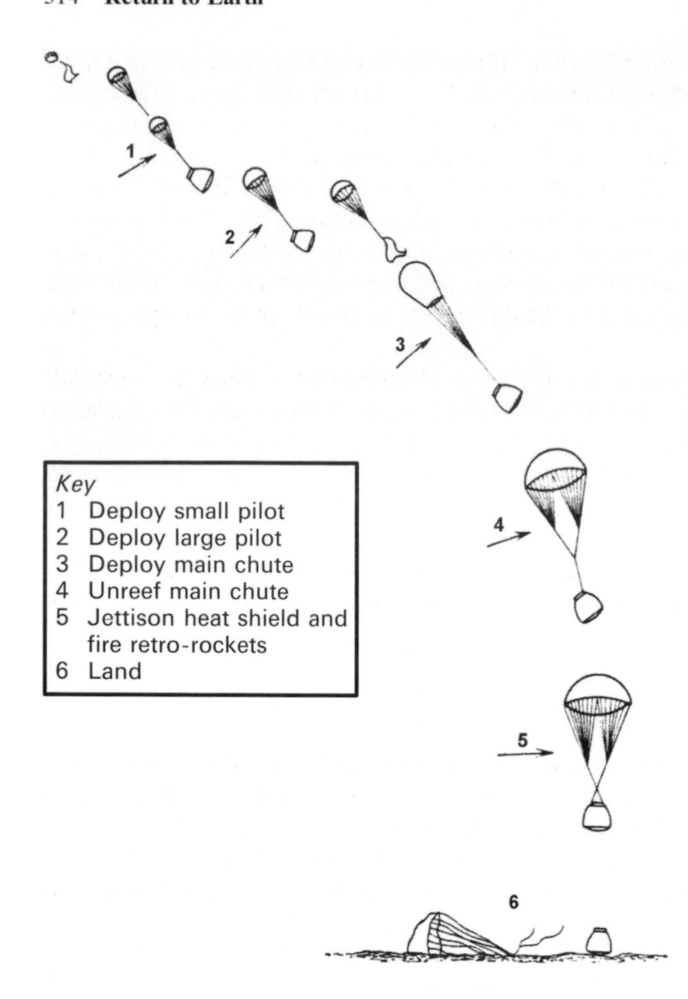

Soyuz parachute sequence.

Seven drop tests were completed as part of a test program during 1965 and 1966 at the Air Force flight test station at Fedosiya. An AN-12 aircraft dropped test capsules from an altitude of 10 km in varying conditions and different terrains. Small problems with leakage of hydrogen peroxide contaminating the parachutes had to be solved before the system was declared operational. Development work continued on qualifying and upgrading the parachutes used on Soyuz.

On Soyuz T (from 1979) the improved Escape Tower design allowed the primary parachute to be used instead of the back-up during the descent of the Crew Module from a higher altitude, and six soft-landing engines were installed on top of the original four units. In the Soyuz TM design (1986) the overall mass of the parachute canopies was reduced by incorporating weight-saving material in the design. This gave up some of the volume in the parachute containers for other improvements in the vehicle. A new altimeter was also incorporated in the soft-landing engine sub-system to help soften landing rates by more accurate firings at altitude. For Soyuz

TMA (from 2002) a series of modifications allowed larger framed crewmembers to be carried comfortably and safely to missions at ISS. This affected some of the subsystems including that of the soft-landing system. In TMA two of the six engines were modified to operate at two different levels of thrust, determined by the landing mass of the DM. This now ranged from 2,980 kg to 3,100 kg. The total landing mass of TMA was 7,200 kg, 200 kg more than that of TM. There were other modifications to the soft-landing sequencer and shock absorbers in the crew couch support frames. A series of ground tests, ground-based drop tests and airborne drop tests were completed between 1997 and 2000 to qualify the new system prior to operational flight.

Other improvements planned for future implementation on Soyuz spacecraft include modifications to the automatic landing control system and the option to deploy parachutes at lower altitudes to improve landing accuracy. By adding two further braking engines, landing accuracy in Russia rather than the traditional Kazakhstan may also be a future option. This is of course dependent on adequate funding (Hall and Shayler, 2003).

Had Zond lunar flights continued, presumably similar recovery and survival equipment would have been carried with amendments to support possible ocean recovery.

Shenzhou

Shenzhou resembles the Soyuz spacecraft not only in appearance but also in its flight profile and hence its landing technique, though the Chinese have developed a main parachute 20% broader than that of Soyuz. The sequence commences at 15 km above the ground when the pilot chute is deployed for 16 seconds and thereby slows the module from a velocity of 180 m/s to 80 m/s. The deceleration chute comes next and further decreases the descent speed to 40 m/s. The main chute is then deployed. It measures 80 m in height and 30 m across when fully deployed. With a surface area of 1,200 m it represents by far the largest parachute installed on a manned spacecraft for recovery. The main chute is suspended below the canopy on one hundred 25 mm diameter cords each one with the capability to withstand a strain of 300 kg. At full deployment the main canopy slows the descent further to 15 m/s. The Chinese have announced that the parachute has been constructed of 1,900 separate pieces of thin but very strong fabric that is able to withstand high loads and temperatures. There is a reserve parachute available in case of main chute failure but this is 60% smaller at 760 m. As with Soyuz, when 1 m off the ground a gamma sensor initiates the soft-landing solid-fuel rocket to reduce the landing impact to 1 m/s. At the same time the parachutes are severed to prevent high winds dragging the capsule across the ground (Harvey, 2004).

Experience in returning payloads from space by parachute has been gained by the Chinese since the early 1960s starting with tests of biological sounding rockets including the T-7A. Some of these tests involved recovering dogs, rats, mice and other biological specimens. The work on recoverable satellites goes back to 1964, and in 1975 China became the third nation (behind the U.S.S.R. and U.S.A.) to recover a

USSR
ВНУТРИ ЛЮДИ!
ОКАЖИ ПОМОЩЬ!
MAN IN SIDE!
HELP!
ВОЗЬМИ
КЛЮЧ
ОТКРОЙ
ЛЮК
KEY HATCH

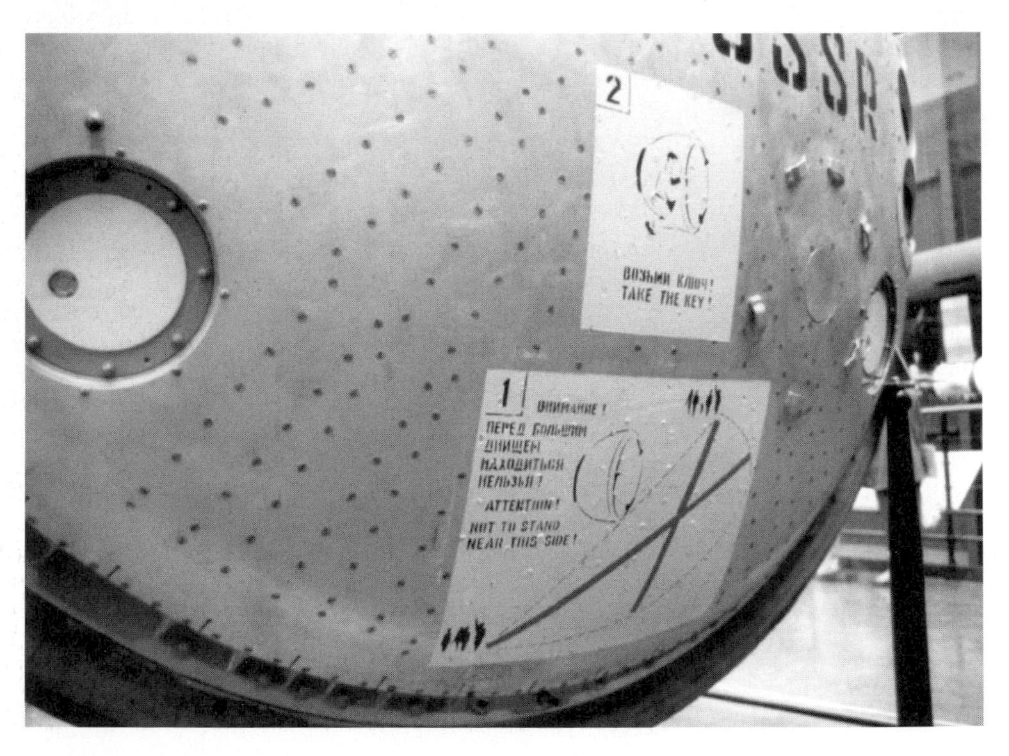
2
ВОЗЬМИ КЛЮЧ!
TAKE THE KEY!
1
ВНИМАНИЕ!
ПЕРЕД БОЛЬШИМ
ДНИЩЕМ
НАХОДИТЬСЯ
НЕЛЬЗЯ!
ATTENTION!
NOT TO STAND
NEAR THIS SIDE!

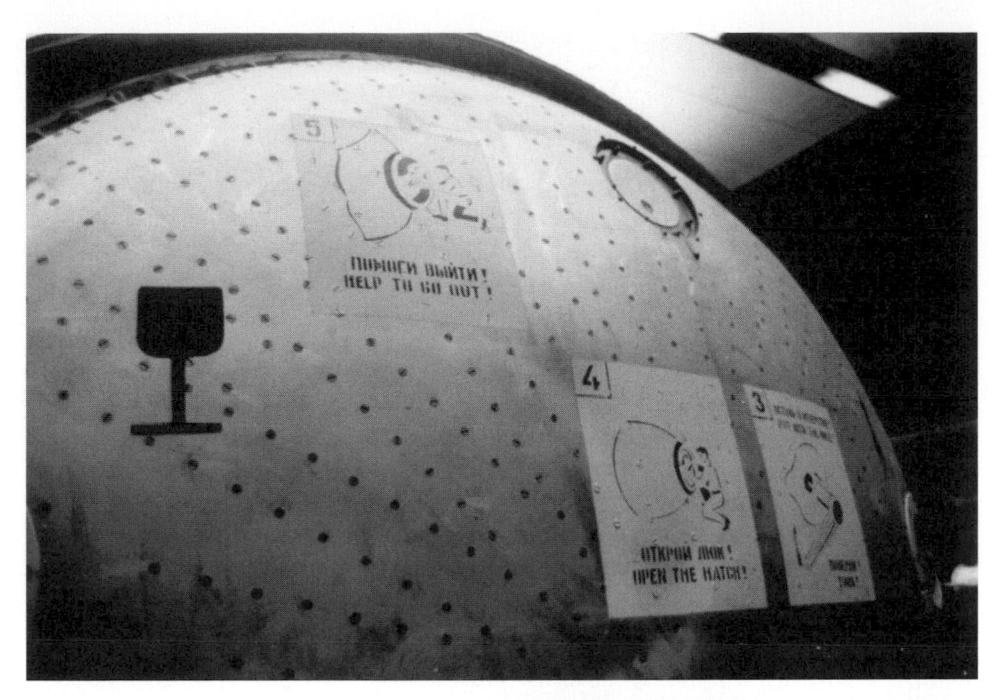

Details of the Soyuz aft-bulkhead revealing soft-landing engines and rescue information.
Courtesy: Andy Salmon

Landing support teams assist in medical and logistical activities at the Soyuz landing site.

Soyuz landing support all-terrain vehicles.

satellite orbit. The experience gained from developing a parachute system and studies of American and Soviet parachute recovery systems helped in the development of a suitable parachute recovery system for what became Shenzhou. After a program of ground and airborne tests the parachute subsystem of Shenzhou was finally qualified by the four unmanned test flights flown between 1999 and 2002.

FLYING HOME

The techniques for returning the Shuttle from orbit were developed over many years, and the terminal phase was demonstrated over the course of the approach and landing tests of the Shuttle in 1977 and similar programs for Buran.

Shuttle

Basically, the go for de-orbit is given about an hour prior to the scheduled landing, depending on weather conditions at the primary landing site (KSC). Should the weather be unfavorable alternative landings at Edwards or White Sands are possible. In addition, there are over 60 emergency landing sites spread across the world that can handle Shuttle landings, though the level of equipment held is not as high as at other primary or abort locations. Not all sites are currently active. Meanwhile, Orbiter re-entry occurs 30 minutes prior to touchdown. Retro-fire, entry and terminal area energy management maneuvers are designed to place an Orbiter in the best position to achieve a landing approach corridor at both the correct altitude and speed. At about 4,000 m, final landing guidance begins at a range of about 12 km and a velocity of about 682 km/h. There is now about 90 seconds to landing with the crew monitoring, supervising and backing up the automated systems. Final approach has a glide slope of about 22° (seven times steeper than a commercial airliner), and a flare is required to align a final glide slope of 1.5°; 22 seconds before touchdown the landing gear is lowered by gravity upon initiation by the crew. Initial touchdown speed of the main gear is approximately 326 km/h, the vehicle rolls about 1,000 meters on wheels with the speed brake deployed and, since 1992, a braking parachute to slow the vehicle down. Modifications to the vehicle have refined these operations over the years but the basic landing profile remains the same. It has proved very successful on all vehicles that have been able to follow the descent to landing (excluding the Challenger and Columbia accidents). For a detailed account of a Shuttle landing phase see Jenkins, 2001, pp. 260–261.

Studies into slowing down the Orbiter have included arrester wires, skids and braking parachutes; tests have determined that the braking parachute was the most efficient method of helping the Orbiter to slow down in the event of a burst tire (which had happened on STS-51D in April 985). Enterprise OV-101 was used to determine whether a landing barrier was effective. The tests in June 1987 at Dulles Airport were successful and, though no arrest system is installed at the primary landing site, they are available at the primary TAL sites should they be required.

A Shuttle Orbiter in glide flight.

Buran

If Buran had flown, it would have demonstrated its capability for early return in its "immediate return" and "early return" modes, and for rescue from a separate Soyuz spacecraft.

An immediate return would have seen a de-orbit any time between 40 minutes and 3 hours after identifying the need to terminate the mission. The 40-minute limit was defined as the minimum time to configure the vehicle and prepare the systems to attempt an early landing. In Hendrickx and Vis (2007) the authors explain that if the ground track occurred over one of the three Soviet landing sites viable to support Buran (the primary one at Baikonur, and the BUp sites in the Soviet Far East and in the Crimea) then landing would be targeted there; if not, the vehicle would be abandoned with the crew ejecting. An important factor in considering exactly when to eject, apart from the state of the vehicle, was the need to ensure the safety of the public (if appropriate, emergency rescue crews were on hand).

Early return would see a return between 3 to 24 hours, offering more time to prepare for re-entry and offering a much better chance of reaching one of the primary landing sites.

The Soviets were very serious about the possibility of rescue by a separate Soyuz (e.g., in this case an inability to de-orbit the Orbiter). Of course, several days would be needed to prepare a suitable Soyuz for the rescue, and compatible docking systems would need to be available on the Buran and Soyuz to effect easy transfer of crew between vehicles. Of course, if the Buran was in range of or attached to Mir or Mir 2, then the station would have acted as a safe haven (as with current Shuttle/ISS operations) until a Soyuz could be brought up to recover the crew. With a one-person Rescue Soyuz launched only two cosmonauts from Buran could be brought home at a time. It was assumed that for the two-person test flights a Soyuz would be readied for launch to support a rescue attempt if required. Multiple Soyuz launches to recover larger crews would not have necessarily placed additional strain on the infrastructure of launch preparations, orbital ground support and emergency recovery teams. Three separate Soyuz missions were launched as early as 1969 on successive days and recovered on successive days. Nothing like that has been attempted since, despite being theoretically possible.

Nominal landing for Buran was possible in automatic mode, flight director mode or manual mode, with automatic mode preferred for manned landings similar to that adopted for the American landings. From Mach 10 to Mach 2 (20 km–4 km) the pre-landing maneuver sequence was completed to align the vehicle with an entry point 14.5 km from the runway, as with the American Shuttle landing, which could be achieved from either end of the runway under the best conditions. A steep glide slope of between 17° and 23° would allow small corrections prior to a pre-flare maneuver at 400 m–500m resulting in a shallow 2° glide slope for landing. The final flare at 20 m led to a touchdown at about 300 km/h–330 km/h and a main brake and drag parachute would be used to help in slowing the vehicle. Sadly, no Buran pilot was able to put the theory and training into practice on a real mission from orbit. However, during the development program a total of 24 approach and landing tests were completed between November 10, 1985 and April 15, 1988 in addition to dozens of ground simulations.

Swift exit from Buran

To support the crew for ejection or in-flight decompression the Buran cosmonauts would have worn the *Strizh* ("swift") pressure garment. The development and testing of this suit took place between 1981 and 1991 and two suits were fitted to manikins for the single unmanned flight of Buran in 1988 to evaluate the performance of their connections to the personnel onboard life support system.

The suit was of the soft type and had an integral helmet. Its operating pressure was 400 hPa in primary mode and 270 hPa in back-up mode. A pressurized cabin with ventilation could support two suits for 24 hours and when using the PLSS decompressed the emergency mode could support the suits for up to 12 hours. The mass was 18 kg, and could sustain a cosmonaut ejecting at altitudes up to 30 km and speeds of Mach 3. In all, 27 models were fabricated for tests and training, but only four flight models were made (Abramov and Skoog, 2003, pp. 352–353).

SURVIVAL EQUIPMENT

To support contingency and off-nominal landings, each manned spacecraft contained a survival package for each crewmember which included rations, protective clothing and accessories, a means of providing drinking water, signaling equipment, medical and first aid kits, tools, and safety or protection equipment. In addition, cannibalizing equipment from the downed spacecraft or landing equipment would have supplemented the survival kit: one of the most useful items being the recovery parachutes. Each spacecraft also included toolkits and housekeeping kits which could be utilized in the event of a protracted wilderness stay awaiting recovery. These were developed from aviation, military or wilderness survival equipment, but adapted for post-spaceflight use. They have changed little in meeting the requirements of the new program and the basics have remained the same, keeping the crewmember alive and well until rescued.

United States

Mercury

Located on the left side of the pilot's couch was a personal parachute which could be used as a third back-up parachute in case of failure of the primary system. The realities of struggling out of the spacecraft to use it though were pretty slim.

The survival kit was a regular Department of Defense kit with added items developed by NASA. This included a one-man raft, a desalting kit, shark repellent, sea dye markers, a first aid kit including medical injectors, distress signals, a signal mirror, portable radio, survival rations, matches, a signal whistle, 3 m of nylon cord, survival flashlight, jack knife, survival knife, pocket waterproof, zinc oxide, soap, water container and SARAH beacon.

Gemini

The survival gear for Gemini was mounted on each astronaut's ejection seat and was attached to the parachute harness by a nylon line. The weight of each pack was 10.4 kg. Provisions for each astronaut included drinking water (1.6 kg); a machete; one-man raft (1.6×0.9 m) with a CO_2 bottle for inflation; sea anchor; dye markers; nylon sun bonnet; survival light (strobe) with flashlight; signal mirror; compass; sewing kit; 4.3 m of nylon line; cotton ball and striker; halazone tablets; a whistle; and batteries for power. There were also a survival radio with a homing beacon and voice reception; sunglasses; and a desalter kit with briquettes (enough to desalt up to eight pints of seawater). The medical kit included stimulant, pain, motion sickness and antibiotic tablets and aspirin, plus injectors for either pain or motion sickness (NASA, 1966).

Apollo

The Apollo program had the potential of returning to various locations on Earth, if an abort was called. For alternative or nominal missions the survival pack was expanded to cover three astronauts for a 3-day (72-hour) period. The items were stowed in two rucksacks. These rectangular rucksacks were stowed in the CM right-hand forward equipment bay; they were fabricated from Armalon, a Teflon-coated glass fabric with a zipper opening and strap handle. This design allowed rapid egress from the stowage location in case of emergency after landing, and its design prevented spilling the contents in the process. Though never called upon during Apollo missions, it fulfilled the fire regulations of onboard equipment; with one person handling and securing the life raft, it was used during training of all crews in its operation and use of contents (McAllister, 1972).

The balance between minimum weight and small volume was the challenge facing designers who had to come up with equipment they hoped would never be used but, if it was, would be good enough to ensure survival. The combination survival light assembly was a good example of this and was actually developed for the Gemini Program, but was adequate for the needs of Apollo as well. It was a hand-held unit primarily used for signaling by means of a strobe light, a flashlight or a signal mirror. The unit also contained a siren, whistle, a compass, fire starter, cotton balls, halogen tablets, a water receptacle, knife blades, needles, nylon cord and fishhooks.

There was also a desalter kit comprising two processing bags, eight chemical packets and mending tape. Each of the chemical packs could produce 1 pint of potable water. The astronaut would gather an amount of seawater and insert one pack of chemicals for a pre-defined period of time. The result was then filtered to produce drinking water. This was a slightly modified DOD unit designed to meet the requirements of human-rated equipment for safe flying in space. There were also three water containers each holding 2 kg of water with access provided by a metal cap. Those used in Mercury and Gemini were made from PVC plastic film with a neoprene-coated nylon restraint added later. Those in Apollo were made of aluminum to meet pressure regulations, meet outgoing and flammability requirements and provide sufficient shock load tests (78*g*). During development it was found that the sides would deflect due to pressure differences; to solve this, the indented "X" was incorporated into the two largest sides. In addition, the cap was at first a press fit stopper that tended to blow out during decompression tests; this was changed to a screw thread. During training, some astronauts complained that the aluminum caused an unpleasant taste to the water. A film liner was suggested to alleviate this problem.

The survival kit also included light-polarizing sunglasses, an aluminum machete and sheath with a cutting edge on one side of the 18 cm long blade and a sawing edge on the other. The flammable nylon handle used on Gemini flights was replaced by an aluminum (non-flammable) one for Apollo. It proved useful in training for cutting or sawing in the jungle, but snapped if used to pry items; a stronger blade was suggested for future use.

A 7.6 m long lanyard system was provided to connect the crew, spacecraft and life raft together during emergency egress. With a mooring lanyard snap-fitted on each end for quick disconnect, it featured three man-lines of 2.4 m for the first crewman to egress and 4 ft for the second and third crewmen. In addition, each crewmember had a life vest for launch abort and post-flight water recovery. Neck and wrist dams located in the pressure garment assembly provided additional water seals during water recovery operations. There was a three-man triangular-shaped life raft designed to be inflated in 25 seconds by two carbon dioxide cylinders to inflate two separate floating tubes. Manual inflation was possible as a back-up. A sea anchor, three bailing buckets and two sea dyes (yellow–green) were installed in the life raft kit. The dye could be seen from an altitude of 1,524 m at a range of 1.6 km.

The signaling radio was a hand-held, dry-cell battery-powered device and could be used connected to the spacecraft antenna, thus meeting the needs for constant emergency signal and communication if a delayed post-landing recovery had occurred. The training frequency was 242 MHz and operationally it was 243 MHz (the international distress frequency). The Gemini unit was used on Apollo 7 through 11, and then a unit specially developed for Apollo was incorporated in the survival kit.

A 152 × 132 cm utility net made of standard fine-mesh nylon was also available from Apollo 12 onwards; primarily used for protection from insects it was also a useful filler for the packed survival kit in the rucksacks. Three pieces of Mylar 152 × 106 cm in size were provided for thermal protection (survival blankets) and for signaling purposes (reflective); this was also a useful filler and reduced the need for a non-functional filler in the already limited mass and volume restrictions. There were three sun bonnets with rear capes to protect the neck and two plastic squeeze bottles each with approximately 56.7 grams of skin protection lotion. Two survival knives completed the kit and were the standard three-bladed all-metal knives used by the U.S. Navy.

The survival equipment carried onboard Apollo between 1968 and 1975—which encompassed all the lunar missions, Earth orbital missions, Skylab and ASTP—was only slightly modified as flights and training continued.

Shuttle

As the Shuttle was designed for land landings there was a reduced requirement for survival equipment for remote location support. There were emergency egress procedures and equipment, and the survival equipment supported this need. Since 1981, the Orbiter has carried a survival pack that supports land or sea capability for all crewmembers (depending on the crew complement of between two and eight) for up to 48 hours and was deployed during emergency escape from the Orbiter. Packaged in a single container it contained an eight-person life raft with a carbon dioxide inflation system and a mooring lanyard assembly. There were two oral inflation tubes, a bellows pump, bailing bucket and sea anchor. There were also 24-hour flotation life vests for each crewmember; these were finally stowed on each seat for ease of access, if required. Signaling equipment consisted of a personal distress signal kit, two smoke/

illumination flares, a sun mirror, two radio beacons with spare batteries and two sea dye markers. The survival pack also included a two-part individual survival kit (based on DOD and earlier kits aboard Mercury-Apollo), a survival blanket, survival knife assembly, desalter bag and chemical packets (NASA, 1981).

From 1988 an additional egress survival pack was incorporated in the lower leg pocket of each suit. One contained a PRC-112 survival radio (with spare antennae, earphone and a signal mirror); there were also Scop/Dex motion sickness pills. The other pack contained chemical lights, a day/night signal flare, a pen gun (with pen gun flares) and a strobe light.

Soviet Union/Russia

Survival packs have been carried on Soviet/Russian craft ever since NAZ portable emergency kits were stored inside the Vostok capsule in 1961. For Soyuz these kits (produced by Zvezda) weigh about 32.5 kg and are located in two triangular carrying cases wedged in between the cosmonaut seats. They contain a large canteen, a soft flask, dried food, a medical kit, signals and flares, a foraging bag, fishing tackle and metal wire garrottes for use as a saw and for hunting. There is also a Makarov pistol with cartridges (TP-82m) and a machete (Hall *et al.*, 2005).

In addition to the Sokol suit, recovery parachutes and the onboard toolkit, the cosmonauts also have the *Granat 6* ("pomegranate") survival pack that includes for each cosmonaut

- A *Forel* ("trout") hydro-suit, which is a one-piece, orange, nylon flotation garment with integral soled feet and hood. The "Neva" inflatable collar has an emergency mouthpiece, emergency beacon, and signal device on its shoulder. Rubberized cuffs, Velcro-closed pockets and a pair of mittens with watertight cuffs and nylon wrist tapes complete the garment. This is an improvement over the Vostok era where cosmonauts had to complete flotation and survival training in their pressure garments or a small one-man dinghy.
- A TZK-14 cold-weather suit with a royal blue, zipped-front anorak and attached mittens. There is also a woolen balaclava, a lined wool knot cap, woolen gloves, one pair of thermal socks and a pair of nylon overboots.

After the TMA-1 landing in 2003 the need for immediate communication with a downed crew outside the nominal landing area was addressed by the inclusion of mobile phone communication equipment in each Soyuz TMA DM.

It is thought that similar search and rescue and survival equipment to the American Shuttle would have been carried on Buran manned missions adapted from those flown on Soyuz.

China

Photos have been released of Chinese taikonauts in training with various elements of survival gear similar to that used by the Russians, including water survival equipment and signaling devices. Therefore, it is reasonable to suggest that the Shenzhou survival and rescue equipment is similar to the Russian Soyuz spacecraft with amendments and adaptations to support Chinese operations.

SEARCH AND RESCUE

Supporting landings, whether normal or contingency, requires a network of personnel and vehicles, normally from the armed services. For water landings this includes significant numbers of surface ships and aircraft adding to the overall mission expense. In addition, a worldwide tracking and communications network is required, and the advent of space-based tracking and data relay satellites has significantly increased the global coverage of communications for all manned missions.

In defining the requirements for recovery, consideration has to be given to off-nominal as well as nominal landing areas. This includes any launch site abort situations prior to the launch of the vehicle, and includes the escape and rescue procedures, equipment and infrastructure to support this phase. The abort landing area needs to be considered prior to safe entry into orbit. This can either be back at the launch site, across the ocean, over land masses or after completing a single orbit around the Earth. Primary landing areas are considered for nominal mission operations, while planned landing areas are considered for alternative landing sites at the end of a normal mission due to unfavorable weather or spacecraft conditions at the preferred primary site (e.g., instead of the Shuttle landing at the Cape in borderline weather conditions it diverts to Edwards instead—it's not an emergency just a better operational alternative). Preferred contingency landing areas are those that are used in an emergency and offer the best facilities to support landing. Alternative landing areas are those offering a wider range of opportunities should a quicker return to Earth be required. The number of recovery personnel is dependent on the category of primary, planned, contingency and alternative landing sites.

United States

Recovery of a crew from space for the United States focused on, between 1961 and 1975, ocean-based recovery teams. From 1981 and the advent of the Space Shuttle, recovery regions were centered more on land-based facilities in the continental United States, Africa and Europe during the launch phase, though a token coverage for contingency landings continues to support Shuttle operations.

Mercury

The above philosophy was developed to ensure recovery of a Mercury astronaut under any conceivable landing conditions that threatened his safety. The recovery itself was divided into location of the landing zone and early notification of the landing in that area by information supplied by the tracking network. Visual observation of the spacecraft would initially be by aircraft. Then, the recovery operation moved into the retrieval phase which completed the recovery of the astronaut and spacecraft from the water by support helicopter, divers and personnel from the primary recovery vessel. This philosophy operated through to ASTP in 1975.

Getting out of Mercury was a little different as the spacecraft was so small. In the early design the astronaut was to be locked inside the capsule and, with as many as 70 torque bolts securing the hatch, an explosive charge was required to release it in an emergency on the pad. After splashdown the astronaut had to crawl through the neck of the spacecraft. This was by no means easy as, first, a panel had to be removed from the instrument unit, and then the primary and emergency parachute container had to be pushed out of the way followed by the life raft and emergency pack before the astronaut could finally squeeze out. Once in the water, the suited astronaut had to inflate the life raft and climb in, securing it to the spacecraft so it did not drift away. The recovery sequence by helicopter was astronaut first then spacecraft. This proved troublesome in training and threatened the life of the astronaut if water filled the spacecraft which would then begin to sink. It was also tiring and hot work. By changing the sequence to allow helicopters to secure a recovery hoop line, the side hatch could be blown giving the astronaut access to a second secure line to the helicopter and allowing him to get into a "horse collar" for winching up into the aircraft. Once inside, the helicopter pulled the spacecraft out of the water. Later a Stullken Flotation Collar was fitted around the spacecraft by para-rescue divers and the astronaut had the choice to stay put and get lifted up inside the spacecraft by the primary recovery craft or jettison the hatch and opt for helicopter rescue.

All Mercury astronauts except Wally Schirra opted for individual recovery. Schirra remained in his craft for pickup. In July 1961 this almost cost the life of astronaut Gus Grissom when the side hatch on his spacecraft "Liberty Bell 7" blew prematurely, resulting in seawater flooding in and loss of the capsule despite valiant attempts by the helicopter crew. Grissom almost drowned, but was recovered safely. Almost a year later in May 1962 Scott Carpenter landed 402 km downrange of the primary recovery range due to late initiation of the retro-rocket. When the main parachute failed to deploy automatically he had to manually instigate deployment. After splashdown, he felt hot and uncomfortable in the confines of the capsule and decided to get out via the neck passageway of Aurora 7, then inflated his life raft and climbed aboard. Realizing it was actually upside down he re-entered the water and turned it over. He spent a "pleasant" couple of hours awaiting helicopter recovery in the life raft, becoming the only astronaut ever to use the survival equipment in the 1961–1975 pioneering period of American manned spaceflight.

Gemini

For Gemini missions, landing areas were divided into the planned landing region and those identified for contingency landings. The planned area included primary sites in the West Atlantic while secondary landings were in the East Atlantic, West Pacific and Mid-Pacific where there were several support ships.

The launch site area for off-the-pad aborts during the initial phase of ascent was an area 65.6 km seawards from Cape Canaveral and 4.8 km towards Banana River inland from LC 19. The launch abort landing area started 65.6 km out at sea from the Cape and extended right across to the west coast of Africa.

Contingency areas were those beneath the ground track of the spacecraft except areas already identified in the planned landing area which required aircraft and para-rescue support divers for recovery within 18 hours after the splashdown, whereupon the crew would rely on the onboard survival and personnel rescue kits, should the need to vacate the spacecraft be required.

Astronauts have praised Gemini as a fine spacecraft but a lousy boat. John Young commented that GT-3 featured the capability to "pitch, heave and roll". The Gemini 6 crew were the only ones to be picked up inside their spacecraft by a recovery vessel, and the Gemini 8 astronauts landed in a secondary landing area due to early termination of their flight.

Apollo

The lunar program adopted the basic principles used on Mercury and Gemini, though returning a spacecraft from the lunar distance required amendments to the deployment of recovery forces and development of speciality tools, procedures and equipment. The support of the U.S. Department of Defense was crucial to the safe and prompt recovery of each crew flying an Apollo spacecraft from 1968 through 1975.

For Earth orbital missions a four-zone concept was employed. There were two in the Atlantic and two in the Pacific, with the Western Atlantic as the primary landing zone. For lunar missions this was made dependent on when the landing was required, either during launch close to the launch site (Sector B—160 km to 5,440 km downrange) or farther east across the Atlantic (Sector A—5,440 km all the way to the west coast of Africa), from Earth parking orbit (West Atlantic or Mid-Pacific), translunar injection to end of mission (mid-Pacific line or mid-Atlantic line). When Apollo 13 aborted then changed back to a free-return trajectory it would have meant an Indian Ocean landing. By adding an after-burn a mid-Pacific line landing was possible. The normal end of mission would be in the mid-Pacific.

Shock attenuation systems did allow a land landing for Apollo in cases of emergency but were never called upon (though the new Orion CM will have land landing capability). Earth studies revealed the design of the CM had two flotation attitudes. Stable I was vertically upright and Stable II was with the vehicle upturned, its heat shield uppermost. Clearly, this second mode was unsuitable for efficient crew recovery; therefore, three inflatable bags were installed on the upper deck of the CM inflated by air compression; this would upright the CM in 5 minutes. Tests were conducted to allow two crewmembers to lower the center of gravity by moving from their couches to the aft of the CM in the event only two bags inflated; this was successful, though never used operationally. Uprighting took 12 minutes; however, if both bag and compressor failed the CM would not be able to self-right. Once upright the para-rescue crews would secure flotation gear around the spacecraft and await helicopter recovery of the three crewmembers before the CM was hoisted aboard the primary carried. The Apollo 7 crew elected to remain in the spacecraft during hoisting. All three Skylab crews were also hoisted in their CM due to the time they had remained in space. Primary recovery sites for Skylab were in the Pacific Ocean, as was the primary landing area for ASTP (NASA, 1975).

Shuttle

Egress from the Shuttle is primarily though the side hatch; a secondary route was via the escape panel system, which originally consisted of the ejection seat panel and the

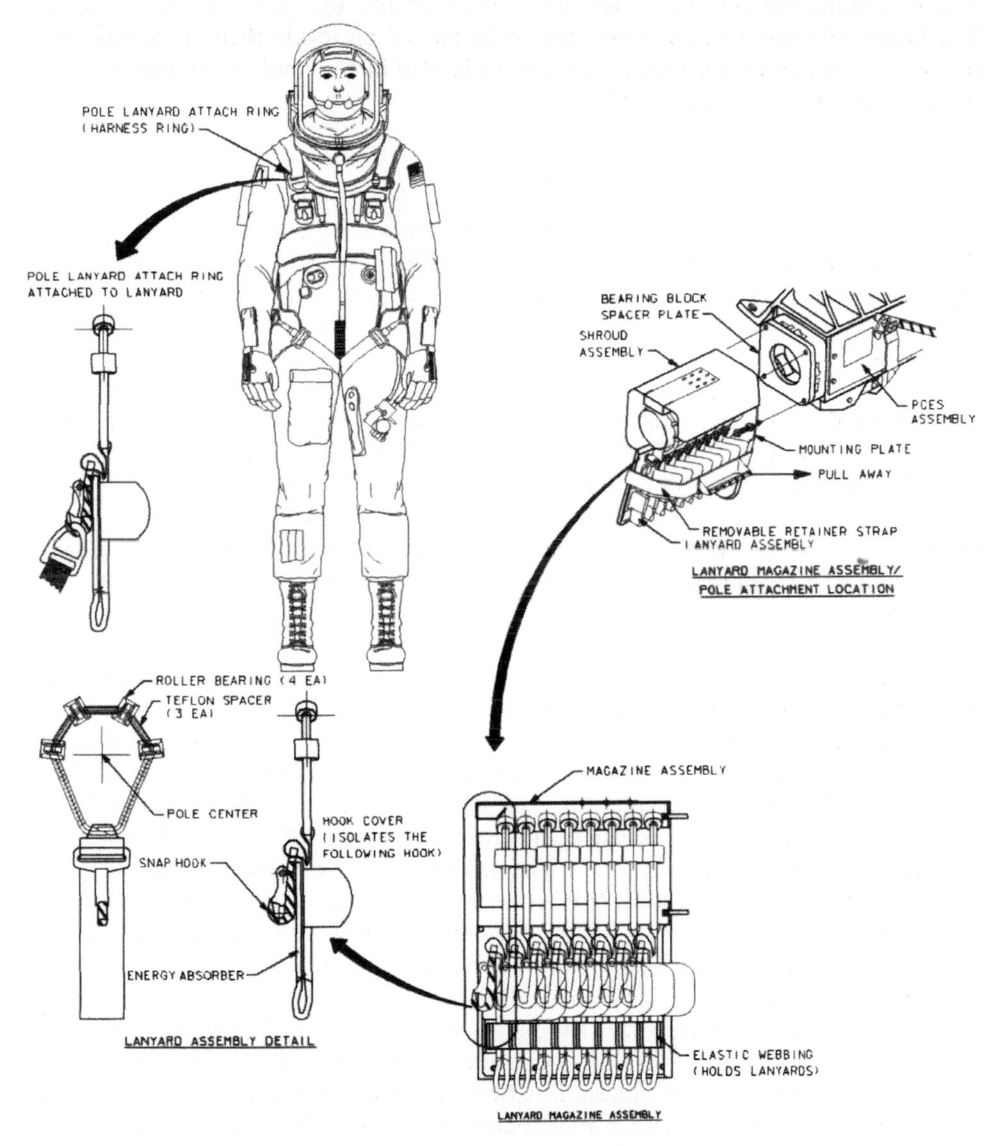

Shuttle escape equipment. Lanyard magazine assembly.

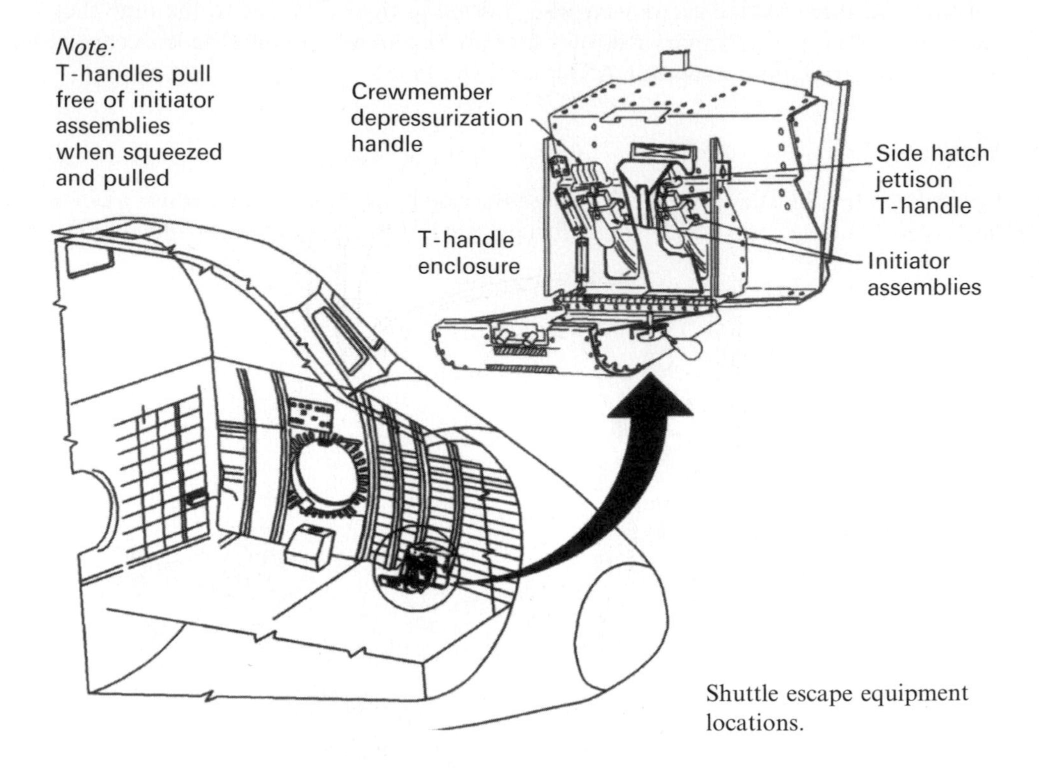

Shuttle escape equipment locations.

left overhead window. When they were removed from Columbia the overhead escape panel system was incorporated in the overhead left window on all subsequent vehicles. Crew training still includes these emergency exits from the vehicle which were originally evaluated by astronaut Jerry Carr in 1977.

An egress bar is permanently attached to the side hatch and serves as a hand hold to assist jumping off. Thermal aprons are used to protect the crewmember from the external surfaces of the Orbiter due to heating as a result of re-entry. A descent "sky-genie" system (there are seven aboard the Orbiter) is provided to lower the astronaut from the overhead escape route or side hatch panel to the ground at a controlled rate of descent. While in orbit, overhead escape is via Window 8 (port—left side overhead window), with the crew using MS2's Seat 4 to climb up through the window

Following Challenger a slide pole arrangement was chosen to provide an in-flight rescue system for the crew in a "controlled gliding flight attitude" from 9,144 m and below. Modifications to the system were required to allow an explosive blow-out of the side hatch to extend the slide pole and allow the astronauts to clear the vehicle cleanly. Wearing pressure garments the mid-deck access hatch from the flight deck had to be enlarged to assist ease of movement between the two levels of the Orbiter. It was detected that a water landing of a Shuttle Orbiter (though evaluated early in the program) was not survivable leaving ditching over water via the slide pole the only option.

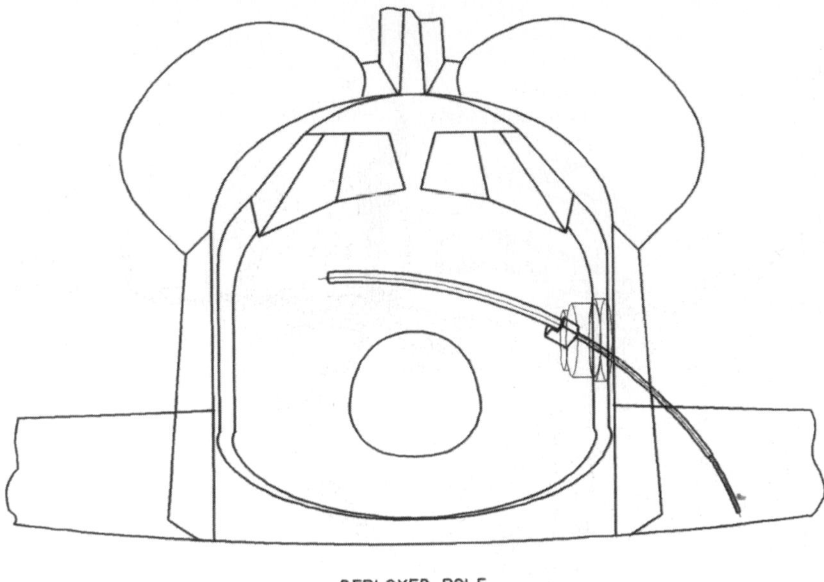

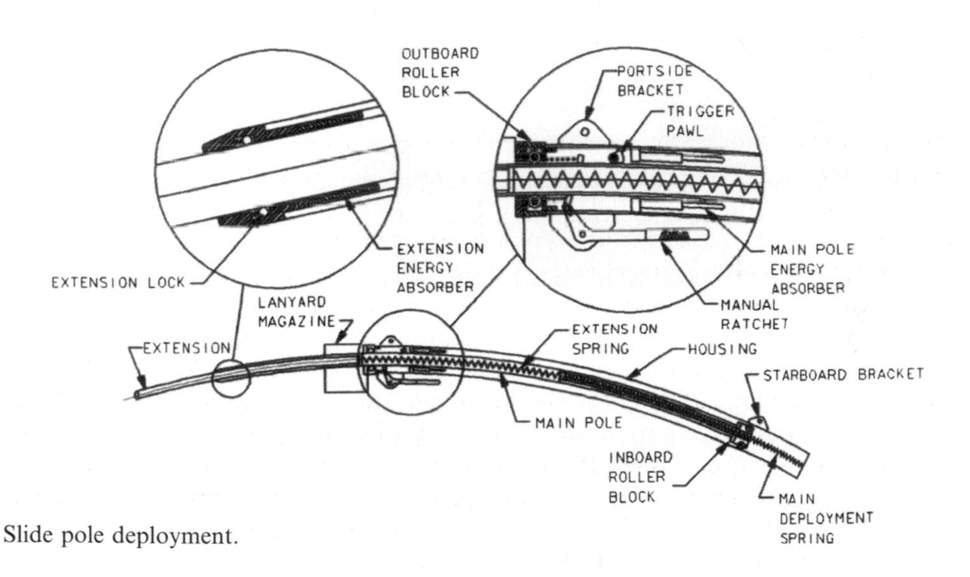

Slide pole deployment.

The emergency ground slide is similar to the inflatable escape chutes provided on commercial aircraft for rapid egress to the ground. It is stowed below the mid-deck and attached to the side hatch or the opening if the hatch had been ejected. It is then inflated by an argon pressurized bottle, and is deployed in 60 seconds; it is designed to remain functional for 6 minutes to allow all astronauts to descend some 3.2 m to the

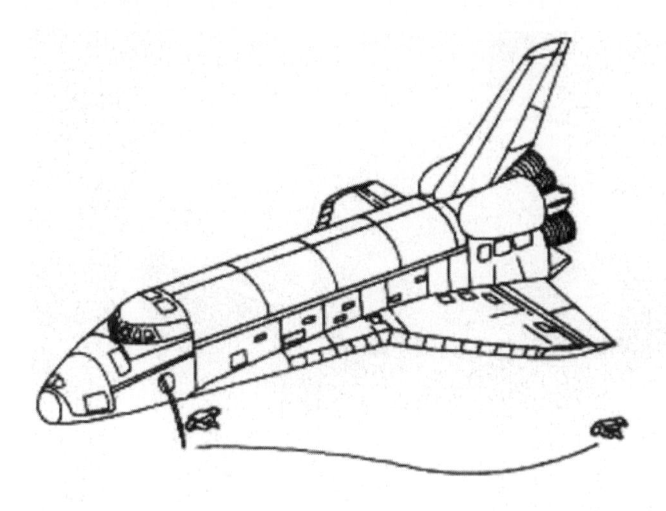

Leaving the Orbiter (top). Testing the tractor rocket escape system (bottom).

ground. It can be separated if a fire or rescue vehicle is required to support an injured astronaut.

The slide pole is a curved, spring-loaded, telescopic steel and aluminum cylinder stowed in the mid-deck on the ceiling in an aluminum housing. Extended it measures 2.7 m beyond the side hatch. In normal operations it is removed from the launch position and reinstalled for re-entry. It has a mass of 108 kg. Eight lanyards for crewmember attachment are stowed in a magazine attached to the end of the pole; they are used to guide the crewmember across the pole, where they slide down and off the end of the pole to escape in the sea.

The system was tested in the spring of 1988; the tests determined that the slide pole was much safer and easier to operate than the proposed tractor rockets that were also considered at the time. It was determined that it would take about 90 seconds for all the crew to exit the vehicle with the astronauts each leaving the vehicle at about 12-second intervals.

Artist's impression cutaway of Shuttle crew compartment.

Operationally, the decision to use the system would have to be taken at about 18,290 m by setting the flight control system to the autopilot mode. At 7,620 m and about 368 km/h a crewmember on the mid-deck (officially MS3 seated next to the side hatch and termed the jump master) would activate the pyrotechnics to equalize the pressure inside the vehicle to that outside. At 7,620 m the vehicle angle of attack is changed by the autopilot to 15° where it has to remain to allow all crewmembers to get out safely. The side hatch is now jettisoned by a crewmember and the jump master activates and extends the slide pole. Each crewmember then attaches themselves to the lanyard system and jumps out of the side hatch, the length of the pole allows for passage beneath Orbiter's left wing. All crew should have exited Orbiter before reaching the 3,050 m altitude for descent by personnel recovery parachute. Installed on Discovery in time for the STS-26 mission in September 1988 it has been standard equipment for all missions since that time.

To support the prospect of launch and entry escape the crews from STS-26 were supplied with partial-pressure suits (Model S1032). Termed Launch Entry Suits (LESs), they were produced by the David Clark Company of Worcester, Massachusetts. A contract was awarded in 1986 for an IVA partial-pressure and crew escape suit, a variant of an existing USAF suit design. The system supported Shuttle flights from 1988 to 1995; then it was replaced by the Advanced Crew Escape

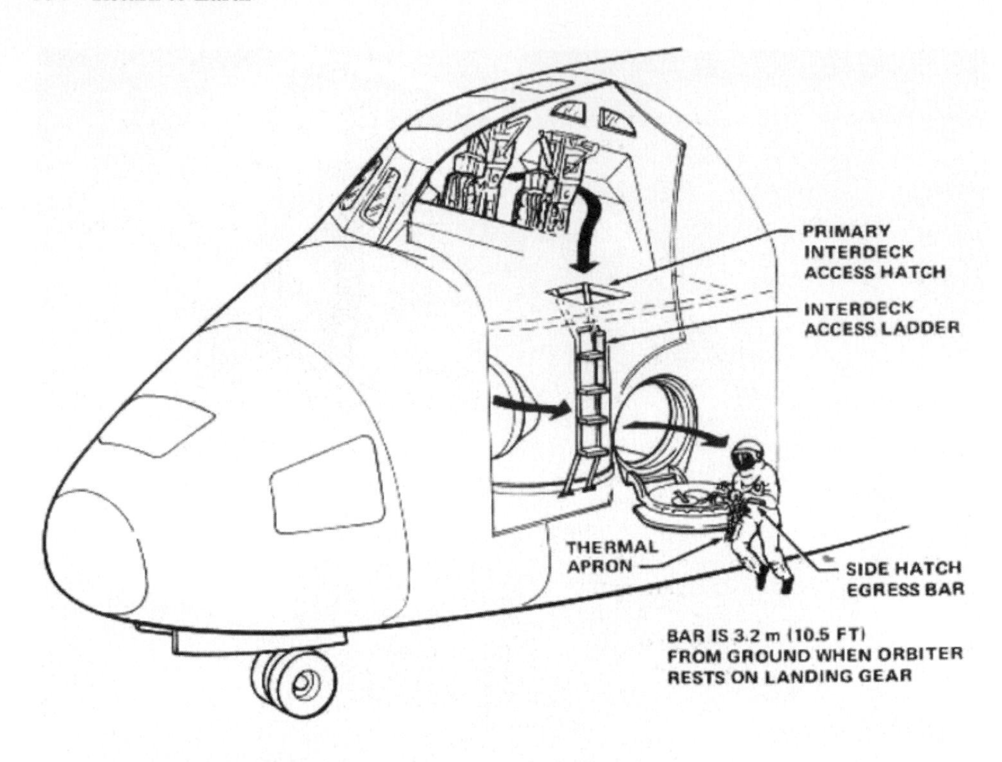

Using the side hatch escape system (top). Though no Shuttle astronaut wishes to land on water, if the Shuttle survives the impact, then this is the equipment provided to keep them afloat until rescue (bottom).

Suit. Operating at 18.6 kPa its weight was about 13.6 kg with 29 kg of support equipment. Primary life support was via the Orbiter with a further 10 minutes provided by a back-up system. There were 49 LESs fabricated for Shuttle operations between 1987 and 1989 (Thomas and McMann, 2006). From 1994 a new suit was introduced called the Advanced Crew Escape Suit (Model S1035), again manufactured by David Clark. This system included parachute and flotation systems to

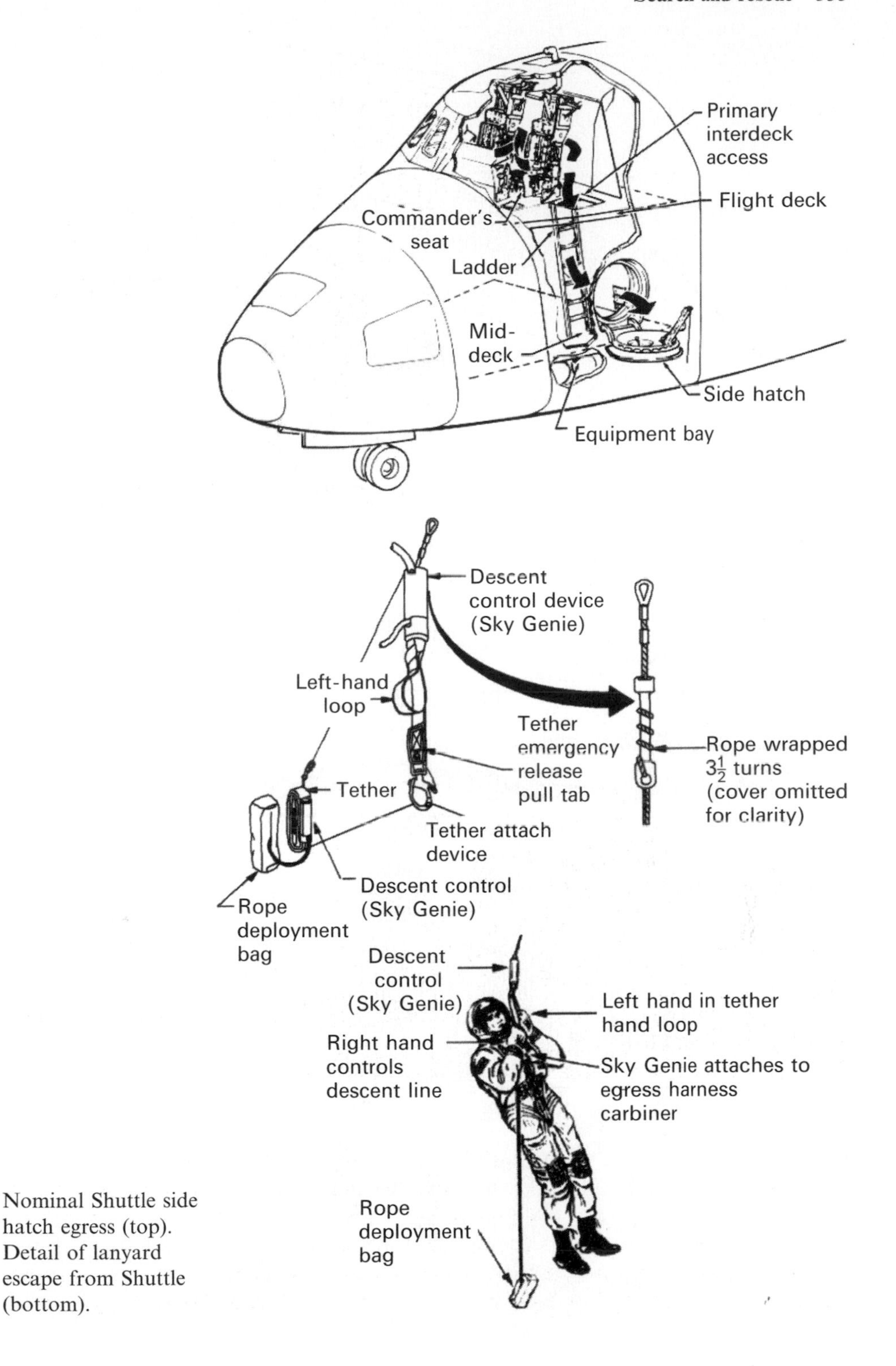

Nominal Shuttle side hatch egress (top). Detail of lanyard escape from Shuttle (bottom).

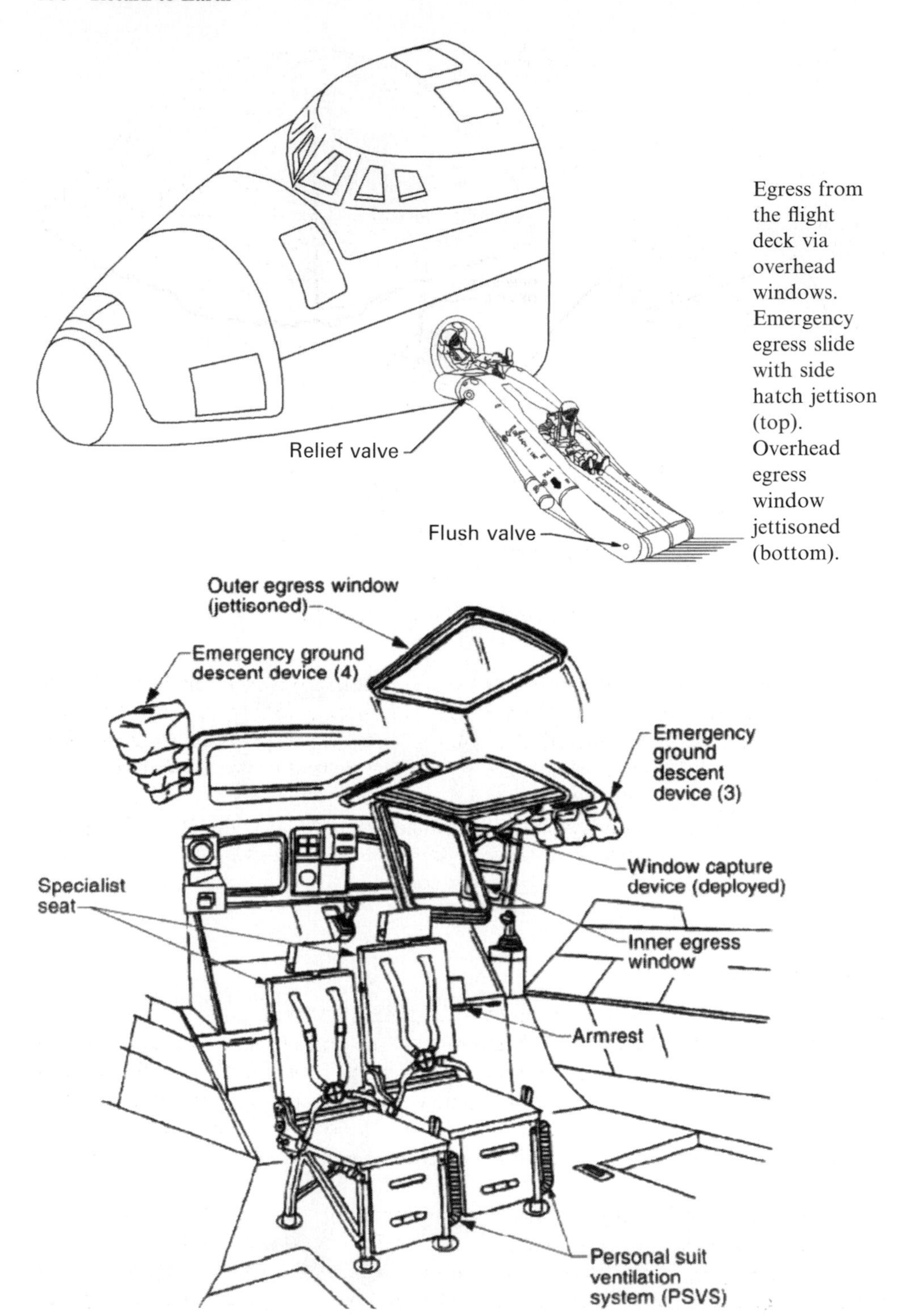

Egress from the flight deck via overhead windows. Emergency egress slide with side hatch jettison (top). Overhead egress window jettisoned (bottom).

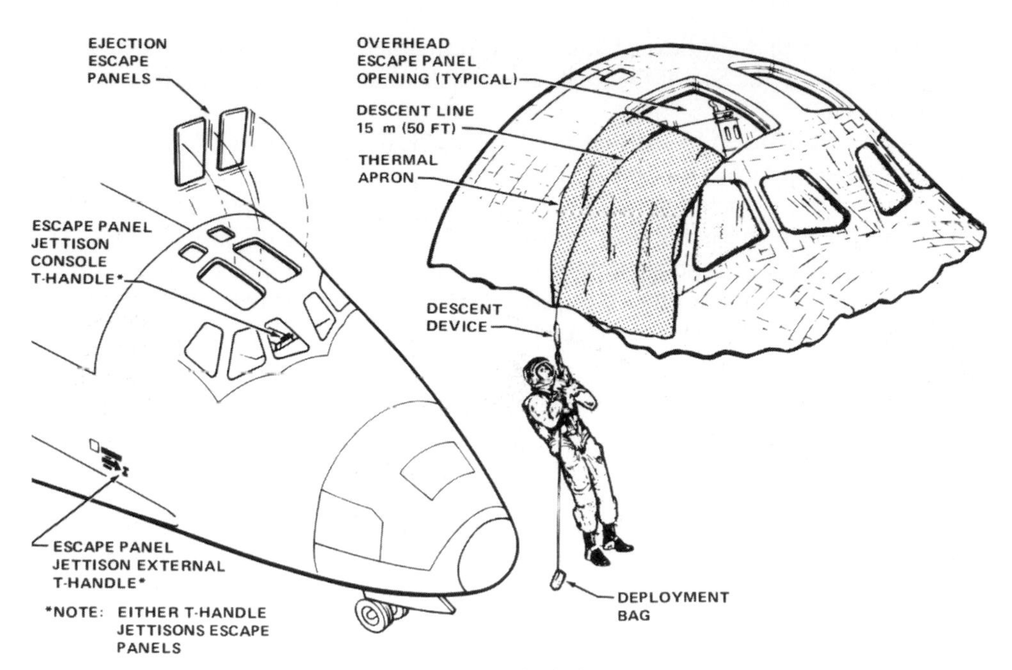

Detail of overhead window egress.

increase survivability. The development of this design commenced in 1990 with the remit to replace the LES; since the end of 1995 it has been worn by each Shuttle flight crew and will remain in operation until the end of the program in 2010. The system entails a full-pressure garment unlike the partial-pressure system on the LES. Operating pressure is 24.1 kPa using the primary supply from the Orbiter; there is also a 10-minute back-up capability. The mass is 12.7 kg with 29 kg of support equipment. In total 70 suits of this type were manufactured for the Shuttle programme.

Soviet Union/Russia

Recovery crews were quickly on hand to support the landing of the Vostok and Voskhod 1 crews, but Voskhod 2 landed outside the primary recovery area and had to wait for rescue crews to reach the craft in difficult circumstances. Some Soyuz craft have experienced delayed recovery but to date no Soyuz has landed outside Soviet/Russian territory, though this is not impossible in some emergency situations. As a result, plans to recover Soyuz from foreign soil or territory, or the ocean, have to be considered.

Contingency and emergency landing sites for Soyuz spacecraft have existed for some time outside the normal Kazakhstan areas. Each night prior to the crew settling down for their sleep period they receive an orbit number and a table which updates the time of a retro-burn to reach designated landing sites in the event of an emergency and return to Earth in the next 24 hours. It appears there are three emergency landing zones for Soyuz in the Sea of Okhotsk, the North American prairies and the steppes of Kazakhstan.

Recovery is handled by the Federal Airspace Search and Rescue Administration who are responsible for the location and return of the DM as well as returning cosmonauts from the landing site. Landings are usually in daylight to aid recovery operations, though this is not always possible due to certain in-flight situations. The recovery forces are equipped with a fleet of aircraft, helicopters and all-terrain vehicles and are directed as quickly as possible to the landing area by radio equipment including direct communications links with the crew as they descend. Landings are normally recorded on film by recovery helicopters, and recovery teams are soon at the landing site to retrieve the crew, taking great care as the spacecraft may still be very hot from its recent re-entry.

Support structures are raised around the upright DM to allow for easier exit by the crew, via a slide, assisted by ground personnel. In winter the crew are wrapped in furs and are quickly examined before being airlifted back to a local airbase, then flown back to Baikonur and thence on to Moscow. The DM is lifted on to one of the all-terrain vehicles after being made safe for transportation back to Energiya for post-flight examination.

China

Images of emergency and nominal spacecraft recovery techniques and equipment resembling that of Soyuz have been published in support of their manned spaceflight program. As the program continues so more information will be forthcoming.

SUMMARY

As with all aspects of human spaceflight, supporting the end of each mission is as complex, involved and hazardous as sending a crew into space or supporting it while there.

As the missions to ISS continue and the Chinese manned program expands so will the need to support a planned return to the Moon and recovery of astronauts from those missions. The equipment, testing and development, and provision of search and rescue as well as survival equipment will be a feature of the Constellation Program and, in time, of the program to send humans to Mars. No matter how long the mission, its target or objectives all crews want to come home to Earth at some point, and to do it safely and efficiently. After all, there is no place like home ...

REFERENCES

Isaak P. Abramov and Å. Ingemar Skoog (2003). *Russian Spacesuits*. Springer/Praxis, Chichester, U.K.

Courtney G. Brooks, James M. Grimwood, Loyd S. Sewnson Jr. (1979). *Chariots for Apollo: A History of Manned Lunar Spacecraft*, NASA-SP-4205. NASA, Washington, D.C.
John Catchpole (2001). *Project Mercury*. Springer/Praxis, Chichester, U.K.
Rex D. Hall and David J. Shayler (2001). *The Rocket Men*. Springer/Praxis, Chichester, U.K.
Rex D. Hall and David J. Shayler (2003). *Soyuz: A Universal Spacecraft*. Springer/Praxis, Chichester, U.K.
Rex D. Hall, David J. Shayler and Bert Vis (2005). *Russia's Cosmonauts inside the Yuri Gagarin Training Center*. Springer/Praxis, Chichester, U.K.
Brian Harvey (2004). *China's Space Program from Conception to Manned Spaceflight*. Springer/Praxis, Chichester, U.K.
Bart Hendrickx and Bert Vis (2007). *Energiya-Buran: The Soviet Space Shuttle*. Springer/Praxis, Chichester, U.K.
D. R. Jenkins (2001). *Space Shuttle*. Midland Publishing, Hinckley, U.K.
Fred A. McAllister (1972). *Apollo Experience Report: Crew Provision and Equipment Subsystem*, NASA-TN-D-6737, March. Manned Spacecraft Center, Houston, TX.
McDonnell Douglas (1961). *Project Mercury Familiarisation Manual*, revised 1 August, Publication Number SEDR 104 Copy, Number 8. McDonnell Douglas, AIS Archive.
NASA (1966). *Gemini Press Kits GT3 (March 1965) through GT12 (November 1966)*. NASA, Washington, D.C.
NASA (1975). *Apollo Program Summary Report*, April, JSC-09423. NASA, Washington, D.C.
NASA (1981). *Space Shuttle News*. NASA, Washington, D.C.
NASA (1999). *Results of the First Manned Orbital Spaceflight 20 February 1962*. Apogee Books, Burlington, Ontario.
David J. Shayler (2001). *Gemini Steps to the Moon*. Springer/Praxis, Chichester, U.K., pp. 301–317.
Asif A. Siddiqi (2000). *Challenge for Apollo*, NASA-SP-2000-4408. NASA, Washington, D.C.
Kenneth S. Thomas and Harold J. McMann (2006). *US Spacesuits*. Springer/Praxis, Chichester, U.K.
John Vince (1966). *Gemini Spacecraft Parachute Landing System*, NASA-TN-D-3496, July. Manned Spacecraft Center, NASA, Washington, D.C.

Summary

Spaceflight is an inherently dangerous mode of transport and method of exploration, and will always be so. It has been clearly demonstrated on more than one occasion that though provisions for escape and rescue can be trained for and provided, this does not mean that the opportunity to use such equipment and procedures is always possible. This must be recognized and accepted more widely if we are ever to venture away from Earth to the planets.

In a recent test of the Orion Parachute Test Vehicle one of the parachutes used in the test failed resulting in heavier landing loads. This was not a recovery parachute but a programmer chute designed to place the mock-up capsule in the correct test conditions. The test on July 31, 2008 at the Yuma Proving Ground was the first-generation design of the Orion Parachute Assembly System and dropped from a C-17 aircraft at 7,620 m. Eighteen parachutes were to be used in the test (ten to align the mock-up and eight on the Orion Test Vehicle). As a result of the failed program the deployment of the drogues, plot and main chutes was delayed imparting heavier loads than expected and resulting in a faster descent and greater land impact. The two drogues deployed and inflated as designed but due to higher loads than expected they separated almost immediately. The three pilot chutes did pull out the three main chutes, but two of them also separated prematurely due to increased loads beyond their design limit. These two chutes probably slowed the test capsule down enough to allow the third to operate as designed (i.e., remaining attached until the vehicle touched down). The test capsule tipped over at impact and sustained some damage, though parts were expected to be reusable.

There are more tests to be conducted on the parachute system before it can become human-rated for operational mission. Alhough the test was a disappointment, it was also part of the learning curve for new equipment and design. While the media considered this a test failure, NASA does not see it that way: to them it was a developmental exercise to evaluate the system, and highlight potential problems and failures in time to redesign and improve the system for manned flights. Though

Landing of the recent Orion abort test.

everyone would like to see hardware and software perform as designed first time and every time, it's clear that this is simply not always possible or realistic. Learning from setbacks is the key to developing improved systems and procedures, and like the programs that preceded it Constellation is no different.

Experiences with launch escape systems, orbital contingency procedures and adequate methods of returning the crew to Earth have been focal points of manned spaceflight for over 50 years and will continue to be so at least for the foreseeable future. In much the same way as with commercial airlines, it is hoped that individual escape systems will not be required, but safety procedures will of course still be paramount. In 2004 NASA predicted the advent in almost 50 years' time of fourth-generation reusable launch vehicles that will be so safe and reliable that no escape system will be required. With an anticipated annual flight rate of 10,000 commercial launches, turnaround will be no greater than a few hours and flight crews will be able to launch the vehicle themselves without any assistance from the ground ... only time will tell.

For the time being escape systems are essential. Having an abort on a vehicle that incorporates its own re-entry capability simplifies the recovery of a crew at speed. Where there is no re-entry capability then a second vehicle has to be dispatched to return the stricken crew; of course, having the time and resources to achieve this is a factor in meeting this goal. There can be safe havens, but these have to be accessible and sustainable. In orbit around Earth return to home does not take that long nor to a degree does return from the Moon compared with the time it would take to rescue a

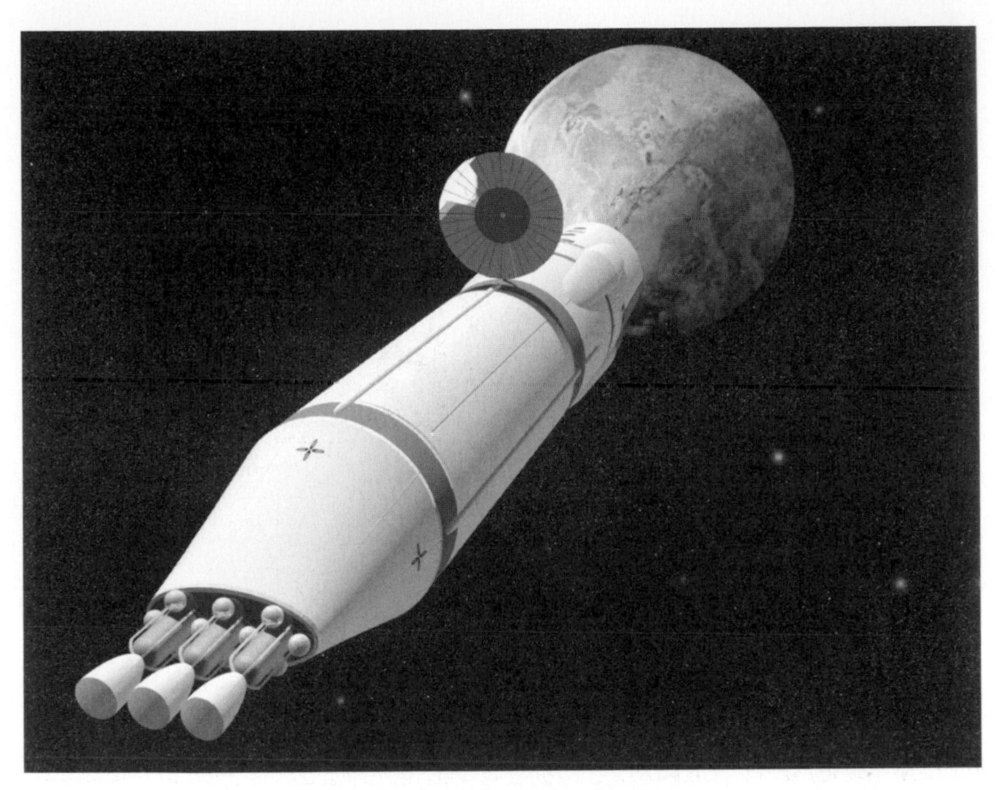

Mars—new challenges for space rescue.

crew from Mars or beyond. Distance and time complicate the provision of rescue and recovery, but this has to be addressed if we are to venture towards the planets.

In these pages I have tried to present the story of how complicated and broad the program is to provide crew rescue and safety at the pad, in leaving Earth, flying in space, exploring a celestial surface and in trying to get home again. These programs of test, failure, retest and success are lengthy, drawn-out processes and certainly not headline makers, unless things go wrong. However, it is much better to encounter problems in development than risk fatality in operation. Crews put their lives on the line flying vehicles into and out of space, relying on safety and rescue systems they hope they will not need and knowing test programs have qualified the equipment. They trust the systems will work if called upon, and this engenders confidence and belief that they will make it home. Otherwise, would they really strap into the vehicle on the pad?

Bibliography

The gathering of information for this book began almost 40 years ago with data on the emergency procedures for the early Apollo manned missions. This interest in safety during Project Apollo was part of a growing interest in the human exploration of space by both the United States and the then Soviet Union. Over the years this expanded into gathering additional data on space accidents and disasters and into the field of space rescue systems, facilities and procedures.

In the compilation of this book numerous sources have been used, far too many to list here, but apart from the cited examples in the text the following books and publications have been of great help and support in narrowing down the data to this present volume. The author would appreciate any further updates and information which could be considered for any future editions of this book and of the earlier book on *Space Accidents and Disasters* (also in this series).

Periodicals

Spaceflight Magazine and *The Journal of the British Interplanetary Society* (British Interplanetary Society).

Aviation Week and *Space Technology* (McGraw-Hill).

Flight International (Reed Business Information).

NASA History Series

1963–2000 *Astronautics and Aeronautics: A Chronology*, NASA (multiple volumes).

1963 *Project Mercury, A Chronology*, NASA SP-4001, James M. Grimwood.

1966 *This New Ocean, A History of Project Mercury*, NASA SP-4201, Loyd S. Swenson Jr., James M. Grimwood, Charles C. Alexander.

1969 *Project Gemini, Technology and Operations: A Chronology*, NASA SP-4002, James M. Grimwood and Barton C. Hacker with Peter J. Vorzimmer.

1969 *The Apollo Spacecraft: A Chronology*, Volume I, NASA SP-4009, Ivan D. Ertel, Mary Louise.

1973 *The Apollo Spacecraft: A Chronology*, Volume II, NASA SP-4009, Mary L. Morse, Jean K. Bays.

1973 *The Apollo Spacecraft: A Chronology*, Volume III, NASA SP-4009, Courtney G. Brooks, Ivan D. Ertel.

1977 *On the Shoulders of Titans, A History of Project Gemini*, NASA SP-4203, Barton C. Hacker, James M. Grimwood.

1977 *Skylab: A Chronology*, NASA SP-4011, Roland W. Newkirk, Ivan D. Ertel.

1978 *The Apollo Spacecraft: A Chronology*, Volume IV, NASA SP-4009, Ivan D. Ertel, Roland W. Newkirk, Coutney G. Brooks.

1978 *Moonport: A History of Apollo Launch Facilities and Operations*, NASA SP-4204, Charles D. Benson, William Barnaby Faherty.

1978 *The Partnership: A History of the Apollo-Soyuz Test Project*, NASA SP-4209, Edward C. Ezell, Linda N. Ezell.

1979 *Chariots for Apollo, A History of Manned Lunar Spacecraft*, NASA SP-4205, Courtney G. Brooks, James M. Grimwood, Loyd S. Swenson Jr.

1980 *States to Saturn, A Technical History of the Apollo/Saturn Launch Vehicles*, NASA SP-4206, Roger E. Bilstein.

1983 *Living and Working in Space: A History of Skylab*, NASA SP-4208, W. David Compton, Charles D. Benson.

1984 *On the Frontier, Flight Research at Dryden 1946–1981*, NASA SP-43093, Richard P. Hallion.

1985 *The Human Factor, Biomedicine in the Manned Space Program to 1980*, NASA SP-4213, John A. Pitts.

1988 *NASA Histroical Data Book, Volume I: NASA Resources 1958–1968*, NASA SP-4012, Jane Van Nimens, Leonard C. Bruno, Robert L. Risholt.

1988 *NASA Historical Data Book, Voulme II: Programs and Projects, 1958–1968*, NASA SP-4012, Linda N. Ezell.

1988 *NASA Historical Data Book, Volume III: Programs and Projects 1969–1978*, NASA SP-4012, Linda N. Ezell.

1989 *Where No Man Has Gone Befroe: A History of Apollo Lunar Exploration Missions*, NASA SP-4214, W. David Compton.

1993 *Suddenly Tomorrow Came: A History of the Johnson Space Center*, NASA SP-4308, Henry C. Dethloff.

1994 *NASA Historical Data Book, Volume IV: NASA Rescources, 1969–1978*, NASA SP-4012, Ihor Y. Gawdiak, Helen Fedor.

1995 *Exploring the Unknown: Selected Documents in the History of the U.S. Civil Space Program, Volume I, Organizing for Exploration*, NASA SP-4407, edited by John M. Logsdon.

1996 *Exploring the Unknown: Selected Documents in the History of the U.S. Civil Space Program, Volume II, Relations with Other Organizations*, NASA SP-4407, edited by John M. Logsdon.

1997 *Wingless Flight: The Lifting Body Story*, NASA SP-4220, R. Dale Reed, Darlene Lister.

1998 *Exploring the Unknown: Selected Documents in the History of the U.S. Civil Space Program, Volume III, Using Space*, NASA SP-4407, edited by John M. Logsdon.

1999 *Exploring the Unknown: Selected Documents in the History of the U.S. Civil Space Program, Volume IV, Accessing Space*, NASA SP-4407, edited by John M. Logsdon.
1999 *Power to Explore: A History of the Marshall Sapceflight Center 1960–1990*, NASA SP-4313, Andrew J. Dunar, Stephen P. Waring.
1999 *The Space Shuttle Decision: NASA's Quest for a Reusable Space Vehicle*, NASA SP-4221, T.A. Heppenheimer
2000 *Challenge to Apollo: The Soviet Union and the Space Race, 1945–1974*, NASA SP-2000-4408, Asif A. Siddiqi.

Other books

1974 *Rescue in Space, Lifeboats for Astronauts and Cosmonauts*, Erik Bergaust (G. P. Putnam's Sons, New York).
2001 *The X-Planes, X1 to X45*, Jay Miller (Midland Publishing, U.K.).
2001 *Space Shuttle: The History of the National Space Transportation System, the First 100 Missions*, Dennis R. Jenkins (Midland Publishing, U.K.).

Praxis Space Exploration Series

(a useful library reference for manned spaceflight history)

1999 *Exploring the Moon: The Apollo Expeditions*, David M. Harland.
2000 *Disasters and Accidents in Manned Spaceflight*, David J. Shayler.
2000 *Challenges of Human Space Exploration*, Marsha Freeman.
2001 *The Rocket Men: Vostok & Voskhod, the First Soviet Manned Spaceflights*, Rex D. Hall, David J. Shayler.
2001 *Skylab: America's Space Station*, David J. Shayler.
2001 *Gemini, Steps to the Moon*, David J. Shayler.
2001 *Project Mercury: NASA's First Manned Space Programme*, John Catchpole.
2001 *Russia in Space: The Failed Frontier?* Brian Harvey.
2002 *Apollo: The Lost and Forgotten Missions*, David J. Shayler.
2002 *Creating the International Space Station*, David M. Harland, John E. Catchpole.
2002 *The Continuing Story of the International Space Station*, Peter Bond.
2003 *Europe's Space Programme: To Ariane and Beyond*, Brian Harvey.
2003 *Russian Spacesuits*, Isaak P. Abramov and Å. Ingemar Skoog.
2003 *Soyuz: A Universal Spacecraft*, Rex D. Hall, David J. Shayler.
2004 *Walking in Space*, David J. Shayler.
2004 *China's Space Program: From Conception to Manned Spaceflight*, Brian Harvey.
2004 *Lunar Exploration: Human Pioneers and Robotic Surveyors*, Paolo Ulivi, David M. Harland.
2004 *The Story of the Space Shuttle*, David M. Harland.
2005 *Russia's Cosmonauts, Inside the Yuri Gagarin Training Center*, Rex D. Hall, David J. Shayler, Bert Vis.
2005 *Women in Space: Following Valentina*, David J. Shayler, Ian Moule.
2005 *Marswalk One: First Steps on a New Planet*, David J. Shayler, Andrew Salmon, Michael D. Shayler.
2005 *Space Shuttle Columbia: Her Missions and Crews*, Ben Evans.

2005 *The Story of the Space Station Mir*, David M. Harland.
2006 *US Spacesuits*, Kenneth S. Thomas, Harold J. McMann.
2006 *Apollo: The Definitive Sourcebook*, Richard W. Orloff, David M. Harland.
2007 *NASA Scientist Astronauts*, David J. Shayler, Colin Burgess.
2007 *Praxis Manned Spaceflight Log*, Tim Furniss, David J. Shayler with Michael D. Shayler.
2007 *Soviet and Russian Lunar Exploration*, Brian Harvey.
2007 *Energiya-Buran: The Soviet Space Shuttle*, Bart Hendrickx, Bert Vis.
2007 *The Rebirth of the Russian Space Program: 50 Years after Sputnik, New Frontiers*, Brian Harvey.
2007 *On the Moon: The Apollo Journals*, Grant Heiken, Eric Jones.
2007 *Animals in Space: From Research Rockets to the Space Shuttle*, Colin Burgess, Chris Dubbs.
2007 *Lunar and Planetary Rovers: The Wheels of Apollo and the Quest for Mars*, Anthony Young.
2007 *Space Shuttle Challenger: Ten Journeys into the Unknown*, Ben Evans.
2007 *The First Men on the Moon: The Story of Apollo 11*, David M. Harland.
2007 *The Story of Manned Space Stations: An Introduction*, Philip Baker.
2008 *How Apollo Flew to the Moon*, W. David Woods.

Website

Mark Wade's Encyclopaedia Astronautix address is *http://www.astronautix.com/*

Index

Printing: Mercedes-Druck, Berlin
Binding: Stein + Lehmann, Berlin